In accordance with the latest Syllabus prescribed by the Council for the Indian Certificate of Secondary Education Examination, New Delhi.

MOST LIKELY

ICSE QUESTION BANK

PHYSICS

CLASS X

By

A PANEL OF AUTHORS

OSWAL PUBLISHERS

1/12, Sahitya Kunj Colony, M.G. Road, Agra - 282002

EDITION
YEAR 2021

DISCLAIMER

With the ambition of providing standard academic resources, we have exercised extreme care in publishing the content. In case of any discrepancies in the matter, we request readers to excuse the unintentional lapse and not hold us liable for the same. Suggestions are always welcome.

ISBN : 978-81-95133-30-7

PUBLISHED BY

OSWAL PUBLISHERS

1/12, Sahitya Kunj Colony, M.G. Road, Agra - 282002

(0562) 2527771-4, +91 7534077222

Contact@oswalpublishers.com

www.oswalpublishers.com

The cover of this book has been designed using resources from Freepik.com

Printed At Upkar Printing Unit, Agra

PREFACE

It is a matter of immense pride for us to present our 'MOST LIKELY ICSE QUESTION BANK' series, especially prepared for students appearing for Board examinations in the oncoming year.

This book is a perfect capsule for building self-confidence during exam. preparation. Based on chunking strategy, the 'Categorywise–Chapterwise' format with its exhaustive set of questions allow the students to cover every category in a chapterwise manner.

Covering easy categories first boosts student's self-esteem and the ascending score braces them to take up challenging categories without fear. This prepares the student to see the exam paper as achievable at all times.

With its simple language and style, it is a one-stop solution for smart study. We are confident that the book will enable the candidates to develop a better understanding of the curriculum and help them organize their learning process. This book shall definitely prove to be a fruitful tool for the students and encourage them towards scholastic excellence.

Constructive suggestions for further improvement of the book are always welcome.

Note: Questions marked with '*' are frequently asked in previous years board examinations.

—**Publisher**

LET'S GO ORGANIC

Growing microgreens can be a good, easy and economical way for a student to start practicing plantation at home.

Microgreens are highly nutritious plants that are harvested after the sprout-stage and just before the maturity phase.

Microgreens may include mustard, coriander, basil seeds, dried peas, beans, broccoli, radish, spinach, beetroot, cauliflower and others.

It is more convenient to start growing microgreens that require low maintenance like mustard, coriander, peas, basil and others.

Start with finding a flat tray or a shoe-box or any tetra-pack carton.

Make holes in the box for easy drainage of water.

Spread two inches of soil in the box. You can get ready-made soil from the garden or nursery. Sprinkle your desired seeds over the soil.

You can spread a mixture of coco-peat and fertile topsoil or a stack of moist tissue papers over it. Remember to keep the tissue paper moist till the seeds start germinating.

Cover the box with newspaper and keep it in a place where there is no direct sunlight.

Spray water twice a day atleast.

On third and fourth day, once the leaves start to appear, start exposing the germinating microgreens to indirect sunlight.

You can try to expose them to direct sunlight for not more than 4 hours.

By ninth or tenth day, your microgreens will be ready, harvest them!

Microgreens can be eaten with sandwiches, wraps, salads, smoothies, juices or as garnishers on pizza, soup and curries.

CONTENTS

Note: Questions marked with '*' are frequently asked in previous years board examinations.

SUGGESTIONS FOR STUDENTS

1. Try to understand the concept and try to apply it to the immediate surroundings.
2. Try to relate the concepts from Mathematics with Physics and vice-versa.
3. Avoid rote learning. Answer in your own words and then verify whether you have covered all points.
4. Watch related videos on YouTube and try to make some projects/models which work on the principles which you have learnt.
5. Pay equal attention to all chapters.
6. Take out time for self-study.
7. Use the reading time of 15 minutes judiciously to make a proper choice of questions from section II by reading the requirements of the question carefully supported with a high degree of concentration.
8. Avoid writing answers which are simply repetition of the question. Instead be specific about the key word in that statement.
9. Candidates must not leave any topic for option. All topics are covered in section I which is compulsory.
10. Avoid changing the order of sequence of questions and numbering system.
11. Handwriting should be neat and legible.
12. Learn the principles, laws and definitions accurately.
13. Ray diagrams and the other diagrams need to be practiced periodically.
14. While writing the answers it is not only important to cover all the points but also to present them in a proper sequence.
15. While solving a numerical it is advisable that the formula needs to be written in the beginning. Essential steps need to be shown and final answer needs to be expressed along with a proper unit. Avoid computation at the first step; let it be plain substitution as the marks are awarded for the correct substitution.
16. Be regular in your study habits. Complete your syllabus well in time. A thorough revision of all topics is all time important.
17. It is advisable to solve previous year's papers in writing.
18. More emphasis should be given on writing rather than memorising.
19. For speed in mathematical calculations; it is advisable to learn tables up to 30, know squares up to 30, cubes up to 15 and basics of fractions and decimal.
20. It is better to inculcate the habit of underlining the important points or key words in the answer.

Word of Advice

Chapter 1. Force, Work, Power and Energy

Topics Found difficult By Students

- Work done when a body moves in a circular path.
- Physical quantity and unit.
- Conversion of units.
- Substitution in a numerical which required some conversion.
- Numerical on moment of force.
- Cases of inter conversion of energy.
- Concept of work. Work done for different angles between force and displacement.
- Work done when displacement is not in the direction of the force.
- Concept of couple.

1. While defining moment of force, some students, instead of writing 'line' of action, they wrote 'point' of application. In many cases, students did not mention *perpendicular distance.* Some even mentioned *displacement.* Some merely wrote *turning effect* and did not mention the axis of rotation or fulcrum. Instead of turning effect of *force,* many wrote turning effect of a *body.*

2. A few students wrote $1\ N\ m = 10^7\ erg$ instead of dyn-cm. Some wrote $1\ J = 10^7\ dyn$. Some related the SI and CGS unit of *force,* instead of relating the SI and CGS unit of *moment of force.*

3. While defining kilowatt hour, some students did not mention that it is the energy consumed / required; in some cases, they mentioned *power.* Some comprehended 1kW h as 1watt energy consumed in 1 hour. Some students used the *concept of kilo* and multiplied by 10^3 for writing the relation between *kilowatt hour* and *joule.* A few students committed errors in converting 1kW h to J, or while converting into MJ.

4. The work done by the satellite at any instant, as it revolves around a planet in a circular orbit, many students mentioned displacement as zero. Students overlooked the words *work at any instant.*

5. Many students were confused between 'cell in use' and 'charging of a cell'.

6. Many of them wrote *microphone* instead of *loudspeaker* as example for-*Electrical energy changes into sound energy.* Some gave technically incorrect answers such as, *bell* instead of an *electric bell.*

7. Many students wrote the unit of *acceleration* instead of writing the unit of *velocity.* A large number of students could not apply the principle of conservation of mechanical energy and thus, could not arrive at the correct answer. Some students did not convert mass into the SI unit 'kg', however in the comparison of kinetic energy and potential energy, that is ½ m v^2 = mgh, as m was cancelled, they got the correct velocity, as the error got nullified.

8. Some only wrote *inside* or *outside* instead of writing *radially inside* or *radially outside the circular path,* w.r.t. *centripetal force.* Some students diagrammatically showed that motion is tangential to the curved path but were unable to make it clear that the *motion of a body moving with a constant speed around a circular path would be accelerated.*

9. A few students wrote vague answers by explaining centrifugal and centripetal forces but did not compare their magnitudes. Some wrote that centripetal force is greater /more than centrifugal force.

10. Some wrote the physical quantity which is measured in calories such as work or energy. Some of them even wrote as joule as they were confused between physical quantity and unit.

11. Students wrote the relation between calorie and the S.I unit of that quantity as 1 J = 4.2/4.186 calorie. Some even wrote 1 cal = 0.4 J/ 1 J/ 41.8 J/ 373 K/ 336 J

12. For the definition of couple, key words/terms such as two, equal, parallel, opposite, not acting along the same line etc. were missing which made the definition incomplete or meaningless. Some of them defined moment of couple or torque instead of couple.

13. Many students wrote the S.I. unit of moment of couple as joule in place of Nm or N or dyn or N/m or N m^{-1}
14. Some students failed to understand that when the forces are same, the moment can be equalised by adjusting their torque arms.
15. Some students wrote either 1 H.P. = 0.746 watt or 1 watt = 746 H.P. Some students were unsure about the exact value of H.P. in watt.
16. Many students failed to convert kW to W. Some wrote the unit of speed as ms^{-2} instead of ms^{-1}.
17. Most students failed to convert kgf to newton and from cm to m that resulted in wrong answers.

Chapter 2. Simple Machines

Topics Found difficult By Students

- Understanding of lever of class 1 with M.A. greater than 1 and class 2 lever.
- Change in mechanical advantage when the load is shifted towards fulcrum in class II lever. Speed of light for all wavelengths in vacuum is the same.
- Advantages of movable and fixed pulley.
- Numerical problems based on p = f.v and machines.

1. Some students did not identify the class of lever correctly.
2. Many students expressed the answer learnt by rote, *i.e.*, "By increasing effort arm". They did not pay attention to the statement *without increasing the length of the lever.* Some students wrote by *increasing load* or *by decreasing effort* without realising that as load changes, effort also changes proportionately. A few students wrote - *change the position of load and effort* without realising that the class of lever would change.
3. For the diagram of clock and tackle system of pulley, students made mistakes as no fixed support was shown, strands drawn were loose, connections shown were incorrect, directions of either/both load and effort were not marked, rough sketches of pulleys were drawn, there were no proper number of pulleys in the fixed and movable block, direction of load and tension were missing, no distinction between fixed and movable block.
4. Most students wrote that V.R. becomes half/double, as weight of block in pulley system is doubled.
5. Most of the students could identify the class of lever but some could not identify the position of fulcrum, hence made a mistake.
6. Majority of the students were unable to answer *what happens to the mechanical advantage of the lever if the load is shifted towards the fulcrum without changing the dimensions of the lever.* They just mentioned mechanical advantage is greater than one without realizing that the mechanical advantage of lever is already > 1 which increases further.
7. Most students wrote E.A. > L.A. but did not mention that load lies between the fulcrum and the effort for lever class II.
8. Some students made conceptual errors by using $L \times L.A. = E \times E.A$, to calculate the distance moved by the effort. Due to calculation errors in the first part, many students got wrong answers in the latter parts. Most students answered the third part correctly.
9. Some students wrote the definitions of M.A., V.R. and efficiency without mentioning V.R. does not change. Some also wrote the definition of V.R. in terms of displacements without stating that these displacements are in the same time.
10. Many students could identify that the movable pulley will require less effort but could not explain it with the M.A. ratio of both pulleys.

Chapter 3. Refraction of Light

Topics Found difficult By Students

- Diagram of refraction from denser to rarer medium to show the formation of an image.
- Completing the path of a ray through 30°, 60° and 90° prism.
- Refraction through any refracting medium when opposite faces are parallel and when they are not parallel.
- Factors affecting critical angle and conditions for total internal reflection.
- Total internal reflection through right angled isosceles prism.

1. Some students took the ray normal at one surface but after refraction, the ray shown was bent at the other surface. The angle with the second surface was marked as 90° by them. Several students did not mark arrows on the rays. Some missed marking arrows on the rays.
2. Some students were confused in F_1 and $2F_1$, F_2 and $2F_2$, between virtual and real image, etc. A few students answered by drawing ray diagram, which was not asked.
3. A few who did not know the concept about colour of light and wavelength wrote *blue* or *red* randomly. Some students treated wavelengths as speed.
4. Some students did not establish a relation between speed and wavelength in a medium and thus wrote incorrect answers.
5. A few students could not establish the relation between the size of the image and change in medium. Some students considered the light ray passing from denser medium to rarer medium, in case of diver in water. Some wrote that the object appears *shorter* while others wrote *of the same size.*
6. In ray diagrams, students made mistakes by not marking arrows on rays, showing partial reflection with refraction, not showing extension of rays and dotted image, showing incorrect image position and incorrect refraction.
7. In ray diagrams, students made mistakes by showing the ray entering the prism as bended, No angle of incidence was marked. After total internal reflection, the ray was shown coming out un-deviated, Refraction was shown at the second surface; The diagram was completed by treating it as right-angled isosceles prism.
8. Many students defined critical angle incorrectly as keywords such as rarer, denser were missing, did not make it clear that it is the angle of incidence, 90° was missing.
9. Many students wrote factors affecting for total internal reflection instead of critical angle, factor as angle of incidence/frequency, additional point/s which were incorrect.
10. Many students wrote separate expression or definition for each ${}_a\mu_w$ and ${}_w\mu_a$ but did not give the relation between the two.
11. Many students could complete the path of first reflected ray correctly but made a mistake in completing the path of the second ray after total internal reflection.
12. A few made following errors in the graph angle of deviation with angle of incidence as prism surface, by giving incorrect relation between the angle of incidence and the angle of deviation, drew diagram of prism instead of graph, incorrect graph showing the variation in the angle of deviation with the angle of incidence, incorrect labelling on the graph.
13. Only a few students wrote the correct mathematical relation between the critical angle and the absolute refractive index of a medium. Many students simply mentioned increases and decreases.
14. Most of the students expressed the conditions required for total internal reflection of light to take place incorrectly. Some wrote that light enters from rarer medium to denser medium. Instead of angle of incidence being greater than critical angle, it was written vice versa and some wrote *incident ray greater than critical angle.*

15. Most of the students drew the ray diagram incorrectly as – the object was not drawn at the focus, the image drawn was not dotted, the arrows were missing before and after refraction, the lens was labelled wrongly.
16. Some were not clear about the concept of critical angle and conditions of total internal reflection.
17. In *factors upon which the heat absorbed by a body depends,* some stated only temperature instead of change in temperature; some mentioned material and specific heat capacity as two different points; some even stated factors such as surface area and black surface which are actually factors affecting the rate of heat absorption
18. Students failed to write that different colours have different refractive index. They were unable to make the point clear that different colours have different speed in a different medium.
19. Some wrote $i > r$ when light enters from rarer to denser medium and $i < r$ when light enters from denser to rarer medium.

Chapter 4. Refraction through Lenses and Optical Instruments

Topics Found difficult By Students

- Numerical problems based on light using sign convention.
- Identification of lenses when virtual images are formed.
- Change in the focal length when the refractive index of the outside medium changes.
- Ray diagrams of lenses.
- Numerical problem on lens formula.
- It is also observed that in ray diagrams in spite of knowing diagrams students loose marks as they don't draw arrows on rays before and after refraction and don't draw dotted line for virtual intersection or image.
- The ability to diverge or converge with respect to the focal length.

1. Some students got incorrect answers due to improper sign convention and incorrect substitution. Some students did not consider the sign convention. A few students wrote unit for magnification.
2. Some students answered that focal length changes with change in medium, but they could not give the correct reason. Several students could not write about the effect on the focal length of the lens due to change in the refractive index of the outside medium.
3. Few wrote convex lens as a thicker lens and concave lens as a thinner lens. Some students wrote thicker lens has greater focal length.
4. Some students were tricked by the words 'object between the optical centre and the principal focus of a lens as well as virtual image' and they did not pay attention to the characteristic that image is virtual and diminished. Those who identified the lens correctly drew correct ray diagram but those who identified the lens incorrectly, draw incorrect ray diagram. Some students did not draw arrows on the rays in the diagram.
5. Several students applied mirror formula in finding the distance of the image from the lens. Many students were not clear about the sign convention. In the final answer unit was also missing.
6. The common flaws observed in most of the answer scripts in numerical were: use of incorrect lens formula; incorrect use of sign convention; incorrect substitutions; incorrect characteristics of the images; incorrect answer with correct unit or correct answer with incorrect unit or without unit.

Chapter 5. Spectrum

Topics Found difficult By Students

- Scattering of light
- Speed of light for all wavelengths in vacuum is the same.
- Most of the students are confused between the concept of dispersion and scattering.

1. For clouds appear white, many students used the term *reflection* or *dispersion* instead of *scattering*. The *size* of the particles and *uniform scattering* was not mentioned by some students. Some students were confused between the white colour of the sky at noon and white colour of clouds.
2. Some students wrote incorrect name of the electromagnetic radiation used for photography in fog. Some wrote *infar* instead of *infra.*
3. Some students wrote the uses/properties of the given radiations instead of source/cause.

Chapter 6. Echoes and Vibrations of Sound

Topics Found difficult By Students

- Correlation between the holes present on the flute and the frequency of sound produced.
- Confusion between the use of $V = \frac{d}{t}$ and $V = \frac{2d}{t}$.
- Numerical based on sound.
- Students were not comfortable in stating the factors affecting the frequency of stretched string.
- Cases of resonance, forced vibrations, free vibrations, etc.
- Safe limit of sound level for humans.
- Mechanical waves.
- Vibrations in a stretched string in relation to the wavelength, amplitude, frequency etc.

1. Many students drew the free vibrations in vacuum graphically in a correct manner, but axes were not labelled. A few students did not take X-axis as the mean position. Some students drew a single curve while several drew a graph of minimum deviation and angle of incidence but labelled the axes as *displacement* and *time.* A few students drew the graph only in the positive quadrant of the graph. It did not have any negative amplitude. Many students drew the graph for damped vibrations.
2. Many students committed the mistake of simply writing "audible in air" or "not audible" without calculating frequency. Some wrote an incorrect unit of frequency. A few used the velocity formula of echo. Several students were not aware of the audible range. Other students calculated 'time', treated it as time period to calculate frequency and finally wrote that the wave is not audible in air. A number of students used speed of sound in water as 336 ms^{-1}.
3. Most of the students were confused between 'SONAR' and 'echo depth sounding'.
4. Many students could not write the reason about the holes present on the flute and its relation with the frequency (pitch) of the sound produced. Some of them even related it to the loudness.
5. In the definition of resonant vibrations following glitches were observed such as keyword such as increase in amplitude was missing, natural frequency matching with forced vibration was written instead of frequency of forced vibration.
6. Many students confused *loudness* with *Quality, pitch, frequency.* Students were unaware about *base drums* and *side drums.*
7. Some common errors were noticed like could not calculate the time taken from the number of oscillations of the pendulum, formula applied v = d/t *in* place of v = 2d/t, did not write the unit, velocity of sound was substituted as 320 m s^{-1} instead of 340 m s^{-1}.
8. In many students, conceptual understanding was missing between free and forced oscillations.
9. A number of students were unable to comprehend the sound waves graph. Many students, instead of calculating proper ratios of amplitudes and wave lengths, just compared them by stating more or less. Some students left the answer in the fractional form without bringing it to the lowest form. A few students stated the values of the quantities for the two waves.
10. Many students did not write the key words such as *without influence of external periodic force, constant frequency and constant amplitude* in the definition for free vibrations of a body.
11. For *the amplitude of a vibrating body continuously decrease during damped vibrations,* most students did not write that external force is resistive in nature or that it causes loss of energy. Some students wrote the

definition of damped vibrations. A few students wrote that damped vibrations are due to the removal of external force.

12. Some used the terms *shrillness* and *loudness* interchangeably, without actually knowing the difference.
13. In the definition of resonance some made no mention about increase in the amplitude and misspelt the unit of loudness.

Chapter 7. Electricity

Topics Found difficult By Students

- Numerical problems based on Electricity, especially in finding the current in parallel branches.
- Specific resistance and its unit.
- Differentiation between resistance and specific resistance.
- In electricity the difference between P.D. and terminal voltage is not clear. In some cases P.D. is misunderstood as only potential drop.

1. Some students were confused between *superconductors* and *semiconductors*.
2. Some wrote random names such as, galvanized rubber, metals, copper, aluminium, tungsten, constantan, mangnanin, nichrome, instead of semiconductor which were incorrect.
3. Several students did incorrect calculation of external resistance. Hence, the calculation for current through the given resistor was also incorrect. A few students either used incorrect formula or made calculation errors.
4. Some students considered the *emf* as the terminal voltage. A few students wrote incorrect unit of the asked physical quantity.
5. Most of the students wrote the effect of an increase in the temperature on the specific resistance of a metal, as does not change or decreases. Some students wrote that with the increase in temperature, the specific resistance of a semiconductor increases or remains the same.
6. Some students were confused in understanding the series and parallel combination of resistors. Some made calculation errors. Some of them wrote an incorrect expression at stage 1 for parallel combination, (instead of $1/R$ as L.H.S they wrote R) hence, in the last step mathematical result was incorrect. Some students left the final answer in improper fraction.
7. Students did not take into consideration the position of the key in the circuit. They did not realise that when the key is closed and when it is open, the resistance present in the circuit differs, so the resistance of the circuit is different in both cases and therefore, the current is also different in both the cases. Many students blindly wrote current is zero when the key k is open. The difference between open circuit and open key was not clear to the students. Some of the students did not write the units for the answers. It was also observed in some scripts that students were not trained in dealing with the internal resistance.
8. Many students showed a straight line instead of a wavy line to depict the resistor in the diagram.
9. Many students showed series combination but used the formula of parallel combination and *vice versa*.
10. Some students misinterpreted voltage drop as potential drop. Many students were unable to score as they did not convert mA to A.
11. Some students could not differentiate between *resistance* and *specific resistance*. Majority of the students wrote about *resistance* instead of *resistivity*.

Chapter 8. Electric Power and Household Circuits

Topics Found difficult By Students

- Connection of live, neutral and earth wire with the appliance.
- Difference between the use of a fuse in a normal circuit and a power circuit.
- Reason for stepping up voltage during power transmission.
- Dual control switch.
- Drawing circuit diagram from the given information.
- In house circuit the series and parallel combination. Also the misconception is ring system has parallel wiring and tree system has series wiring.

1. Some students wrote the reason for not using a copper wire as a fuse wire was the low melting point of copper/copper melts quickly/low specific heat capacity of copper. Some wrote the characteristics of a fuse only without comparing it with copper.
2. Students wrote advantage of connecting electrical appliances in parallel combination as advantages of series combination, current received is 220 volts, work on the same current in place of voltage, wrote about advantages of ring system of wiring, same thickness of wire can be used, monthly Power bill is reduced.
3. A large number of students wrote characteristics of a fuse wire as high melting point and low resistivity. Some even expressed about good conductance or low specific heat capacity which was irrelevant.
4. Students were confused between live wire and neutral wire in the diagram given with the questions.
5. Some students drew incorrect diagram of an A.C. generator as they drew battery in place of load, in place of slip ring they showed split rings, drew transformer diagram, drew slip rings correctly but labelled them as split ring, drew DC motor diagram.
6. Most of the students were unaware of the word *phase wire*. Some students wrote the answer in terms of colours of wire. A few students wrote hot wire or live/earth or neutral/earth instead.
7. Some were unable to comprehend the meaning of the term *fuse rating*.
8. Many students could not write about the absolute necessity of the *fuse* in power circuits correctly.
9. Students wrote the colour code for all the three wires which implies that they did not know which wire is connected to the metallic body of the appliance. While some did not write the colour code but identified the wire as earth wire.
10. Most students did not draw the diagram of dual switch control correctly. Vague diagrams were drawn by few students, such as physical appearance of a switch. In some cases, load was connected directly to the live wire. Proper labelling was not done by many students.
11. Most of the students wrote about the maximum limit of current through the wire but failed to write about the melting of wire in case the current exceeds.
12. Many students wrote about minimising the power losses but failed to write about the decrease of current due to increased voltage.
13. Students appeared confused due to the internal resistance. Some did not understand the meaning of terminal voltage and made vague calculations. To calculate energy in joule minutes were not converted to seconds. Some divided by 60 and converted the time in hour and wrote unit kWh without dividing by 1000.

Chapter 9. Megnetic Effect of Current

Topics Found difficult By Students

- The diagram of AC generator and DC motor.
- Difference between the number of turns of the coil and just coils.
- Identification of magnetic poles in electromagnets.
- Causes of energy losses in the transformer.
- Students were also confused about the working and the characteristics of transformers.
- Similarities between A.C generator and D.C motor.
- Factors affecting induction in transformer

1. Some students were not able to write the angles between the magnetic field and the length of the conductor for the asked forces.

2. A few students wrote that the magnetic polarity of the loop that faces us as *south to north* or *north to south*, which was incorrect.
3. Some students did not pay attention to the *number of turns remains the same* and wrote that the strength of the magnetic field produced by this loop could be increased by increasing the number of turns of the coil. A few students wrote that the strength of the magnetic field can be increased by increasing the magnetic strength.
4. Many students did not connect the coil to the terminals shown in the diagram. Some showed the coils in the incorrect arms or showed incorrect number of turns of the coil i.e. more turns for less voltage and vice-versa. A few students labelled it correctly.
5. Many students incorrectly drew the diagram of a DC motor as they drew diagram of AC generator, did not label the diagram or labelled incorrectly, drew slip rings instead of split rings, drew the diagram with the load, did not draw the field magnet.
6. Many students incorrectly answered for ways to increase speed of rotation of DC motor by use of incorrect technical terms. Some got confused with the induction, while some wrote number of coils in place of number of turns of a coil
7. Majority of the students could not apply right hand thumb rule correctly and instead of writing towards east, various incorrect answers for example, North-South, South-East, West, North, clockwise etc. were observed.
8. Many students identified magnetic effect of current but quite a number of students wrote that current is flowing through the conductor, about electromagnetic induction, about Lorentz Force, about electric field instead of magnetic field.
9. Some showed lack of conceptual clarity on the working of a transformer.

Chapter 10. Specific Heat Capacity and Latent Heat

Topics Found difficult By Students

- Graphs on calorimetry.
- Numerical problems based on Heat.
- Ratio concept in solving numerical.
- Heat capacity and specific heat capacity.
- Problems on calorimetry.
- Calorimetry numerical were found difficult.
- Difference between heat capacity, SHC

1. Almost all the students gave partially correct answers by relating the difference in the absorption of heat or to heat capacity instead of specific heat capacity of the stone and water. Some students wrote that the stone takes less time since it is a good conductor instead of writing about the specific heat capacity.
2. Some students compared heat capacities instead of specific heat capacities. Substitution for masses was incorrect. Several students substituted for masses correctly but could not express the specific heat capacities in the ratio form. Some wrote the relationship between heat capacity and specific heat capacity of a substance in the form of symbols without giving the meaning of the symbols.
3. Numerical related to melting of ice, many students made errors by not taking into consideration the heat absorbed by ice after melting to reach equilibrium temperature, not substituting the change in temperature correctly, not writing the unit for the final answer.
4. Many were unable to write the reason for naming one of the given substances as a good conductor of heat on the basis of their specific heat capacities.
5. Many students got confused with latent heat and specific heat capacity and between specific heat capacity and specific resistance.
6. Several students in the definition of calorimetry wrote the key word study in place of measurement. In many scripts students stated the principle of calorimetry.
7. Majority of the students could not express a temperature in degree Celsius in S.I. unit of temperature. Several students wrote 1°C = 273 K. Some students wrote t°C-273 or T = 273 – t°C or t = 273+K, 1°C=1 K.

8. Some students wrote hydrogen or kerosene, or petrol or diesel can be used as a coolant in car radiators.
9. Some students solved the specific latent heat of fusion of the metal equation using incorrect expressions $Q = mc\Delta t$ only or $Q = ml$ only or using $Q = ml + mc\Delta t$. A few students wrote the incorrect unit.
10. Many students were confused in the use of symbols such as C, C′ and H. Several students deviated from the usual symbols but did not explain them.
11. Some of them confused it with the definition of specific heat capacity and the same was observed with the unit. It was also observed that some students wrote $°K$ which is the incorrect way of expressing the unit of temperature.
12. Most of the students wrote about prevention of conduction instead of prevention of transference of heat energy by radiation.
13. For ice, most of the students used the formula mL only and used the value of latent heat capacity of ice (336 J/g) instead of the given value 330 J/g. Some students did not involve calorimeter in the heat exchange. Calculation errors were also observed.
14. Some used the value of S.H.C. of water in a wrong unit.
15. While writing the equation many students did not consider increase in temperature of melted ice to 50°C. It was also observed that students were unable to solve numerical involving mixtures with change of state. It was also observed that they did not pay attention in writing the units correctly. For eg in many places, gram was written as gm instead of g.

Chapter 11. Radioactivity and Nuclear Energy

Topics Found difficult By Students

- Completing the nuclear reactions in reverse order.
- Nuclear fusion reaction as thermonuclear reaction
- Beta emission during carbon dating.
- Concept of isobar.
- Conditions required for a nucleus to be radioactive.
- Mass number and atomic number.
- Energy sources especially nuclear energy.

1. Some students mixed up the definitions of *fission* and *fusion*. The words *bigger* and *smaller* were used in place of *heavier* and *lighter*, respectively. Vague statement was written by a few students such as; *large energy is given out during nuclear fusion* without comparing the quantity of nuclear material.
2. Some students gave precautions to be taken while handling the radioactive substances for their disposal.
3. A few students did not write number of neutrons. Some students used neutrons as atomic mass for beta emission. Some forgot to add the number of protons and neutrons and some wrote only the number of protons as mass number.
4. Various reasons were observed hydrogen nucleus to emit an alpha particle such as Hydrogen is not a radioactive substance, after Alpha emission atomic number decreases by 2 and mass number decreases by 4, Hydrogen has one proton and one neutron or hydrogen has one proton and one electron. Some students gave irrelevant explanation. Very few students attempted it to explain with a nuclear reaction.
5. Majority of the students wrote about large amount of heat energy being released during the nuclear reaction but did not talk about the reaction taking place at very high temperature. Some students explained nuclear fusion reaction in terms of heat released.
6. Some students could balance the reaction correctly but majority of them could not do so. They could not apply the principle of conservation of mass number and atomic number in completing the reaction. Some students wrote He in place of H.
7. Many students defined nuclear fission incorrectly as they related it to the size of the nucleus instead of the mass of the nucleus, splitting of an atom instead of nucleus, defined nuclear fusion.
8. It was observed that majority of the students found the atomic number correctly, but while finding the mass number, did not take into consideration mass of neutrons.
9. Many students, instead of writing lighter and heavier nuclei, wrote smaller and bigger atoms/nuclei for nuclear fusion. Some students wrote the definition of nuclear fission.

10. A few students were confused between isotopes and isobars and thus wrote examples with incorrect mass number and atomic number wrote the elements without atomic number and mass numbers or gave examples of isotopes.
11. Most of the students wrote the two factors upon which the rate of emission of thermions depends incorrectly by mentioning *melting point* instead of *temperature;* others wrote *area* instead of *surface*.
12. Some students wrote *neutron electron ratio* instead of *neutron proton ratio.* Some wrote Instead of n:p, p:n is greater than 82 and in some cases, instead of neutron, electron was written, Some got confused between *atomic number* and *atomic* mass. Some gave the definition of radioactivity rather than the condition.
13. Some confused thermionic emission with radioactive phenomenon and gave the answer as α, β and γ.
14. Most of the students were confused with household voltage and hence wrote 220V or vague values such as 1000V.
15. When beam passes though electric field, students wrote partial answers, such as: deflects towards screen, deflects and forms parabolic path, but direction was missing, deflects towards positive terminal.
16. Most of the students had a vague idea of free electrons. They wrote that all valence electrons are free electrons.
17. Some students wrote that free electrons do not possess any energy. In some answers, the word *sufficient* was missing. Lack of clear idea of why they can't leave the metal surface was observed.
18. A few found it difficult to work backward to find the atomic number and mass number. They either made calculation errors or interchanged the atomic no. and mass
19. Students committed errors in the order of radiations. They were confused about ascending or descending and also in mathematical signs i.e. > or < used by them.
20. They were not sure about energy change in CRT.

Miscellaneous

- Some did not express their answers in ratio form and left it in fractional form. A few did not get the inverse proportion between power and time for same work. Some students wrote their answer in the form of > and < and not in ratio form.
- Some students made the errors while solving numerical, such as wrote incorrect formula, took shift as apparent depth, made calculation errors, did not write the unit in the final answer, errors in copying the given data, error in calculation , incorrect unit given etc.
- Several students did not convert mass from g to kg
- Many students could not do substitution in the ratio form. Some students did not express the final answer in the ratio form with lowest term as asked in question.
- Most of the students could name the unit of physical quantity obtained by the given formula correctly; however, a few students were unable to name the unit.

❑

Chapter at a Glance

Set 1

Chapter 1. Force, Work, Power and Energy

Concept	Example	Formulae	Units	
			SI	CGS
Moment of force/Torque turning effect of force about fulence/axis of rotation (a vector qty.)	Turning a wrench	τ = F × ($\perp^r$ distance of line of action from axis of rotation)	1 Nm = 10^7 dyn cm 1 kgfm = 9.8 Nm 1 gfcm = 980 dyrence	
Moment of couple (pair of equal, opposite and parallel forces not acting along same line)	Turning a steering wheel	τ_c = F × Couple Arm ($\perp^r$ distance between 2 forces)	same as τ	
Equilibrium: State of body (rest or motion) under action of two or more forces remains unchanged	book lying on table, revolution of Earth around Sun	$F_{resultant} = 0$ $\tau_{resultant} = 0$		
Principle of Moments: alg. sum of clockwise moments = alg. sum of anti-clockwise moments *anti-clockwise is considered +ve, clockwise as –ve	A beam balance	$W_1 \times l_1 = W_2 \times l_2$		
Centre of Gravity: Point on a rigid body at which entire weight is assumed to act	C.G. of rod lies at midpt. of its axis	$\Sigma\tau = 0$ (alg. sum of moments of weights of particles)		
Work: said to be done when there is displacement in direction of force applied (a scalar qty.)	Lifting a load	W = Fcos θ × S θ = angle between F and S S = displacement	(Joule)	1J = 10^7 eg. (or dyn cm)
*W.D. by force of gravity	apple falling from a tree	W = mgh		
*W.D. against force of gravity.	ball thrown veritically upwards	W = – mgh		
Power: rate of doing work per unit time (scalar qty.)	energy consumed by bulb in one day	$P = \frac{W}{t}$ P = F × Velocity	1W (J/s) = 10^7 erg s^{-1}	
Energy: capacity to do work *When W.D. by a body, energy ↓ *When W.D. on a body, energy ↑ (scaler qty.)	When a striker hits a corner coin, energy is transferred from striker to coin	E = W	1 HP (horsepower) = 746 W same as W 1 KWh = 3.6×10^6 J 1 cal = 4.18 J 1 eV = 1.6×10^{-19} J	

Potential Energy: energy possessed by virtue of position or configuration	Water stored in tank/ dam			
Kinetic Energy: energy possessed by virtue of motion	A vehicle in motion	$K = \frac{1}{2} mv^2$ $K = \frac{p^2}{2m}$, p = momentum		
Gravitational Pot. Energy (due to gravity)		$U = mgh$ $U = mg(h_2 - h_1)$		
Work Energy Theorem: W.D. = Increase in K.E.		$W = \frac{1}{2} m (v^2 - u^2)$		
Law of Conservation of Energy: Total energy in a system remains constant K + U = constant	ball thrown vertically upwards Simple pendulum released from rest	$u = \sqrt{2gh}$ $v = \sqrt{2gh}$		

Chapter 2. Simple Machines

Concepts	Symbol	Formulae
Load: Opposing or resistive force against machine	L	
Effort: External force on Machine to overcome load	E	
Mechanical Advantage (depends on friction and can vary for a machine)	M.A.	$M.A. = \frac{\text{Load}}{\text{Effort}} = \frac{L}{E}$ M.A. > 1 ⇒ force multiplier
Velocity Ratio (depends only on design and is constant for a machine)	V.R.	$V.R. = \frac{\text{Vel. of E}}{\text{Vel. of L}} = \frac{V_E}{V_L}$ $V.R. = \frac{d_E}{d_L}$, d = displacement V.R. < 1 ⇒ Vel. multiplier
Input: Work done on machine	W_{input}	$W_{input} = E \times d_E$
Output: Work done by machine on load	W_{output}	$W_{output} = L \times d_L$
Efficiency *Ideal machine, η = 1 (M.A. = V.R.) *Actual machine, η < 1 (M.A < V.R.)	η	$\eta = \frac{W_{output}}{W_{input}} \times 100\%$ $\eta = \frac{M.A.}{V.R.}$
Principle of a lever: (a lever works on principle of moments) Moment of load about fulcrum = Moment of Effort about fulcrum		Load × Load Arm = Effort × Effort Arm $\frac{L}{E} = \frac{\text{Effort arm (E.A.)}}{\text{Load arm (L.A.)}}$ $M.A = \frac{E.A.}{L.A.}$: law of levers
Single fixed pulley		Ideal: M.A. = V.R. = 1 η = 100% *i.e.,* Actual: M.A. < 1, V.R. = 1 η < 100%, *i.e.,*

Single movable pulley		Ideal	$E = T$, T = tension $L = 2T$ $M.A. = \frac{2T}{T} = 2$ $V.R. = 2$ $\eta = 100\%$
		Actual	$M.A. < 2$ $V.R. = 2$ $\eta < 100\%$
Combination of Pulleys (i) One fixed pulley with several movable pulleys		Ideal	$M.A. = 2^n$, n is the number of movable pulleys $V.R. = 2^n$ $\eta = 100\%$
		Actual	$M.A. < 2^n$, $V.R. = 2^n$ $\eta < 100\%$
(ii) **Several pulleys in block and tackle system:** two sets of pulleys, upper set attached to a rigid support at the top is called block, lower set (movable) carrying load is called tackle.		Ideal	$L = nT$, n is the total no. of pulleys in both the blocks $E = T$ $M.A. = \frac{nT}{T} = n$ $E = \frac{L}{n}$ (Force multiplier) $V.R. = n$ $\eta = 100\%$
		Actual	$E = T$, $L + W = nT$ (W = weight of the lower block along with pulleys) $M.A. = n - \frac{W}{E}$ $M.A. < n$ $V.R. = n$ $\eta = \left(1 - \frac{W}{nE}\right) \times 100$ $\eta < 100\%$

Chapter 3. Refraction of Light

Wave Speed (C) = Frequency (υ) × Wavelength (λ)

↓

As light travels from one medium to another:
*υ remains constant
*C ↑ or ↓ ⇒ λ ↑ or ↓

↓

Refraction of Light (bending of light travelling from one optical medium to another)

→ Rarer to denser → C ↓ ⇒ bends towards normal; $\angle i > \angle r$

→ Denser to rarer → C ↑ ⇒ bends away from normal; $\angle i < \angle r$

→ Incident normally → no deviation; $\angle i = \angle r = 0$

↓

Laws of Refraction

↓

$\frac{\sin i}{\sin r} = {}^1\mu_2$

${}^1\mu_2$ = Refractive index of medium 2 *w.r.t* medium1

←

$\mu = \frac{c}{v}$

μ = absolute refractive index of a medium

c = speed of light in vacuum/air

v = speed of light in medium

→ ${}^1\mu_2 = \frac{v_1}{v_2} = \frac{\mu_2}{\mu_1}$

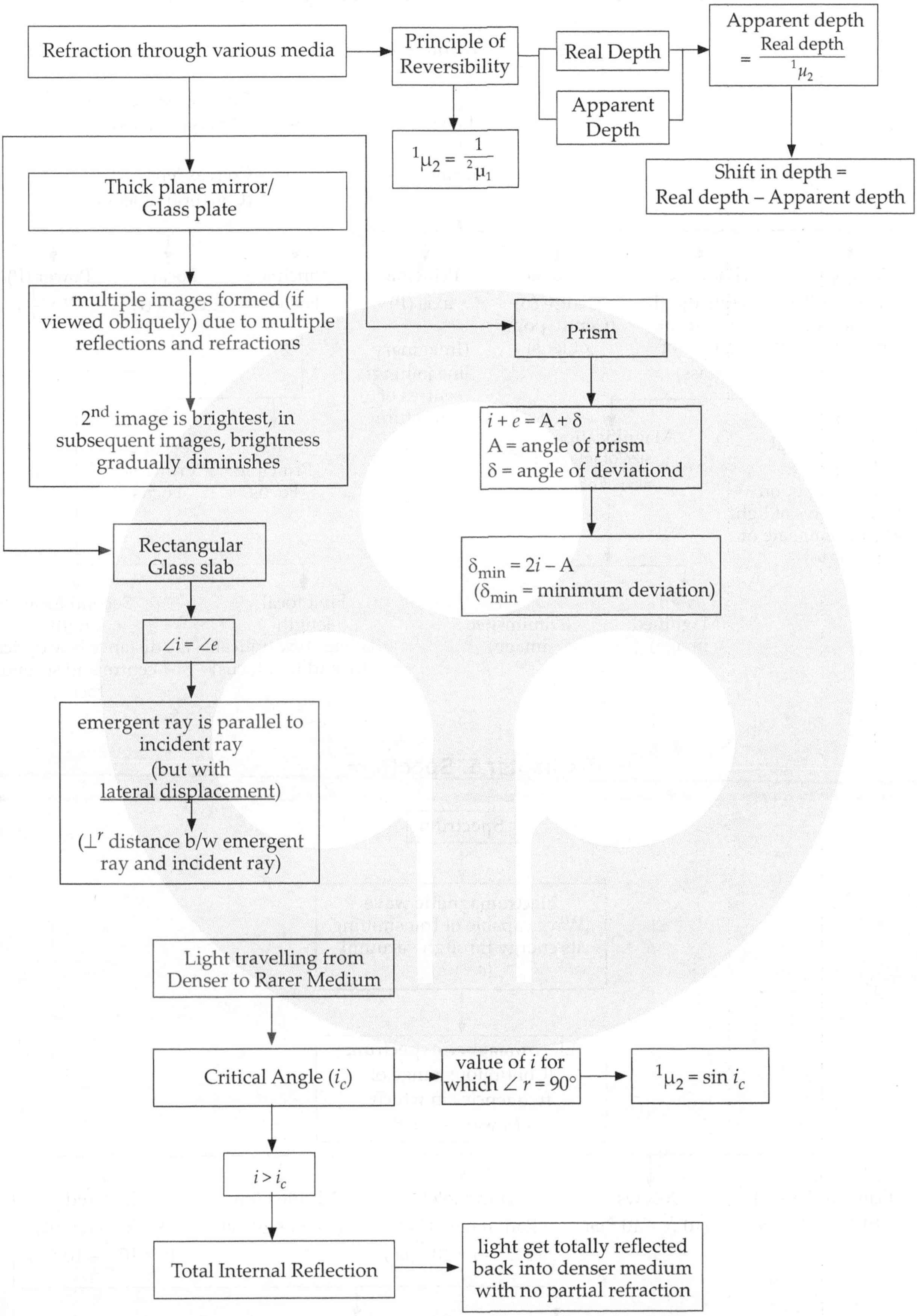
Refraction through various media
Principle of Reversibility
$^1\mu_2 = \frac{1}{^2\mu_1}$
Real Depth
Apparent Depth
Apparent depth $= \frac{\text{Real depth}}{^1\mu_2}$
Shift in depth = Real depth – Apparent depth
Thick plane mirror/ Glass plate
multiple images formed (if viewed obliquely) due to multiple reflections and refractions
2nd image is brightest, in subsequent images, brightness gradually diminishes
Prism
$i + e = A + \delta$
A = angle of prism
δ = angle of deviationd
$\delta_{min} = 2i - A$
(δ_{min} = minimum deviation)
Rectangular Glass slab
$\angle i = \angle e$
emergent ray is parallel to incident ray (but with lateral displacement)
($\perp^r$ distance b/w emergent ray and incident ray)
Light travelling from Denser to Rarer Medium
Critical Angle (i_c)
value of i for which $\angle r = 90°$
$^1\mu_2 = \sin i_c$
$i > i_c$
Total Internal Reflection
light get totally reflected back into denser medium with no partial refraction

Chapter 4. Refraction through Lenses and Optical Instruments

Recraction Through Lenses

- Lens
  - Concave lense (Diverging lens)
  - Convex lens (Converging lens)
- Terms
  - Centre of curvature (C) (Centre of sphere of glass)
  - Radius of Curvature (R) (radius of sphere of glass)
    - Real and virtual images (depending upon whether rays of light actually emanate or appear to)
    - Magnification ($m = \frac{\text{size of image}}{\text{size of object}}$)
      - $m > 1$ (Magnified image)
      - $m < 1$ (Diminished image)
  - Optical Centre (o) (Centre point of lens)
  - Principle axis (P) (Imaginary line joining centres of curvature)
  - Principle Focus
    - First Principle Focus
    - Second Principle Focus
  - Focal Length (f)
    - First focal length (distance b/w optical centre and first focus)
    - Second focal length (distance b/w optical centre and second focus)
  - Power (P) ($P = \frac{1}{f}$)

Chapter 5. Spectrum

Spectrum

- **Electromagnetic wave** {Wave capable of transmitting its energy through vacuum} ($c = \upsilon\lambda$)
  - **Electromagnetic spectrum** {Continuous range of frequencies in which EM waves exist}
    - Gamma Rays (γ) $(10^{-16} - 10^{-12})m$
    - X-Rays $(10^{-12} - 10^{-8})m$
    - Ultraviolet Radiations (UV) $(10^{-8} - 4 \times 10^{-7})m$
    - Visible Light $(4 - 8) \times 10^{-7}m$
    - Infrared Radiations (IR) $(8 \times 10^{-7} - 10^{-3})m$
    - Microwaves $(10^{-3} - 3)m$
    - Radiowaves $(3 - 10^{5})m$

Visible Light
(VIBGYOR)

- Scattering of Light (re-radiation of light in different directions)
  - Rayleight Scattering (Scattering of light by by molecules of air)
    $I \propto \frac{1}{\lambda^4}$
- Violet → Indigo → Blue → Green → Yellow → Orange → Red
  Increasing order of λ
- Dispersion (Splitting of white light into light into VIBGYOR)
  - Spectrum (Band of different colours) obtained from dispersion) eg. rainbow

Chapter 6. Eches and Vibrations of Sound

Sound Waves

- Longitudinal waves in solid, liquid and gas, can propagate as transverse waves only in solid and on the surface of liquid
- A form of mechanical waves that produces sensation of hearing
- Propagation
  - Compression (hight press, low vol.)
  - Parefaction (low press, high vol.)
- Terms related to Wave Motion
  - Amplitude (A)
  - Time Period (T)
  - Frequency (v)
  - Wavelength (l)
  - Wave Velocity (V)
- Formulae
  - $\nu = \frac{1}{T}$ ↔ $v = \sqrt{\frac{E}{\rho}}$
    E = Elasticity
    ρ = Density
    $v_{solids} > v_{liquids} > v_{gases}$
  - $v = \nu\lambda$
- **Reflection of Sound Waves**
  - ECHO
    - Sound heard after reflection from distant object
    - $v = \frac{2d}{t}$
      v = Speed of sound in air
      d = Distance b/w source and reflecting surface
      t = Time after which echo is heard
    - SONAR
    - RADAR

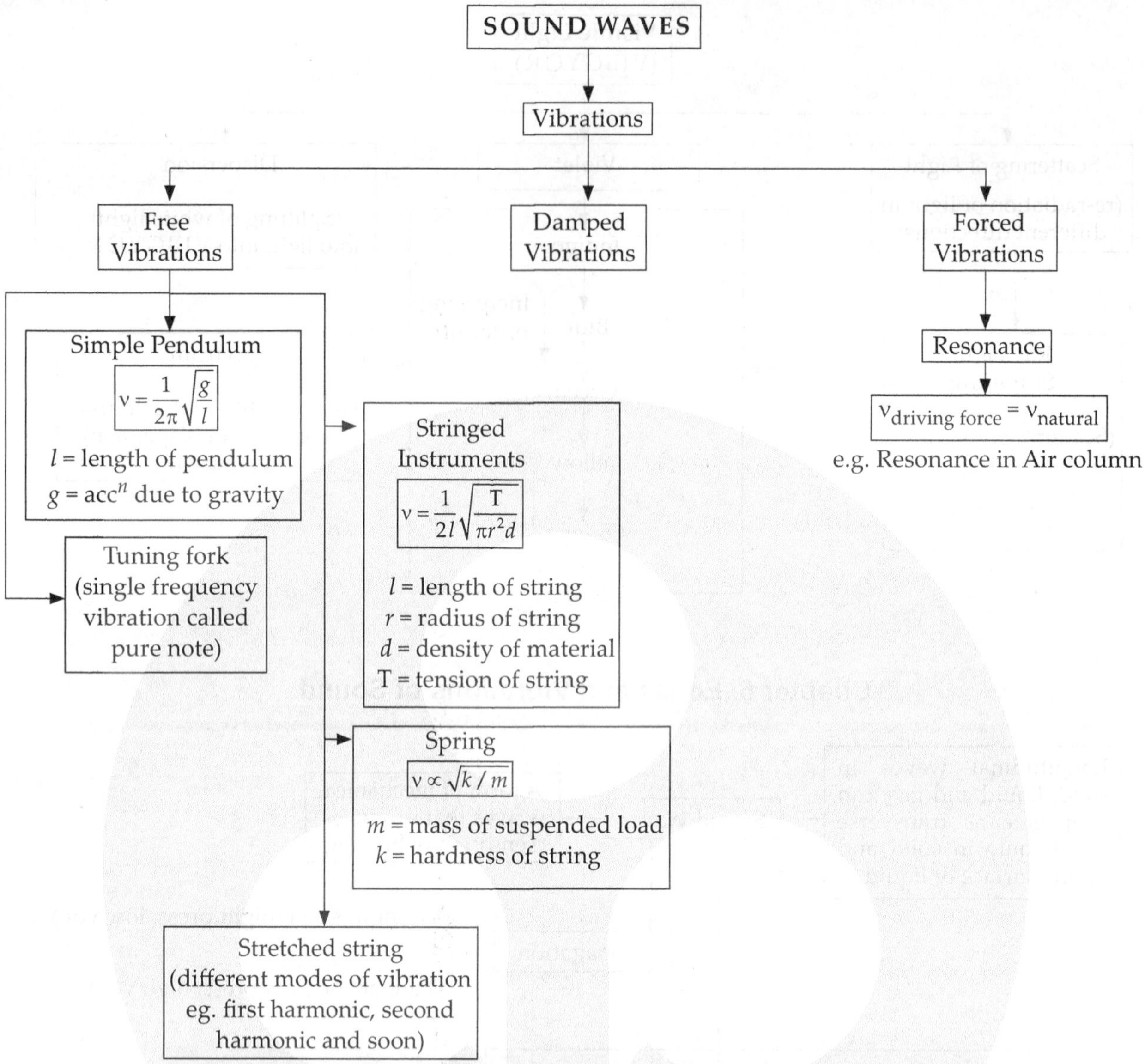

Chapter 7. Electricity

Concepts	Simbol	Formula	SI Unit
Current: Rate of flow of charge	I	$I = \frac{Q}{t} = \frac{ne}{t}$	Ampere (A)
Potential: electrical state of a conductor which determines the direction of flow of charge when 2 conductors are connected	V_A (potential at point A)		Volt (V)
Potential difference (p.d.): Work done in moving a unit +ve charge	V	$V = \frac{W}{Q}$ $V = V_A - V_B$	Volt (V)
Ohm's law: $V \propto I$	V = IR R = resistance		

Resistance: Obstruction offered to the flow of current	R	$R = \frac{V}{I}$ $R = \rho\frac{l}{A}$ r = resistivity l = length of wire A = area of cross-section	Ohm (Ω)
Conductance: A measure of how quickly current can flow within conductor	G	$G = \frac{1}{R}$ $G = \frac{\sigma A}{L}$ Conductivity $\sigma = \frac{1}{\rho}$	Ohm^{-1} $(\Omega)^{-1}$
Resistors in series	R_S	$R_S = R_1 + R_2 + R_3 + ...$	
Resistors in Parallel	R_P	$\frac{1}{R_P} = \frac{1}{R_1} + \frac{1}{R_2} + \frac{1}{R_3} + ...$	Volt (V)
Electromotive force of a cell (emf): p.d. b/w terminals of cell when it is in open circuit	ε	$\varepsilon = \frac{W}{q_0} = V + v = 1\,(R + r)$	Volt (V)
Terminal voltage of a cell: p.d. b/w electrodes of cell when it is in closed circuit	V	$V = \varepsilon - V$ $V + Ir = \frac{\varepsilon r}{R + r}$	Volt (V)
Internal Resistance of cell: resistance offered by electrolyte inside the cell	r	$r = \left(\frac{\varepsilon}{V} - 1\right) R$	Ohm (Ω)

Chapter 8. Electric Power and Household Circuits

Electrical Energy

Total work done by current through a conductor in a given time

$W = IQ = VIt = I^2Rt$

- Electric Power
  - rate of energy consumption
  - $P = \frac{W}{t} = VI = I^2R$
- Power Rating of Appliances
  - Resistance of Filament (R) $= \frac{V^2}{P}$
  - Current through Filament (I) $= \frac{P}{V}$
- Household Consumption of Electrical Energy
  - Joule's Law of Heating
  - $H = I^2Rt$
  - heat produced in a wire

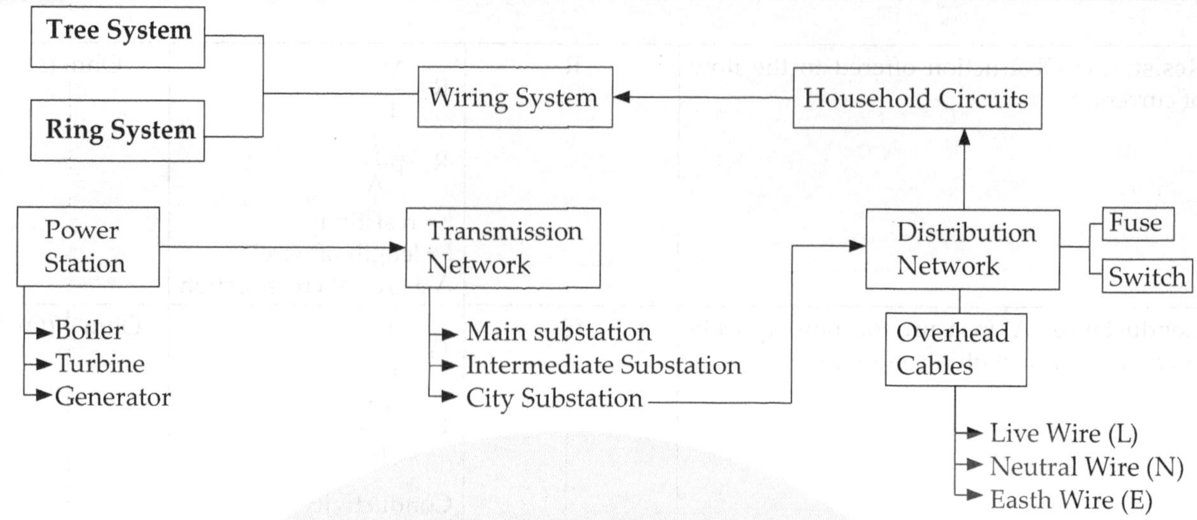

Chapter 9. Magnetic Effect of Current

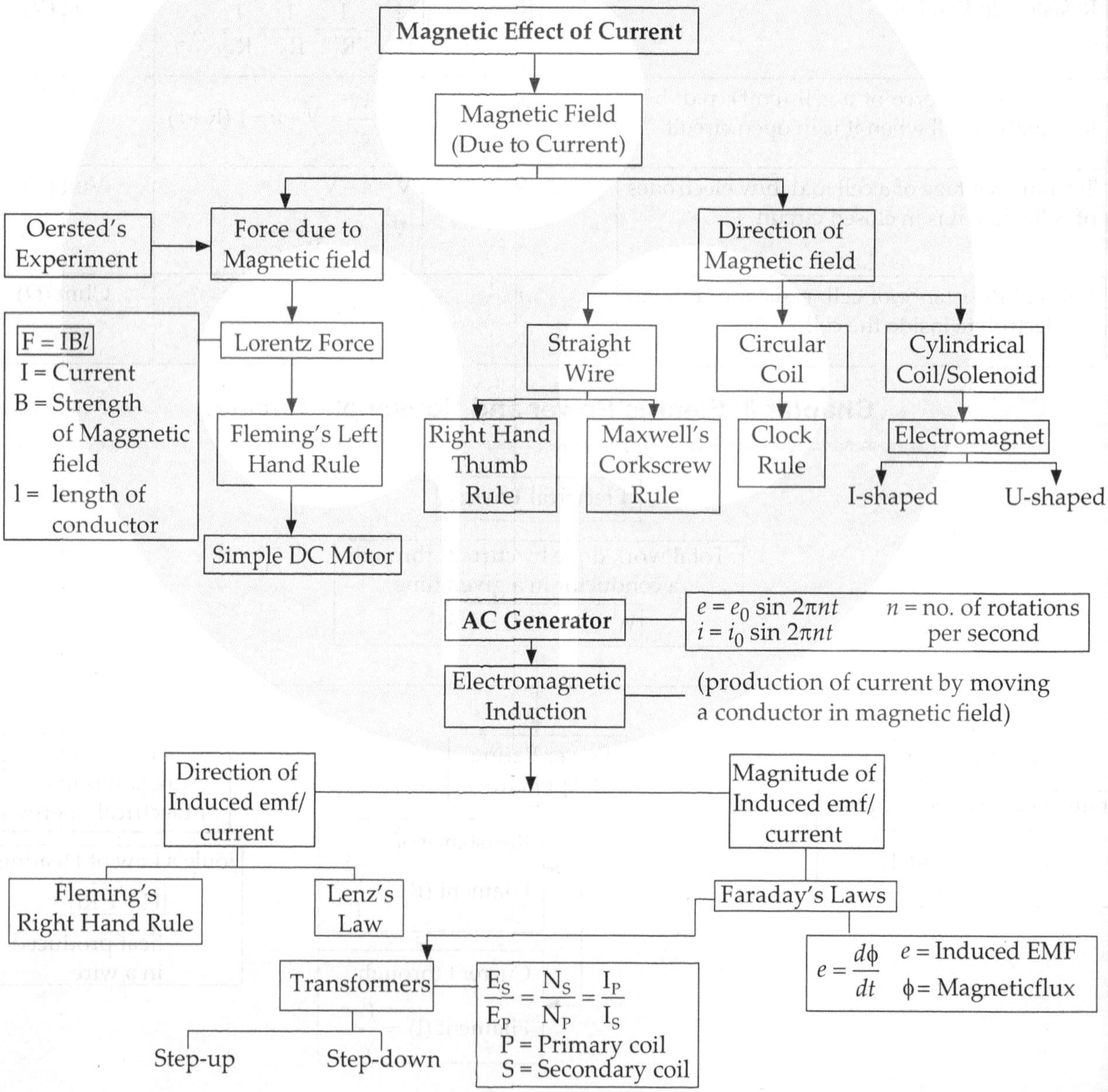

Chapter 10. Specific heat Capacity and Catent Heat

Concept	Definition	Symbol	Formulea	S.I./Unit
Heat	Total internal energy of a body	Q	$Q = cm\Delta t$ M= mass Δt = rise (or change) in temperature c = Specific heat capacity	Joule (J) 1 calorie = 4.2J (cal)
Heat (or thermal capacity)	heat required to raise temperature of a given mass by 1°C/1K	C′	$C' = \frac{Q}{\Delta t}$	$JK^{-1}/J°C^{-1}$
Specific Heat Capacity	heat required to raise temperature of unit mass by 1°C/1K	c	$c = \frac{Q}{m\Delta t}$ $c = \frac{C'}{m}$	$JKg^{-1}K^{-1}/JKg^{-1}°C^{-1}$
Calorimeter	Instrument used for measuring heat gained/lost			
Principle of calorimetry (or Principle of Mixture)	based on law of conservation of energy *i.e.,* Heat lost by hot body = Heat gained by cold body		$m_1c_1(t_1 - t) = m_2c_2(t - t_2)$ t = final temperture	
Latent Heat (Melting, freezing, vaporization, condensation, sublimation, solidification)	heat absorbed or released during change of state without change in temperature	Q	Q = Lm L = Specific latent heat	Joule (J)
Specific Latent Heat	latent heat required by unit mass to undergo phase change	L	$L = \frac{Q}{m}$	Jkg^{-1}

Chapter 11. Radioactivity and Nuclear Energy

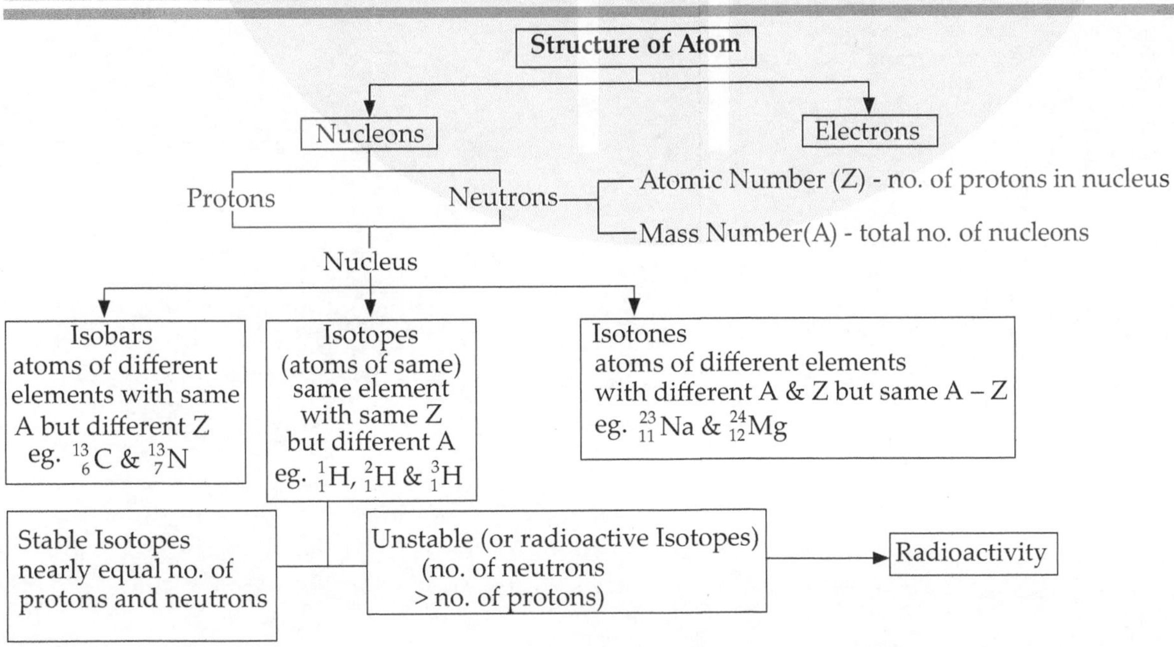

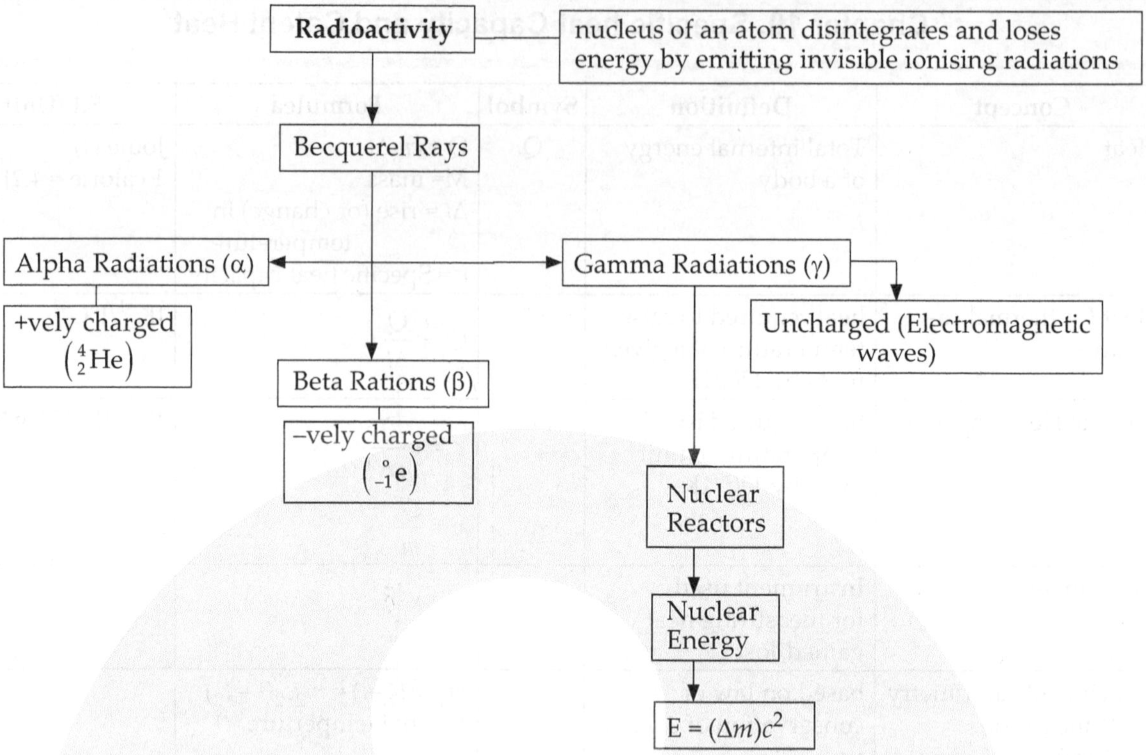
Radioactivity
nucleus of an atom disintegrates and loses energy by emitting invisible ionising radiations
Becquerel Rays
Alpha Radiations (α)
+vely charged ($^4_2$He)
Gamma Radiations (γ)
Uncharged (Electromagnetic waves)
Beta Rations (β)
–vely charged ($^0_{-1}$e)
Nuclear Reactors
Nuclear Energy
E = (Δm)c^2

Short Answers

Set 2

Chapter 1. Force, Work, Power and Energy

1. State the condition when on applying a force, the body has:
(i) translational motion
(ii) rotational motion

Ans. (i) When the body is free to move.
(ii) When the body is pivoted at a point.

2. A body is acted upon by two unequal forces in opposite directions, but not in same line. What type of motion will the body have?

Ans. The body will have rotational as well as translational motion.

3. Define translational motion.

Ans. When a force acts on a rigid body which is free to move, the body starts moving in a straight line in the direction of force. This is called translational motion.

4. (i) Define moment of force.*
(ii) Write the relationship between the SI and CGS unit of moment of force.

Ans. (i) The turning effect on the body about an axis is due to the moment of force applied on the body and is equal to the product of the magnitude of the force and the perpendicular distance of the line of action of the force from the axis of rotation.
(ii) 1 Nm = 10^7 dyne cm.

5. State two factors on which moment of force about a point depends.

Ans. Moment of force depends on two factors:
(i) Magnitude of force applied.
(ii) Distance of line of action of the force from the axis of rotation.

6. State the units of moment of force.

Ans. In CGS system, unit of moment of force is dyne cm and in MKS system, unit of moment of force is newton-metre. If force is measured in gravitational units then the unit of moment of force in CGS and MKS system are gf-cm and kgf-m respectively.

7. State when moment of force is positive and when it is negative.

Ans. If the couple has a tendency to rotate the body in anticlockwise direction then its moment is taken positive and if the tendency of rotation is clockwise then the moment is negative.

8. What is meant by the term 'moment of force'? If the moment of force is assigned a negative sign then will the turning tendency of the force be clockwise or anticlockwise?

Ans. It is equal to the product of the magnitude of the force and the perpendicular distance of the line of action of force from the axis of rotation. If moment of force is assigned a negative value, it the turning tendency of the force is clockwise.

9. (i) Define couple.*
(ii) State the S.I. unit of moment of couple.

Ans. (i) Two equal and opposite parallel forces, not acting along the same line forms a couple. A couple is always needed to produce rotation.
(ii) S.I. unit of moment of couple is Newton × metre (Nm).

10. Name two kinds of equilibrium.

Ans. (i) Static equilibrium
(ii) Dynamic equilibrium

11. State two conditions for a body to be in equilibrium.

Ans. (i) The resultant of all the forces acting on the body should be equal to zero.
(ii) The algebraic sum of moments of all the forces acting on the body about the point of rotation should be zero.

12. What do you mean by equilibrium of a body?

Ans. When a number of forces acting on a body produce no change in its state of rest or motion, the body is said to be in equilibrium.

13. Give an example of static equilibrium.

Ans. A book lying on a table.

14. Give an example of dynamic equilibrium.

Ans. A body moving in a circular path with a uniform speed.

* Frequently asked previous years Board Exam Questions.

15. State the principle of moments.

Ans. According to the principle of moments, for a body to be in rotational equilibrium, the algebraic sum of all clockwise moments is equal to the algebraic sum of all anticlockwise moments about the same fulcrum.

16. Give two examples of couple action in our daily life.

Ans. (i) Driving the pedal of a bicycle.

(ii) Winding a clock with the key.

17. Can centre of gravity be situated outside the material of the body? Give an example.

Ans. Yes, it can be situated outside the material of the body.

Example: Centre of gravity of a ring.

18. (i) Where is the centre of gravity of a uniform ring situated?

(ii) The position of the centre of gravity of a body remains unchanged even when the body is deformed. State whether the statement is true or false.

Ans. (i) The centre of gravity of a uniform ring is at its geometric centre.

(ii) False.

19. Where does the position of centre of gravity lie for:

(i) a circular lamina

(ii) a triangular lamina

Ans. (i) **Circular lamina:** Geometric centre of the lamina.

(ii) **Triangular lamina:** Point of intersection of medians.

20. Where is the centre of gravity of a uniform ball lies?

Ans. At its geometrical centre.

21. Explain why standing passengers are not allowed on the upper deck of a double decker bus.

Ans. The centre of gravity is raised and on an inclined road the bus may topple.

22. Name the force required for uniform circular motion. State its direction.

Ans. Centripetal force is required for uniform circular motion. It is always directed towards the centre of circle at each point of its path.

23. Which of the following remains constant in uniform circular motion, speed or velocity or both?

Ans. In uniform circular motion, speed remains constant.

24. Is it possible to have an accelerated motion with a constant speed? Name such type of motion.

Ans. Yes, it is possible.

Example: Uniform circular motion.

25. Define uniform circular motion.

Ans. The motion of a body around a circular path with uniform speed but variable velocity such that it is acted upon by uniform acceleration is called uniform circular motion.

26. Define centrifugal force.

Ans. A force acting on a body away from the centre of circular path is called the centrifugal force.

27. What is the work done by the gravitational force on the moon that revolves around the earth?

Ans. The moon is revolving around the earth. In this process, the work done by the gravitational force is zero because the displacement of the moon is at right angle to the gravitational force.

28. Define centripetal force.

Ans. The force acting on a body moving in a circular path which is directed towards the centre is called centripetal force.

29. Where is the centre of gravity of a hollow cone situated?

Ans. At a height of $h/3$ from the base on its axis.

30. Why is centrifugal force a virtual or a fictitious force?

Ans. The centrifugal force is not a real force, but it is fictitious (virtual) force because it really does not exist, but is considered to describe or understand a certain motion.

31. With reference to their direction of action, how does a centripetal force differ from a centrifugal force?

Ans. The direction of centripetal force is towards the centre of the circle whereas centrifugal force is opposite to centripetal force.

32. Can the C.G. be situated outside the material of the body? Give an example.

Ans. The position of C.G. of a body depends on the distribution of mass in the body. It can be situated outside or inside the material of the body.

For example : The C.G. of a wire is at its mid-point but C.G. of a ring lies at the centre where there is no material.

33. Explain why one leans forward while climbing up a hill.

Ans. A person going uphill has to bend forward but while coming down has to bend backwards, this is to keep the C.G. at the centre of the body and between his feet.

34. Explain why a tight rope-walker often holds a long pole in his hands when in action.

Ans. A tight rope-walker carries a long pole in his hand to maintain the C.G. such that the vertical line passes through the rope.

35. Explain why a ship loaded with light goods is more liable to be over turned than the one loaded with heavy goods.

Ans. A ship loaded with light goods is more liable to be over turned than the one loaded with heavy goods. This is because the C.G. in the former case is high and so a slight displacement due to water current may cause the vertical line passing through the C.G. to fall outside the base, thereby overturning the ship.

36. Explain why when we carry a weight on one hand, we bend on the other side.

Ans. This is to let our C.G. fall within the base of our support.

37. A coolie, with a load of 30 kg on his head, walks on the platform. If he walks a distance of 60 m, what is the work done by him?

Ans. The work done by the coolie is zero because there is no displacement of the load in the direction of force, which he is applying vertically upwards.

38. How much work is done by a body when it moves in a circular path? Give reason.

Ans. The work done by the body is zero because the force (centripetal force) is normal to the displacement.

39. State the relation between the S.I. and C.G.S. units of work.

Ans. S.I. unit is joule (J), C.G.S. unit is erg.

$$1J = 10^7 \text{ erg.}$$

40. What do you understand by work ? Is work a scalar or a vector quantity?

Ans. Work is said to be done when a force displaces a body through some distance, in its own direction,

i.e., Work = Force × Displacement

$$W = F \times S$$

Therefore, work is a scalar quantity.

41. In the following cases write yes, if the work is being done and no, if no work is being done.

(i) A man trying to push a wall (ii) A man moving on the rough ground

(iii) A coolie standing with a heavy load on his head (iv) A boy climbing up a staircase

Ans. (i) No (ii) Yes (iii) No (iv) Yes

42. State two conditions when the work done by a force acting on a body is zero.

Ans. (i) When there is no displacement by the force on the body.

(ii) When the displacement is normal to the direction of force.

43. State the condition when the work done by force is (i) positive and (ii) negative.

Ans. (i) Work done by a force is positive when the displacement is in the direction of the force.

(ii) Work done by a force is negative when the displacement is in the direction opposite to the force.

44. A force F when acts on a body, displaces it by a distance d in a direction at an angle θ with the direction of force. Write down the expression for the work done by the force.

Ans. Work done by the force, $W = Fd \cos \theta$.

45. When does a force do work ? What is the work done by the moon when it revolves around the earth?

Ans. Work is said to be done if a body undergoes displacement due to applied force.

Work done = Force × Displacement

Work done is zero by the moon, as there is no displacement since it is moving in a circular path.

46. State the amount of work done by an object when it moves in a circular path for one complete rotation. Give a reason to justify your answer.

Ans. Amount of work done is equal to zero.

Work is said to be done only when there is displacement produced. In case of a body moving in a circular path, the body comes to its original place, therefore, there is no displacement and hence work done is zero.

47. State two conditions for work to be done.

Ans. The two conditions are :

(i) Application of force on the body.

(ii) Displacement of the body in the direction of force.

48. The work done by a fielder when he takes a catch in a cricket match, is negative. Explain.

Ans. The work done is negative, because force required to catch the ball and displacement are in a opposite direction to each other (*i.e.,* $\theta = 180°$).

49. What is the S.I. unit of work? Define it.

Ans. The S.I. unit of work is newton-metre or joule (J).

$$1 \text{ joule} = 1 \text{ N} \times 1 \text{ m}.$$

1 J of work is said to be done by a force of 1 N if it displaces a body by 1 m in the direction of force.

50. Is it possible that no transfer of energy may take place even when a force is applied to a body?

Ans. Yes, it is possible when force is perpendicular to the direction of displacement.

51. What physical quantity does electron volt (eV) measure? How is it related to S.I. unit of that quantity?

Ans. The physical quantity measured by electron volt is energy

$$1 eV = 1{\cdot}6 \times 10^{-19} J$$

52. What do you understand by the term energy? Is energy a scalar or vector quantity? How is energy possessed by a body measured?

Ans. Energy is defined as the capacity of a body to do work. Energy is a scalar quantity. The energy possessed by a body is measured by the amount of work done by a body while changing its state of rest or of motion.

53. State the S.I. and C.G.S. units of energy. How are they related ?

Ans. The S.I. unit of energy is joule (J) and C.G.S. unit is erg.

$$1 J = 10^7 erg.$$

54. (i) Define power.

(ii) Is power a scalar or vector quantity?

Ans. (i) Power is defined as the rate of doing work per unit time or work done per second,

i.e., $$\text{Power} = \frac{\text{Work done in joule}}{\text{Time in second}}$$

or $$P = \frac{W \text{ (in joule)}}{t \text{ (in second)}}$$

(ii) Power is a scalar quantity.

55. Can two agents, doing the same work, have different power?

Ans. Yes, since power = $\frac{\text{work}}{\text{time}}$, the two agents can have different powers if they take different amounts of time to do the same work.

56. State S.I. unit of power and define it.

Ans. S.I. unit of power is J/s or Js^{-1} (or watt W). The power consumed by a body is said to be 1 watt when 1 J of work is done by it in 1 sec.

57. Justify the statement, "Power can be expressed as the product of force and velocity."

Ans. $$\text{Power} = \frac{\text{Work}}{\text{Time}} = \frac{\text{Force} \times \text{Displacement}}{\text{Time}}$$

$$= \text{Force} \times \frac{\text{Displacement}}{\text{Time}}$$

$$= \text{Force} \times \text{Velocity}$$

(Since rate of change of displacement is velocity)

58. State larger units of power.

Ans. The larger units of power are :

$$1 \text{ kilowatt} = 1000 W = 10^3 W$$

$$1 \text{ megawatt} = 10{,}00{,}000 W = 10^6 W$$

59. State the practical unit of power mainly used in mechanical engineering and write down its equivalent in watt.

Ans. The practical unit of power used in mechanical engineering is 'Horse Power' (H.P.)

$$1 \text{ H.P.} = 746 W = 0{\cdot}746 kW$$

60. What happens to the power if the time for doing a work is reduced from t second to $t/10$ second?

Ans. The power becomes 10 times of its initial value. This is because power = work/time. Thus, power $\propto \frac{1}{\text{time}}$ for a given work. Reducing the time by a factor of 10 thus increases the power 10 times.

61. Distinguish between work and power.

Ans. Work does not depend on time, while the power depends on time as power is the work done in unit time.

62. State the energy changes in the following devices while in use,

(i) A loudspeaker

(ii) A glowing electric bulb

Ans. (i) **Loudspeaker:** Electrical energy to sound energy.

(ii) **Glowing electric bulb:** Electrical energy to heat and light energy.

63. Write five forms of energy.

Ans. (i) Mechanical energy

(ii) Chemical energy

(iii) Solar energy

(iv) Heat energy

(v) Light energy.

64. What do you understand by mechanical energy? What are the different forms of mechanical energy?

Ans. The energy possessed by a body due to its state of rest or its state of motion is called mechanical energy. Mechanical energy is of two forms : (i) potential energy, and (ii) kinetic energy.

65. What do you understand by the potential energy of the body?

Ans. The potential energy of a body is the energy possessed by a body by virtue of its state of rest or position.

66. What do you understand by the kinetic energy of the body ? Give three examples of bodies possessing kinetic energy.

Ans. The kinetic energy of a body is the energy possessed by the body by virtue of its state of motion. (i) A moving bullet (ii) A flowing river (iii) Wind.

67. How is kinetic energy possessed by a moving body measured?

Ans. The kinetic energy possessed by a moving body is measured by the amount of work done by an opposing force in bringing the body to rest from its present state of motion.

68. A man climbs a slope and another walks same distance on level road. Which of the two expends more energy and why?

Ans. The man moving along the slope expends more energy. It is because, while moving along slope, he gains height and hence has extra potential energy.

69. Is it possible that a body may possess energy even when it is not in motion?

Ans. Yes, the body may possess potential energy.

70. What kind of energy is possessed in the situations when?

(i) A cocked up spring of air gun.
(ii) A stone lying on the top of roof.
(iii) A fish moving in water.
(iv) A horse running along a level road.
(v) Water stored in dams.
(vi) An electron spinning around nucleus.
(vii) A shooting arrow.
(viii) A stone in a stretched catapult.

Ans. (i) P.E. (ii) P.E.
(iii) K.E. (iv) K.E.
(v) P.E. (vi) K.E.
(vii) K.E. (viii) P.E.

71. Obtain an expression for the potential energy of a body of mass m at a height h above the ground.

Ans. Force due to gravity on the body F = mg

Work done in lifting the body against the force of gravity to a height h above the ground

$$W = F \times S = mg \times h = mgh$$

Thus, potential energy of body = mgh.

72. When an arrow is shot from its bow, it has kinetic energy. From where does it get the kinetic energy?

Ans. A stretched bow possessed potential energy on account of a change in its shape. To shoot an arrow, the bow is released. The potential energy of the bow is converted into the kinetic energy of the arrow.

73. What happens to the K.E. when (i) the mass of the body is doubled at constant velocity and (ii) the velocity of the body is doubled at constant mass?

Ans. We have, $K.E. = \frac{1}{2}mv^2$.

Hence,

(i) On doubling the mass, at constant velocity, the K.E. becomes double of its original value.

(ii) On doubling the velocity, at constant mass, the K.E. becomes $(2)^2$ *i.e.*, 4 times its original value.

74. Give three examples of bodies possessing potential energy.

Ans. (i) An object kept at a height above the ground, (ii) A compressed spring, and (iii) A stretched catapult.

75. What is dissipation of energy?

Ans. In transforming the mechanical energy, a part of it gets converted into another form of energy which cannot be utilised for useful work, this loss of energy is called the dissipation of energy.

76. State the energy changes which takes place when electricity is obtained from solar energy.

Ans. Heat energy from sun is converted into electricity.

77. What is the main energy transformation that occurs in :

(i) Photosynthesis in green leaves,
(ii) Charging of a battery.

Ans. (i) Light energy to chemical energy..
(ii) Electrical energy to chemical energy.

78. State the law of conservation of mechanical energy.

Ans. According to the law of conservation of mechanical energy, the sum of kinetic energy and potential energy in a system always remains constant.

79. Name two kinds of potential energy.

Ans. (i) Gravitational potential energy
(ii) Elastic potential energy

80. Name three kinds of kinetic energy.

Ans. (i) Translational kinetic energy
(ii) Rotational kinetic energy
(iii) Vibrational kinetic energy

81. Name the form of mechanical energy, which is put to use.

Ans. Kinetic energy.

82. Name the form of mechanical energy which is the stored form.

Ans. Potential energy.

83. A body of mass 50 kg and another body of mass 100 kg have equal momentum. Which body will have more kinetic energy?

Ans. The body of mass 50 kg will have more kinetic energy. Since, $K = p^2/2m$ where 'K' is kinetic energy, 'p' the momentum and 'm' the mass, thus for equal 'p', when 'm' decreases, K increases.

84. A body is thrown vertically upwards. Its velocity keeps on decreasing. What happens to its kinetic energy as its velocity becomes zero?

Ans. The kinetic energy changes to potential energy.

85. A moving car is suddenly stopped by applying the brakes. What happens to the kinetic of the car?

Ans. On stopping the car, the kinetic energy of the car changes into heat and sound energies.

86. Name the most commonly used unit of heat energy and state how it is related to the S.I. unit.

Ans. Most commonly used unit of heat energy is calorie.

1 calorie = 4·18 joule

87. A body falls freely under gravity from rest. Name the kind of energy it will possess :

(i) at a point from where it falls.

(ii) while falling.

Ans. (i) Potential energy.

(ii) Potential energy and kinetic energy.

88. State the work-energy theorem.

Ans. According to the work energy theorem, the work done by a force on a moving body is equal to the increase in its kinetic energy.

Work done = Increase in kinetic energy.

89. State the principle of conservation of energy.

Ans. Principle of conservation of energy : It states that energy can neither be created nor be destroyed but can be transferred from one form to another form.

90. State the energy changes which takes place when :

(i) A body is released from rest from the top of a building.

(ii) A body is thrown vertically upwards from the ground.

(iii) In a heat engine.

(iv) A bulb glows, when torch light is switched on.

(v) A car moves up a hill.

(vi) A toy car spring is wound and car is made to run on level floor.

(vii) Water stored in dams rotates turbine connected to a dynamo.

(viii) An air gun is loaded and then fired.

(ix) A magnesium ribbon burns in air.

(x) A stone projected vertically upwards, returns back to the thrower.

(xi) Water freezes in the freezing chamber of a fridge.

(xii) Photographic film is exposed to sunlight.

(xiii) Food is digested by animals.

(xiv) During electrolysis.

(xv) Burning of coal.

(xvi) Petrol engine of a running motor car.

(xvii) A battery lighting up a bulb.

(xviii) Electricity obtained from nuclear energy.

(xix) Electricity obtained from solar energy.

(xx) Electricity obtained from wind energy.

(xxi) Electricity obtained from hydro energy.

Ans. (i) The potential energy changes to kinetic energy.

(ii) The kinetic energy changes to potential energy.

(iii) The heat energy changes to mechanical energy.

(iv) The chemical energy of cell changes to electrical energy. The electrical energy first changes to heat and finally to light energy.

(v) The K.E. of moving car changes to potential energy.

(vi) The work done in winding the spring gets stored in it in the form of elastic potential energy. This potential energy is then released in the form of kinetic energy.

(vii) P.E. of stored water changes to K.E. of flowing water. The K.E. of flowing water rotates turbine. The turbine in turn rotates the coil of generator and changes into electrical energy.

(viii) During loading, the mechanical energy changes to potential energy of spring. On firing, the P.E. of spring changes to K.E. of bullet.

(ix) During burning of magnesium, chemical energy changes to heat and light energy.

(x) The kinetic energy changes to potential energy when stone rises up. The potential energy then changes back to kinetic energy, when stone comes down.

(xi) The kinetic energy of water molecules is released in the form of heat energy.

(xii) The light energy changes to chemical energy.

(xiii) The chemical energy changes into heat energy.

(xiv) The electrical energy changes to chemical energy.

(xv) The chemical energy changes into heat energy.

(xvi) The chemical energy changes into mechanical energy.

(xvii) The electrical energy changes into heat energy and light energy.

(xviii) The heat energy changes into electricity.

(xix) The heat energy changes into electricity.

(xx) The K.E. changes into electricity.

(xxi) The kinetic energy changes into electricity.

91. Give one example when :

(i) Heat energy changes to kinetic energy.

(ii) Kinetic energy changes to heat energy.

(iii) Sound energy changes to electrical energy.

(iv) Electrical energy changes to sound energy.

(v) Light energy changes to chemical energy.

(vi) Chemical energy changes to light energy.

(vii) Electrical energy changes to mechanical energy .

(viii) Mechanical energy changes to electrical energy.

(ix) Potential energy changes to electrical energy.

(x) Electrical energy changes to potential energy.

(xi) Mechanical energy to heat energy.

(xii) Kinetic energy to potential energy.

(xiii) Electrical energy to heat energy.

(xiv) Chemical energy to electrical energy.

Ans. (i) In an automobile engine, heat energy changes to kinetic energy.

(ii) When two stones are rubbed against each other, kinetic energy changes to heat energy.

(iii) When we speak in front of microphone, sound energy changes to electrical energy.

(iv) When a loudspeaker works, the electrical energy changes to sound energy.

(v) During photosynthesis, light energy changes to chemical energy.

(vi) During burning of magnesium, chemical energy changes to light energy.

(vii) In an electric motor, electrical energy changes to mechanical energy.

(viii) In an electric generator, the mechanical energy changes to electrical energy.

(ix) The P.E. of stored water in dams, changes to electrical energy in generators.

(x) When electrolysis is carried out, the electrical energy changes to potential energy.

(xi) Rubbing of palms.

(xii) Throwing up a pebble from the ground.

(xiii) Electric heater while in use.

(xiv) Drawing current from an electric cell.

92. Define a kilowatt hour. How is it related to joule?

Ans. One kilowatt hour (kWh) is the energy spent or work done by a source of power 1 kW in 1 hour.

$$1 \text{ kWh} = 3.6 \times 10^6 \text{ J}$$

93. A satellite revolves around a planet in a circular orbit. What is the work done by the satellite at any instant ? Give a reason.

Ans. The work done by the satellite at any instant is zero because the force required (centripetal force) to go around the planet is perpendicular to the displacement at any instant of its motion.

Chapter 2. Simple Machines

1. What is a machine?

Ans. A machine is a device by which we can overcome a resistance by applying a comparatively small force at a convenient point and in a desired direction or to obtain a gain in speed.

2. State the principle of machine.

Ans. According to the principle of machine, work done by a machine or output work can never be greater than the work done on the machine or input work. In an ideal machine, "the output work of a machine is equal to the input work".

3. Give an example where a machine is used to obtain gain in speed.

Ans. In automobiles, gear arrangement system (machine) is used to obtain gain in speed.

4. What is the efficiency of an ideal machine?

Ans. The efficiency of an ideal machine is 1 (or 100%).

5. Write an expression to show the relationship between mechanical advantage, velocity ratio and efficiency for a simple machine.

Ans. Efficiency $= \dfrac{\text{Mechanical advantage}}{\text{Velocity ratio}} \times 100$

6. Name a machine which can be used to multiply force and change the direction of force applied.

Ans. Nut cracker and Hand pump.

7. Define mechanical advantage (M.A.) and velocity ratio (V.R.) of machine.

Ans. Mechanical advantage : The ratio between load and effort is called mechanical advantage (M.A.) :

$$\text{M.A.} = \frac{\text{Load (L)}}{\text{Effort (E)}}$$

Velocity ratio : The ratio between displacement of effort and displacement of load is called velocity ratio (V.R.) :

$$\text{V.R.} = \frac{d_E}{d_L}$$

Where d_E and d_L are the displacement of effort and load respectively in time t.

8. Comment on the statement "The mechanical advantage of a machine is greater than 1". Give an example of a lever whose mechanical advantage is :

(i) Greater than 1, (ii) Equal to 1.

Ans. The statement means that the effort required for the machine to overcome a load is less than the load.

(i) The crowbar,

(ii) The beam of common balance.

9. There is no gain in mechanical advantage in the case of a single fixed pulley. Explain, why the pulley is then used?

Ans. It is used only to change the direction of effort to a more convenient direction and to use one's own weight for the effort.

10. In the case of a block and tackle arrangement, the mechanical advantage increases with the number of pulleys. Explain.

Ans. With the increase in number of pulleys in the arrangement of block and tackle, the number of strands of string supporting the load increases and so its mechanical advantage increases.

11. State two ways of increasing the efficiency of a block and tackle system of pulleys.

Ans. Efficiency of a block and tackle system of pulleys can be increased by :

(i) making the lower movable block as light as possible, and

(ii) minimizing the friction in the bearings of pulleys by using lubricants.

12. Write a relation expressing the mechanical advantage of a lever.

Ans. Mechanical advantage of a lever $= \dfrac{\text{Effort arm}}{\text{Load arm}}$.

13. What is the relationship between the mechanical advantage and the velocity ratio for :

(i) An ideal machine,

(ii) A practical machine.

Ans. (i) Mechanical advantage = Velocity ratio as efficiency is 1.

(ii) Mechanical advantage is less than velocity ratio as efficiency is less than one.

14. State the principle of a lever. What is its M.A.?

Ans. The principle of a lever is derived from the principle of moments. It states that, "The moment of the load about the fulcrum is equal to the moment of the effort about the effort arm", *i.e.*,

$$\text{Load} \times \text{Load arm} = \text{Effort} \times \text{Effort arm}$$

$$\text{M.A. of a lever} = \frac{\text{Load}}{\text{Effort}} = \frac{\text{Effort arm}}{\text{Load arm}}$$

15. Which type of levers have mechanical advantage always more than 1? Give reason.

Ans. The mechanical advantage of second order lever is always more than 1. This is because the effort arm is always longer than the load arm of these levers.

16. Give three examples for levers of I order.

Ans. Levers of Ist order :

(i) Railway signal

(ii) Handle of water pump

(iii) Motor car foot brake

(iv) Nail cutter

17. Give three examples for levers of II order.

Ans. Levers of II order :

(i) Lock and key

(ii) Opener of soda water bottle

(iii) Closing a door

18. Give three examples for levers of III order.

Ans. Levers of III order :

(i) Cutting bread with knife

(ii) A boy writing on a piece of paper

(iii) Forceps

(iv) A fishing rod.

Short Answers

19. Which class of lever found in the human body is being used by a boy:

(i) When he holds a load on the palm of his hand.

(ii) When he raises the weight of his body on his toes.

Ans. (i) Class III lever.

(ii) Class II lever.

20. What change can be made in a class II lever to increase its mechanical advantage?

Ans. A class II lever has mechanical advantage always greater than one. To increase the mechanical advantage of class II lever, length of effort arm must be increased and distance between fulcrum and load should be decreased.

21. State the class of levers and the relative positions of load (L), effort (E) and fulcrum (F) in each of the following cases :

(i) A bottle opener

(ii) Sugar tongs.

Ans. (i) **A bottle opener:** Class II lever, here the load L is in between the effort E and the fulcrum F.

(ii) **Sugar tongs:** Class III lever, here the effort E is in between the fulcrum F and the load L.

22. What is a single fixed pulley? Give its one use.

Ans. A single fixed pulley is a wooden disc with a grooved rim. It is used to lift (or raise) a water bucket from the well.

23. State two reasons, why the efficiency of a pulley system is not 100 percent?

Ans. The two reasons are:

(i) Some effort is used up in overcoming friction between the string and the grooves of the pulleys.

(ii) Some effort is used up in lifting up the movable block along with the load.

Since output work is less than input work, efficiency is always less than 100%.

24. Give two reasons, why the efficiency of a single movable pulley system is not 100%?

Ans. (i) Due to friction in the movable parts of the pulley.

(ii) Due to weight of the movable pulley.

25. The following belongs to which class of lever?

(i) The Physical balance

(ii) A see-saw

(iii) An oar of a boat

(iv) Human arm

(v) Pliers Tools

(vi) A claw-hammer

(vii) Rowing of a boat

(viii) A fire tongs

(ix) Sugar tongs

(x) A pair of scissors

(xi) Wheel barrow

(xii) Nut cracker

(xiii) Lemon crusher

(xiv) Knife

(xv) Fore-arm

Ans.

(i) Class I	(ii) Class I
(iii) Class II	(iv) Class III
(v) Class I	(vi) Class I
(vii) Class II	(viii) Class III
(ix) Class III	(x) Class I
(xi) Class II	(xii) Class II
(xiii) Class II	(xiv) Class III
(xv) Class I	

Chapter 3. Refraction of Light

1. A pond appears to be 2.7 m deep. If the refractive index of water is $\frac{4}{3}$, find the actual depth of the pond.

Ans. Apparent depth = 2.7 m

$$\mu_w = \frac{4}{3}$$

$$\because \quad \mu_w = \frac{\text{Actual depth}}{\text{Apparent depth}}$$

$$\Rightarrow \quad \frac{4}{3} = \frac{\text{Actual depth}}{2.7}$$

$$\Rightarrow \quad \text{Actual depth} = \frac{4}{3} \times 2.7 \text{ m} = 3.6 \text{ m}$$

2. The wave lengths for the light of red and blue colours are nearly 7.8×10^{-7} m and 4.8×10^{-7} m respectively.

(i) Which colour has the greater speed in a vacuum?

(ii) Which colour has a greater speed in glass?

Ans. (i) Both colours of light have same speed in vacuum.

(ii) In glass, speed of red light is more than that of blue light.

3. What is meant by refraction of light ?

Ans. It is a phenomenon in which a ray of light deviates from its original path, while travelling from one optical medium to another optical medium.

4. Name the phenomenon responsible in the following cases:

(i) Apparent bending of a stick in water

(ii) Twinkling of stars.

Ans. (i) Refraction of light.

(ii) Refraction of light.

5. Why do the faces of persons sitting around camp fire appear to shimmer?

Ans. The rays coming from the face of a person, sitting across the camp fire, suffer refraction on passing through hot air. Thus, we see an apparent image of the person. Since hot air is in motion, therefore, its optical density changes, which in turn shifts the image rapidly. This rapidly shifting image gives a shimmering effect.

6. Why does a fisherman aims at the tail of fish during spear fishing?

Ans. Due to refraction, when the rays travel from water to air, the apparent image of a fish is formed at a higher level and ahead of the actual position of fish. Thus, when the fisherman aims at the tail of the apparent image of fish, his spear hits the head of actual fish.

7. During sunset and sunrise, the sun is seen when it is slightly below the horizon. Give reason.

Ans. Light rays from sun while coming towards earth suffers successive refractions from a rarer to a denser medium and so it bends towards the normal at each refraction. These rays on earth appear to be coming from an apparent position of sun which is higher than its actual position. Thus, the sun is seen even when it is slightly below the horizon during sunset and sunrise.

8. Glass is transparent in nature. Why does glass powder look opaque? When water is poured over it, it again becomes transparent. Why?

Ans. In case of powdered glass, irregular reflection takes place from various particles. Thus, light is sent back into the same medium. Therefore, it looks opaque. On putting water over the powder, incident light gets refracted through water. That is why it looks transparent.

9. (i) A ray of light passes from water to air. How does the speed of light change?

(ii) Which colour of light travels fastest in any medium except air?

Ans. (i) When light passes from water to air *i.e.,* from denser to rarer, its speed increases.

(ii) Red light travels fastest.

10. State two conditions for no change in direction of light ray on refraction.

Ans. (i) When angle of incidence at the boundary of two media is zero (*i.e.,* $\angle i = 0$).

(ii) When refractive index of medium 2 is same as that of medium 1.

11. Does the depth of a tank of water appear to change or remain the same when viewed normally from above?

Ans. Depth of the tank remains the same when viewed normally from above.

12. Name one main factor on which the direction of bending of a ray of light depends.

Ans. It mainly depends upon the optical density of the two media through which a ray of light travels.

13. How does a ray of light bend when it travels from:

(i) Rarer to denser medium.

(ii) Denser to rarer medium.

(iii) When it is normal to the interface of the two media.

Ans. (i) When a ray of light travels from rarer to denser medium, it bends towards the normal.

(ii) When a ray of light travels from denser to rarer medium, it bends away from the normal.

(iii) The ray goes straight into second medium without any bending.

14. What is the principle of reversibility of light?

Ans. The principle of reversibility of light states that light follows the same path if the direction of the travel of light is reversed.

15. A fish swimming in a pond seems nearer than it really is. Explain.

Ans. This is because a light ray coming from the fish inside water when suffers refraction at the water-air interface, it bends away from the normal at the point of incidence. The refracted ray appears to be coming from a point at a depth less than the actual depth of the fish. That is why, the fish seems to be nearer than it really is.

16. (i) State the Snell's laws of refraction of light.

(ii) State the unit of refractive index.

Ans. (i) **Laws of refraction of light:**

(a) Incident ray, refracted ray and normal at the point of incidence all lie in the same plane.

(b) Ratio of sine of angle of incidence to the sine of angle of refraction is a constant quantity known as refractive

Short Answers

index of second medium with respect to the first medium.

(ii) Refractive index has no unit as it is a ratio of two similar quantities.

$$\mu = \frac{\sin i}{\sin r}$$

17. A glass slab is placed over a page on which the word VIBGYOR is printed with each letter in its corresponding colour.

(i) Will the image of all the letters be in the same place?

(ii) If not, state which letter will be raised to the maximum. Give a reason for your answer.

Ans. The image of all the letters of the word VIBGYOR in their corresponding colours will not be at the same place. In glass, red colour has the higher speed than violet colour, so red will be raised to the maximum while violet will be at the bottom and the other colours will be raised in the order I-B-G-Y-O.

18. What is a prism?

Ans. A prism is a piece of glass or any transparent medium formed by two triangular and three rectangular surfaces.

19. A lighted candle is placed infront of a thick plane mirror and is viewed obliquely.

(i) How many images will be formed?

(ii) Which image is the brightest and why?

Ans. (i) Theoretically, infinite number of images are formed but practically a few (or multiple) images can be seen.

(ii) The second image formed will be the brightest due to a strong first reflection of the incident light at the silvered surface of the mirror.

20. State two advantages of using right angle prism as a reflector, rather than a plane mirror.

Ans. (i) There is no absorption of light in a prism and hence image is brighter.

(ii) Image is sharper in case of prism.

21. State three actions that a total reflecting prism can produce.

Ans. (i) A total reflecting prism can deviate the path of a ray of light by 90°.

(ii) A total reflecting prism can turn a ray of light through 180°.

(iii) A total reflecting prism can erect an inverted image without deviation.

22. How does the angle of deviation produced by a prism depend on the colour of light used? Which colour of white light is deviated (i) most, (ii) least, by a prism ?

Ans. The angle of deviation produced by a prism increases with the decrease in wavelength of the light used.

(i) Violet light is deviated most by a prism.

(ii) Red light is deviated least by a prism.

23. What do you understand by the deviation produced by a prism? Define the angle of deviation.

Ans. When a ray of light passes through a prism, it bends towards the base of the prism. Thus, the path of the emergent ray differs from the path of incident ray. This is called the deviation produced by a prism.

The angle of deviation is defined as the angle between the incident ray and the emergent ray.

24. Mention one difference between reflection of light from a plane mirror and total internal reflection of light from a prism.

Ans. In reflection of light from a plane mirror, light reflects partially. Some part reflects and some part transmits, whereas in total internal reflection through a prism, 100% reflection of light occurs.

25. Can light be 'piped' like sound in a doctor's stethoscope?

Ans. Yes, light can be piped from one end to the other with negligible loss by allowing it to enter one end of a rod of transparent material. The light will undergo total internal reflection at the wall of the rod and will proceed along its axis. This is actually being achieved by using a bundle of glass fibres.

26. Images formed by totally reflecting prisms are brighter than the image formed by ordinary reflected light, why?

Ans. This is because, in total internal reflection 100% of incident light gets reflected. On the other hand, in ordinary reflection, some part of incident light is lost. So, images formed in the first case are brighter than the images formed in the second case.

27. State the factors on which the angle of deviation depends.

Ans. The angle of deviation depends upon:

(i) The refracting angle of prism

(ii) The material of prism

(iii) The colour of light

(iv) The angle of incidence

28. Name the material for which refractive index is found to be maximum.

Ans. The refractive index is maximum for diamond.

29. State the relation between the refractive index μ and the velocity of light (v_m) in that medium.

Ans. Refractive index:

$$\mu = \frac{\text{Velocity of light in vacuum (or air)}}{\text{Velocity of light in the medium}}$$

$$\mu = \frac{c}{v_m}$$

$$v_m = \frac{c}{\mu}$$

30. What is the refractive index of :

(i) Vacuum (ii) Water

Ans. (i) 1 (ii) 1·33

31. For which colour of white light the refractive index of a transparent medium is: (i) the least, (ii) the most? Give reason to your answer.

Ans. (i) The refractive index of a transparent medium is least for red light.

(ii) The refractive index of a transparent medium is most for violet light.

The reason is that the speed of violet light is least in a transparent medium while the speed of red light is most in transparent medium.

32. What should be the ratio of the speed of light through the liquid to the speed through glass so that there is no refraction of light at the boundaries of the glass block when the system is illuminated by light of one colour?

Ans. Since there is no refraction of light at the boundaries of the glass block, it is possible only when the speed of light through the liquid as well as through the block is the same. Hence, the ratio between the speed of light through the liquid to the speed of light through glass must be 1 : 1.

33. Comment on the statement "The refractive index of glass is 3/2."

Ans. It means that the speed of light in glass is 2/3rd the speed of light in air (or vacuum).

34. Define the term refractive index of a medium in terms of velocity of light.

Ans. **Refractive Index:** It is defined as the ratio of velocity of light in medium 1 to the velocity of light in medium 2.

35. How is the refractive index of a medium related to its real depth and apparent depth?

Ans. $\text{Refractive index} = \frac{\text{Real Depth}}{\text{Apparent Depth}}$

36. Express the refractive index μ of a medium in terms of the velocity of light.

Ans. The refractive index 'μ' of an optical medium is defined as the ratio between the velocity of light in vacuum and the velocity of light in the given optical medium. Thus,

$$\mu = \frac{\text{Velocity of light in vacuum}}{\text{Velocity of light in the given optical medium}}$$

37. Express the refractive index μ of a medium in terms of the angle of incidence *i* in air and the angle of refraction *r* in a denser medium.

Ans. If light travels from air into another denser medium, the refractive index 'μ' of the medium is,

$$\mu = \frac{\text{Sine of the angle of incidence in air}}{\text{Sine of the angle of refraction in the denser medium}}$$

i.e. $\mu = \frac{\sin i}{\sin r}$

38. If a ray of light passes from medium I to medium II without any change of direction, what can be said about the refractive indices of these media ($< i \neq 0$)?

Ans. When a ray of light travels from medium I to medium II without any change of direction, it is possible under either of the following two conditions:

(i) A ray of light is incident normally on the surface separating two media.

(ii) The refractive index of medium I is equal to the refractive index of medium II.

39. Explain briefly what causes the twinkling of stars at night?

Ans. Stars are very far from us. The light ray coming from a star undergoes successive refractions on entering earth's atmosphere and bends towards the normal. Hence, an apparent image of the star is seen. Simultaneously, due to movement of air, the refractive index of layers of atmosphere keep on changing, *i.e.*, the apparent position of star keeps changing. Thus, the star appears to twinkle.

40. An object placed in one medium when seen from the other medium, appears to be vertically shifted. Name two factors on which the magnitude of shift depends.

Ans. (i) The refractive index of the medium.

(ii) The thickness of the first medium.

41. What is total reflecting prism? Name two instruments in which it is used.

Ans. A total reflecting prism is a right angled prism made of glass with other two angles each equal to 45°. It is used in (i) periscope and (ii) binoculars.

42. Why upper surface of water contained in a beaker and above eye level appears silvery?

Ans. Critical angle for water is 48·6°. The rays of light entering in water from below, suffer refraction. If these rays strike the water-air surface at an angle which is greater than 48·60°, they get totally internally reflected. These rays on emerging out of water, appear to come from the upper surface of water, which in turn appears silvery.

43. Why do diamonds sparkle?

Ans. Critical angle for diamond is quite less and is only 24°. Besides this, diamonds are cut very sharply, making large number of refracting surfaces. Hence when a ray of light enters a diamond, it suffers a series of total internal reflections on account of very small critical angle. This makes a ray of light literally trapped within the diamond for a little time and this trapped light energy makes a diamond sparkle.

44. Write down the relation between the critical angle and the refractive index of the medium.

Ans. $$\sin i_c = \frac{1}{\mu}$$

where i_c is the critical angle and μ is the refractive index of the medium.

45. How is critical angle related to the refractive index of the medium?

Ans. $n = \dfrac{1}{\sin i_c}$ where, $n \rightarrow$ refractive index,

$ic \rightarrow$ critical angle.

46. Can the absolute refractive index of a medium be less than one?

Ans. No, the absolute refractive index of a medium can not be less than one because speed of light in any medium is always less than that in vacuum.

47. Why does sun appear bigger during sunset or sunrise?

Ans. The sun appears bigger during sunrise or sunset due to atmospheric refraction which leads to some optical distortion.

48. State one factor on which critical angle for a given pair of media depends. The critical angle for glass-air interface is 45° for yellow light. Will it be equal to, less than or greater than 45° for (i) red light, (ii) blue light?

Ans. The value of critical angle for a given pair of media depends on the colour of light.

(**Hint:** Remember VIBGYOR—Violet has minimum i_c whereas Red has maximum)

(i) For red light, the critical angle will be greater than 45°.

(ii) For blue light, the critical angle will be less that 45°.

49. Name the factors affecting the critical angle for the pair of media.

Ans. Factors affecting the critical angle for a pair of media are:

(i) the colour (or wavelength) of incident light.

(ii) temperature of the medium.

50. What is meant by the statement, "The critical angle for diamond is 24°"? How is the critical angle of a material related to its refractive index?

Ans. (i) It means that when a light ray is incident inside a diamond at an angle greater than 24°, the light will be totally internally reflected.

(ii) Relation of critical angle and refractive index :

$$\mu = \frac{1}{\sin i_c}$$

Where μ = refractive index of material and c = critical angle.

51. What is meant by the term 'critical angle'?

Ans. **Critical angle:** It is the angle of incidence in the denser medium corresponding to which the angle of refraction in the rarer medium is 90°.

52. Fill in the blanks to complete the sentences.

(i) An object placed in a denser medium when viewed from a rarer medium, appears to be at a .... depth.

(ii) An object placed in a rarer medium when viewed from a denser medium, appears to be at a......... depth.

Ans. (i) lesser,

(ii) greater.

Chapter 4. Refraction through Lenses and Optical Instruments

1. What is lens?

Ans. A lens is a portion of transparent refracting medium having at least one of its two surfaces spherical. One of the surfaces may also be plane.

2. What is the difference between a double convex and a bi-convex lens?

Ans. In both cases the two surfaces are convex towards air. In case of double convex lens the two surfaces may have any values for the two radii of curvature of the two surfaces while in bi-convex lens the radii of curvature of the two surfaces, has the same value. Thus, every bi-convex lens is a double convex lens but every double convex may not be bi-convex.

3. We can burn a piece of paper by focussing the sun rays by using a particular type of lens. Name the type of lens used for the above purpose.

Ans. Convex lens.

4. A lens forms an erect, magnified and virtual image of an object. Name the lens.

Ans. Convex lens.

5. Which lens can produce a real and inverted image of an object?

Ans. Convex lens.

6. Point out difference between a convex lens and a concave lens.

Ans.

	Convex Lens	Concave Lens
(i)	It is thicker at the centre and thin at the ends.	It is thinner at the centre and thick at the ends.
(ii)	It converges rays of light.	It diverges rays of light.
(iii)	It has a real focus.	It has a virtual focus.

7. How will you decide whether a given piece of glass is a concave lens, convex lens or a plane glass plate?

Ans. Hold the given piece of glass over some printed number:

(i) If the letters appear magnified, the given piece is a convex lens.

(ii) If the letters appear diminished, the given piece is a concave lens.

(iii) If the letters appear to be of the same size, then it is a plane glass mirror.

8. Out of the two lenses—one concave and the other convex—state which is a convergent or a divergent type of a lens. Give a reason of your answer.

Ans. The convex lens is a convergent type of a lens since it converges the light rays to a point while the concave lens is a divergent type of lens since it spreads the light rays.

9. Name the type of lens that has, (i) a real focus, (ii) a virtual focus.

Ans. (i) Convex lens.

(ii) Concave lens.

10. State the condition when a lens is called an equi-convex or equi-concave.

Ans. A lens is called an equi-convex or equi-concave, when radii of curvature of the two surfaces of lens are equal.

11. An object is placed infront of a lens between its optical centre and focus. The formed image is virtual, erect and diminished. Name the lens used.

Ans. The lens used is concave lens.

12. Ranbir claims to have obtained an image twice the size of the object with a concave lens. Is he correct? Give a reason for your answer.

Ans. No, he is not correct. Because concave lens always forms virtual, erect and diminished image.

13. In sun glasses, both of its surface are curved, yet their behaviour is neither like a convex lens nor like a concave lens. State the reason.

Ans. In sun glasses, the outer surface is convex and the inner surface is concave, hence the power of one surface is positive and that of the other is negative. By making both these powers equal, the effective power of the lens becomes zero. Hence, these lenses are simply like a thin glass having no power.

14. Can a normal convex lens behave like a concave lens and vice-versa?

Ans. Yes. If a lens is kept in a medium whose refractive index is more than that of the material of the lens, the normal convex lens starts behaving like a concave lens and vice-versa.

15. What will be the behaviour of a lens, if it is placed in water instead of air?

Ans. Focal length depends upon the refractive index of the material of the lens with respect to the medium surrounding it. Since the surrounding medium changes from air to water, the refractive index decreases and so the focal length of the lens increases.

16. (i) If the lens is placed in water instead of air, how does its focal length change?

(ii) Which lens, thick or thin has greater focal length?*

Ans. (i) The focal length of the lens will increase in water (Focal length of the lens depends on the refractive index of the material of lens relative to its surrounding medium)

(ii) Thin lens will have greater focal length.

17. How are the distance of object (u), distance of image (v) and focal length (f) of a lens related?

Ans. The distance of object (u), the distance of image (v) and focal length (f) of a lens are related as

$$\frac{1}{v} - \frac{1}{u} = \frac{1}{f}$$

18. A concave mirror and convex lens are held in water. What changes, if any, do you expect in their focal length?

Ans. The focal length of a mirror does not depend upon the nature of the medium in which it is placed whereas the focal length of a lens depends upon the medium in which it is placed. Thus there will be no change in the focal length of the concave mirror whereas the focal length of the convex lens will increase.

19. Where should an object be placed in front of a convex lens in order to get:*

(i) an enlarged real image

(ii) enlarged virtual image?

Ans. (i) Object must be placed between first focal point (F_1) and the centre of curvature ($2F_1$) of the lens.

(ii) Object must be placed between the first focal point (F_1) and the lens.

20. What are conjugate foci?

Ans. Two points lying on the principal axis such that image of an object kept at such point is formed on the other point, are called conjugate foci.

21. State the position of object, position of image, nature of image when: Convex lens is used as objective lens of astronomical telescope.

Ans. Object is at infinity.
Image is formed at F_2 on the other side of lens. Image is real, inverted and almost diminished to a point.

22. State the position of object, position of image, nature of image when: Convex lens is used as objective lens of photographic camera.

Ans. Object is in between infinity and $2F_1$. Image is formed between F_2 and $2F_2$. Image is real, inverted and diminished.

23. State the position of object, position of image, nature of image when : Convex lens is used as an erecting lens in terrestrial telescope.

Ans. Object is at $2F_1$. Image is formed at $2F_2$. Image is real, inverted and same size as object.

24. State the position of object, position of image, nature of image when: Convex lens is used in cine projector.

Ans. Object is between F_1 and $2F_1$. Image is between $2F_2$ and infinity. Image is real, inverted and enlarged.

25. State the position of object, position of image, nature of image when: Convex lens is used as in search light.

Ans. Object is at F_1. Image is formed at infinity. Image is real, inverted and highly enlarged.

26. State the position of object, position of image, nature of image when: Convex lens is used in observing biological specimens.

Ans. Object is between F_1 and O. Image is formed on the same side of lens. Image is virtual, erect and magnified.

27. Write the three characteristics of the image formed by a convex lens of focal length 20 cm for the object at distance (i) 10 cm, (ii) 30 cm, (iii) 40 cm, (iv) 60 cm from the lens.

Ans. (i) For the object at distance 10 cm, the image is virtual, upright and enlarged.

(ii) For the object at distance 30 cm, the image is real, inverted and enlarged.

(iii) For the object at distance 40 cm, the image is real, inverted and same size.

(iv) For the object at distance 60 cm, the image is real, inverted and diminished.

28. A converging lens is used to obtain an image of an object placed in front of it. The inverted image is formed between F_2 and $2F_2$ of the lens. Where is the object placed ?

Ans. Object is beyond $2F_1$.

29. What happens to the image formed by a convex lens if its lower part is blackened?

Ans. Every part of a lens forms a complete image. If the lower part of the lens is blackened, the complete image will be formed but its intensity will decrease.

30. State the nature and position of the object on the principal axis to obtain a real and magnified image.

*** Frequently asked previous years Board Exam Questions.**

Ans. Lens is convex; position of the object is between F and 2F.

31. State the nature and position of the object on the principal axis to obtain a virtual and magnified image.

Ans. Lens is convex; object to be placed between F and the lens.

32. State the nature and position of the object on the principal axis to obtain a real image of the same size.

Ans. Lens is convex; object to be placed at 2F.

33. State three characteristics of the image of an extended source formed by a concave lens.

Ans. Virtual, upright and diminished.

34. State the nature and position of the object on the principal axis to obtain a virtual and diminished image

Ans. Lens is concave; object may be at any place on the principal axis between the lens and infinity.

35. When does a ray of light falling on a lens pass through it undeviated?

Ans. A ray of light falling on the lens passes through it undeviated when it passes through optical centre.

36. Do we expect any change in the position, nature and size of the image, formed by a concave lens, with a change in the position of the object?

Ans. No, the image formed by a concave lens is:

(i) Always formed between the optical centre and the focus of the lens.

(ii) Is always virtual, erect and diminished in size. This holds good for all positions of the object.

37. Define the term magnifying power of a simple microscope. How does it depend on the focal length of the lens used?

Ans. The magnifying power of a simple microscope is defined as the ratio of the angle subtended by the image at the eye to the angle subtended by the object at the eye if it is assumed to be placed at the least distance of distinct vision.

Shorter the focal length of the lens used, larger is its magnifying power.

38. Name the lens used as a magnifying glass.

Ans. Convex (or converging) lens is used as a magnifying glass.

39. Why is the focal length of the lens used as magnifying glass kept short?

Ans. The focal length of the lens used as magnifying glass is kept small to increase its magnifying power.

40. Will the image formed by a magnifying glass be real or virtual?

Ans. The image formed by a magnifying glass will be virtual.

41. State one practical use of a magnifying glass.

Ans. A magnifying glass is used by the watch makers to see small screws, etc.

42. Name the lens for which magnification is always less than 1.

Ans. Concave lens.

43. A convex lens produces on a screen an image twice the size of the object. If the position of image and object be interchanged, what will be the magnification then?

Ans. On interchanging the position of image and object, the magnification will be 1/2.

44. What information we seek by linear magnification of the image when it is positive or negative?

Ans. If magnification is positive, then image will be virtual and erect whereas if it is negative, then image will be real and inverted.

45. Define magnification produced by a lens. State its unit.

Ans. The magnification produced by a lens is defined as the ratio of the size of the image to the size of the object, *i.e.*,

$$\text{Magnification } (m) = \frac{\text{Size of the image}}{\text{Size of the object}}$$

It has no unit.

46. Name the lens for which magnification can be 1. For what position of the object will the magnification be 1?

Ans. The lens for which magnification can be 1 is convex lens. For this, the object must be at 2F (*i.e.*, at distance twice the focal length, in front of the lens).

47. How is the magnification (m) produced by a lens related to the object distance (u) and the image distance (v)?

Ans. The magnification produced by a lens $= \frac{v}{u}$.

48. State two applications each of (i) convex lens, and (ii) concave lens.

Ans. (i) Convex lens is used (a) as a magnifying glass, and (b) in spectacles to correct hypermetropia.

(ii) Concave lens is used (a) in spectacles to correct myopia, and (b) as eye lens in Galilean telescope.

Short Answers

Chapter 5. Spectrum

1. An electromagnetic radiation is used for photography in fog.
   (i) Identify the radiation.
   (ii) Why is this radiation mentioned by you, ideal for this purpose?*

Ans. (i) Infrared radiation.

(ii) They have low frequency, the energy associated with them is also low so they do not scatter much and can penetrate appreciably through it.

2. Light of different colours is deviated through different angles, by a prism. Explain the reason.

Ans. The angle of deviation produced by a prism depends on the refractive index of the material of prism. The refractive index of material of prism is different for different colours of light because the speed of different colours is different inside the prism. Thus, the light of different colours is deviated through different angles, by a prism.

3. Give reason, why violet colour of white light deviated most and red colour of white light deviated least.

Ans. The reason is that the speed of violet light in glass is least while that of red light is most. Since, refractive index of glass = speed of light in air/ speed of light in glass, so the refractive index of glass (the material of prism) is most for violet light and least for red light. Hence, the violet light is deviated most and red light is deviated least.

4. In the atmosphere which colour of light gets scattered the least?

Ans. In atmosphere, red colour scatters the least.

5. Name any two electromagnetic waves which have a frequency higher than that of violet light. State one use of each.

Ans. The two electromagnetic waves and their uses having frequency higher than that of violet light are:

X-rays: X-rays are very extensively used for medical purposes, *e.g.*, in surgery to detect bone fractures by studying X-rays photographs.

Gamma rays: Gamma radiations are used in the cure of cancer.

6. Write approximate wavelength for (i) blue, (ii) red light. Which colour has the higher speed in vacuum? Which colour has the higher speed in glass?

Ans. (i) The wavelength of blue light is from 4640 Å to 5000 Å.

(ii) The wavelength of red light is from 6200 Å to 8000 Å.

Both colours have the same speed in vacuum. Red light has the higher speed in glass.

7. What is understood by the term dispersion of light? What is its cause?

Ans. The splitting of white light into its constituents colours when white light passes through a prism, is called dispersion of light.

The cause of dispersion is the deviation of light of different colours through different angles, by a prism.

8. What is the region beyond the red end of spectrum called?

Ans. Infrared region.

9. How will you investigate the existence of the radiation beyond the red and violet extremes of the spectrum?

Ans. If a blackened bulb thermometer is moved from violet end towards the red end, first a steady rise in temperature is observed, but as the thermometer goes beyond the red end, there is rapid rise in temperature. This shows the existence of some kind of radiation producing the heating effect beyond the red end of the spectrum.

If the radiations from the red end to the violet end are made to fall on the silver chloride solution, it almost remains unaffected. Beyond the violet end of the spectrum it first turns violet and finally it becomes dark brown (or black). This shows the existence of some kind of radiation which are chemically more active than visible light.

10. What do you understand by spectrum? Name the seven prominent colours in order, as obtained in the spectrum of white light.

Ans. The spectrum is the colour band obtained on a screen when white light passes through a prism.

The seven prominent colours obtained in the spectrum of white light, in order are: Violet, Indigo, Blue, Green, Yellow, Orange, and Red.

* Frequently asked previous years Board Exam Questions.

11. Distinguish between an impure spectrum and a pure spectrum.

Ans. A spectrum in which there is overlapping of different colours is called an impure spectrum. In other words, in an impure spectrum, the colours are not distinct and independent but they overlap. On the other hand, a spectrum with each colour distinct and independent is called a pure spectrum.

12. Name the radiations whose wavelength range is from 100 Å to 4000 Å (or 10 nm to 400 nm).

Ans. UV radiations.

13. Name two sources of ultraviolet radiations.

Ans. (i) Sun, (ii) Electric arc.

14. How can the presence of ultraviolet radiations be detected?

Ans. In presence of ultra-violet radiations, the silver chloride solution becomes dark brown or black.

15. Name two sources of infrared radiations.

Ans. (i) Infrared lamp, (ii) Fire.

16. How can the presence of infrared radiations be detected?

Ans. In presence of infrared radiations, the temperature recorded by a blackened bulb thermometer rapidly increases.

17. State two properties of infrared radiations which differ from visible light.

Ans. (i) The wavelength of infrared radiations is longer than that of visible light.

(ii) The infrared radiations have more heating effect than that of visible light.

18. State three properties of infrared radiations similar to that of visible light.

Ans. (i) Infrared radiations and visible light both travel in vacuum with a speed 3×10^8 ms^{-1}.

(ii) Both the infrared radiations and the visible light obey the laws of reflection and refraction.

(iii) Both the infrared radiations and the visible light are electromagnetic waves.

19. Infrared radiations are used as signals during war. Explain.

Ans. Infrared radiations, because of their longer wavelength are scattered less by the fog or mist. So they are used as signals during war.

20. State two properties of ultraviolet radiations which differ from visible light.

Ans. (i) Ultraviolet radiations are of wavelength shorter than that of visible light.

(ii) Ultraviolet radiations are more chemically active than the visible light.

21. State two uses of ultraviolet radiations.

Ans. (i) In producing vitamin D in plants.

(ii) For sterilising purpose.

22. The photographic dark rooms are provided with infrared lamps. Explain the reason.

Ans. The infrared radiations do not affect the ordinary photographic film, so they are used in the photographic dark room.

23. State three properties of ultraviolet radiations similar to visible light.

Ans. (i) Ultraviolet radiations can travel in vacuum as the visible light can do, with the speed 3×10^8 ms^{-1}.

(ii) Both the ultraviolet radiations and the visible light are electromagnetic waves.

(iii) Both the ultraviolet radiations and the visible light obey the laws of reflection and refraction.

24. Give one use of microwaves

Ans. Microwaves are used for communication.

25. Give one use of infrared radiations.

Ans. Infrared radiations are used as signals during war.

26. Write the range of wavelength of: (i) the visible spectrum, (ii) the ultraviolet spectrum, and (iii) the infrared spectrum.

Ans. (i) The wavelength range for the visible spectrum is from 4000 Å to 8000 Å.

(ii) The wavelength range for the ultraviolet spectrum is from 100 Å to 4000 Å.

(iii) The wavelength range for the infrared spectrum is from 8000 Å to 10^7 Å.

27. Name the radiations used for detection of fracture in bones.

Ans. X-rays.

28. What is the region beyond the violet end of spectrum called?

Ans. Ultraviolet region.

29. Name the radiation of wavelength longer than 8×10^{-7} m.

Ans. Infrared radiations.

30. What effect do the ultraviolet radiations have on the human body?

Ans. The ultraviolet radiations cause skin diseases on the human body.

31. Explain, why quartz prism is required for obtaining the spectrum of ultraviolet light?

Ans. The reason is that glass absorbs the ultraviolet radiations while the quartz does not absorb them.

32. Rock salt prism is used instead of glass prism to obtain the infrared spectrum. Give reason.

Ans. The reason is that glass absorbs the infrared radiations while the rock salts does not absorb them.

33. Which characteristic property of light is responsible for the blue colour of the sky?

Ans. Scattering of light.

34. Give three effects of scattering of sunlight by the earth's atmosphere.

Ans. Three effects of scattering of sunlight by the earth's atmosphere are:

(i) The sun appears reddish in colour at sunrise and sunset.

(ii) The colour of the sky is blue.

(iii) The sky appears black in the absence of atmosphere.

35. Define scattering of light.

Ans. When light passes from one optical medium to another, then a part of the light is absorbed by particles of the medium and then subsequently re-radiated in different directions. This phenomenon is termed as scattering of light.

36. When sunlight enters the earth's atmosphere, state which colour of light is scattered (i) the most and (ii) the least?

Ans. (i) Violet light is scattered the most.

(ii) Red light is scattered the least.

Chapter 6. Echoes and Vibrations of Sound

1. State the two laws of reflection of sound.

Ans. Similar to light, sound obeys the laws of reflection :

(i) The angle of incidence is equal to the angle of reflection.

(ii) The incident wave, the reflected wave and the normal, all lie in the same plane.

2. State two practical applications of reflection of sound.

Ans. (i) Megaphone or speaking tube.

(ii) Ear trumpet or hearing aid.

3. What is the requirement for the reflection of a sound wave?

Ans. The requirement for the reflection of a sound wave is that the size of the reflecting surface must be bigger than the wavelength of the sound wave.

4. What is meant by an echo?

Ans. The sound heard after reflection from a distant object, after the original sound dies off, is called an echo.

5. What is sound ranging?

Ans. The process of detecting the presence of an obstacle by hearing the echo of the sound reflected from it called sound ranging.

6. Point out the waves from the given list, *i.e.*, light, sound, radio-waves, ultrasonic waves and thermal radiation, in which the energy is propagated by mechanical waves.

Ans. The energy in sound and ultrasonic waves is propagated by mechanical waves.

7. How does army make use of echoes?

Ans. Army makes use of echo to locate the artillery positions of the enemy by the method of sound ranging.

8. State two medical uses of echo.

Ans. (i) Ultrasonography—for imaging the human organs.

(ii) Echocardiography—to obtain image of human heart.

9. State three conditions for the formation of echoes.

Ans. (i) The minimum distance between the source of sound and the reflecting surface should be 17 metres.

(ii) The size of reflector must be larger than the wavelength of sound.

(iii) The intensity of sound should be sufficient such that it can be heard after reflection.

10. Why will an echo not be heard when the distance between the source of sound and the reflecting surface is 10 m?

Ans. Echo will not be heard because minimum distance between the source and the reflecting surface should be 17 m.

11. State two applications of echo.

Ans. Echoes find many applications such as :

(i) In RADAR, to detect the presence and find the range of obstacles, and

(ii) Medical use in ultrasonography etc.

12. State two differences between light waves and sound waves.

Ans. Two differences between light waves and sound waves:

Light waves	Sound waves
(i) Light waves do not require a medium for propagation.	Sound waves require a medium for propagation.
(ii) Light waves travel with a speed of 3×10^8 m/s.	Sound waves travel with a speed of 332 m/s.

13. Give one use of sound ranging.

Ans. Sound ranging can be used in 'RADAR' to detect the presence of an obstacle and also to find its range.

14. State the relation between speed of sound (v), distance (d) between the reflecting surface and source of sound, and time (t) for echo to be heard.

Ans. $$v = \frac{2d}{t}$$

15. Write the full form of: (i) SONAR, (ii) RADAR. Also, state the waves that are used in both.

Ans. (i) SONAR–Sound navigation and ranging.

(ii) RADAR–Radio detection and ranging.

SONAR makes use of ultrasonic waves and a RADAR makes use of electromagnetic waves such as radio-waves or micro-waves.

16. What is SONAR? Give the principle of sonar.

Ans. SONAR is a technique that makes use of ultrasonic waves for sound ranging. It is equipped to measure even short time intervals quite accurately. A sonar works on the principle of echoes. We send a strong and short (ultrasonic) signal towards the bottom of the ocean. The reflected signal is then detected by it. From the knowledge of the time after which the reflected sound (echo), we can calculate the depth of the ocean.

$$\text{Depth of ocean} = v \times t/2$$

where v is the velocity of wave in water and 't' is the time after which the echo is received.

17. What do you understand by 'Echo-location system of Dolphins' ?

Ans. It is similar to that of bats. It enables the dolphin to navigate among its companions and larger objects. It also enables the dolphin to detect fish, squid and even small shrimp by producing sound and receiving its echo.

18. What are fishing boats?

Ans. These are the boats provided with SONAR. Ultrasound is produced, and their echo is received which enables the boatman to pin point the location of fish in deep water.

19. What are reverberations?

Ans. If the reflected sound is heard along with the original sound because of a series of reflections from various reflectors one after another in an empty room thereby forming a continuous rolling sound is called reverberation.

20. Give one example of reverberation of sound.

Ans. If we speak in an empty hall, a continuous rolling sound is heard due to reverberation of sound.

21. What do you mean by free or natural vibrations?

Ans. The vibrations produced in a body, on being slightly disturbed from its mean position, in the absence of any external force on it, are called free or natural vibration.

22. What do you mean by natural frequency of vibration of a body?

Ans. The number of vibrations executed per second by a freely vibrating body is called the natural frequency of the body.

23. Why are free vibrations practically not possible?

Ans. Theoretically, free vibrations are possible only in vacuum. Practically they are not possible, because of the presence of medium which decreases the amplitude and frequency of vibration.

24. What is meant by echo depth sounding?

Ans. The method by which the position of an obstacle hidden in water is located by producing a sound and hearing its echo is known as echo depth sounding.

25. How does the presence of medium affect the amplitude of free vibrations of a body?

Ans. The presence of medium decreases the amplitude of free vibrations of the body since it exerts a frictional force on the body which has a tendency to resist the motion.

26. State two factors on which natural frequency of a body depends.

Ans. (i) The shape of the body, and (ii) The size of the body.

27. A person walking past a railway line, at the middle of night hears a ringing sound along with the sound of his foot steps. Explain.

Ans. When the vibrations produced by the feet of the person are impressed on the rails, they vibrate with forced vibrations, thereby producing a ringing sound along with the sound of his foot steps.

28. How is it possible to detect the filling of a pitcher under a tap by hearing its sound at a distance?

Ans. As the water level in the pitcher increases, the length of the vibrating air column above the water surface in it decreases and hence the frequency of sound produced increases *i.e.*, the sound gets shriller.

29. A bucket kept under a running tap is getting filled with water. A person sitting at a distance is able to get an idea when the bucket is about to be filled.

(i) What change takes place in the sound to give this idea?

(ii) What causes the change in the sound?

Ans. (i) As the water level in the bucket rises, the sound becomes shriller and shriller.

(ii) As the length of air column, above the water surface in the bucket decreases, frequency of the sound produced increases and the sound becomes shriller and shriller (*i.e.*, of high pitch).

30. A metallic blade is made to vibrate at its one end by keeping its other end clamped. The vibrations of the blade die away after some time. Explain, how you lower the frequency of the vibration of the blade?

Ans. The vibrations of the blade die away after some time due to air resistance (or damping).

The frequency of vibration of the blade can be lowered by:

(i) Increasing the vibrating length of the blade.

(ii) Increasing the thickness of blade.

31. Explain why are soldiers asked to walk out of step while crossing bridges?

Ans. When the soldiers walk in step, they produce some fixed frequency. If this frequency corresponds to the natural frequency of bridge, then resonance can take place. This in turn will make the bridge to vibrate violently, and hence make it collapse. To avoid such situation, the soldiers are asked not to march in step.

32. What do you understand by the damped vibration? Give one example.

Ans. The periodic vibrations of a body of amplitude decreasing with time, due to the presence of medium, are called damped vibrations.

Example: The vibrations of a simple pendulum in air.

33. What are forced vibrations? Give two examples.

Ans. When an oscillatory system is made to oscillate under the action of an externally applied periodic force, it is said to execute a 'forced vibration'.

Examples:

(i) The vibrations produced in the board of a guitar when its string is made to vibrate.

(ii) The vibrations produced in the table top when the stem of a vibrating tuning fork is pressed against it.

34. What do you mean by resonance? When does resonance occur?

Ans. Resonance is a special case of forced vibration, when a body vibrates under the influence of an external periodic force, with a very large amplitude. Resonance occurs when the frequency of external force is either equal to or is integral multiple of the natural frequency of the vibrating body.

35. Explain why does a wine glass start rattling, when a note of some particular frequency is struck by a piano?

Ans. When the glass rattles, at that moment, its natural frequency corresponds with frequency of piano note, thus, resonance takes place, which makes the glass to vibrate violently.

36. When acoustic resonance takes place, a loud sound is heard. Why does this happen? Explain.

Ans. When acoustic resonance takes place, a loud sound is heard. This is because the natural frequency becomes equal to the frequency of external applied force, hence a loud sound is heard.

37. State two ways in which resonance differs from forced vibrations.

Ans. (i) In resonance it is necessary that frequency of externally applied force should be equal to natural frequency of the body whereas, it is not necessary for forced vibrations.

(ii) In forced vibrations the amplitude of vibrations is small whereas, in resonance the amplitude of vibrations is large.

38. The rear view mirror of a motor bike starts vibrating violently at some particular speed of the motor bike. What is the name of the phenomenon taking place? Why does this happen? What could be done to stop the violent vibrations ?

Ans. This phenomenon is called 'resonance'.
The frame of a motor cycle along with the rear view mirror is found to vibrate violently when it is driven to a particular speed. This happens when the frame has the natural frequency of vibration equal to that of the piston when the engine is driven at that particular speed. The violent vibrations can be stopped by changing the speed of the motor bike.

39. State two differences between free vibrations and damped vibrations.

Ans.

Free Vibrations	Damped Vibrations
(i) The amplitude of free vibrations remain constant.	The amplitude of damped vibrations gradually decreases with time.
(ii) There is no loss of energy in free vibrations.	There is some loss of energy in damped vibrations in the form of heat.

40. State two differences between free vibrations and forced vibration.

Ans.

Free Vibrations	Forced Vibrations
(i) The frequency of free vibrations remain constant with time.	The frequency of forced vibrations changes with the change in the frequency of the applied force.
'(ii) The frequency of free vibrations depend on the shape and size of the body.	The frequency of forced vibration is equal to the frequency of the applied force.

41. With which of the following frequencies does a tuning fork of 256 Hz resonate?
288 Hz, 314 Hz, 333 Hz, 512 Hz.

Ans. Tuning fork will resonate with 512 Hz.

42. A vibrating tuning fork is placed over the mouth of a burette filled with water. The tap is opened and the water level gradually falls. It is observed that the sound becomes the loudest for a particular length of air column. What is the name of the phenomenon taking place when this happens ? Why does the sound become the loudest?

Ans. 'Resonance' in air column. Frequency of vibrating tuning fork becomes equal to the frequency of vibrating air column. As a result, amplitude of the sound produced increases. Since loudness is proportional to square of amplitude, so sound is loudest.

43. What is persistence of audibility or persistence of hearing?

Ans. The sensation of any sound persists in our ears for about 0·1 second after the exciting stimulus ceases to act. This time is called persistence of audibility or persistence of hearing.

44. Name the unit used for measuring the sound level.

Ans. Unit for measuring sound level is decibel (dB).

45. Name the subjective property of sound related to its frequency.

Ans. The subjective property of sound related to its frequency is pitch.

46. Name the characteristics of a musical sound.

Ans. There are three main characteristics: (i) Loudness, (ii) Pitch, and (iii) Quality.

47. Which characteristics of sound will change if there is a change in its amplitude and its wave form.

Ans. Due to change in amplitude, the intensity changes and due to change in wave form the quality of sound changes.

48. How are bats able to fly in the dark?

Ans. Sound waves of frequency higher than 20,000 hertz are not audible to human ears. Those are known as ultrasonics. During flight in complete darkness, a bat produces ultrasound in series of impulses. The impulses on reflection from the obstacle under observation are received by the bat. The bat interprets the distance of the obstacle from the time interval between the transmission and reception of the impulses. The phase difference between the echoes received by the ears enables the bat to locate the exact position of the obstacle.

49. Explain why stringed musical instruments, like the guitar, are provided with a hollow box.

Ans. Stringed instruments are provided with a hollow box so that resonance may take place and a loud sound can be produced.

50. What do you mean by Quality or timbre of sound?

Ans. Quality or timbre of sound is that characteristic which distinguishes two sounds of the same loudness and same pitch, but emitted from two different sources.

51. Explain why, when a tuning fork (vibrating) is held close to ear, one hears a faint sound, but when the same vibrating tuning fork is placed on table, such that its handle is in contact with table, one hears a loud sound.

Ans. When the tuning fork is held close to ear, then small amount of air is disturbed and hence sound is faint. When the handle of the vibrating tuning fork is held against table, it makes the table top to vibrate with forced vibrations. As the table top has a larger surface area, therefore, large volume of air is set into vibrations, thereby producing a loud sound.

52. What change, if any would you expect in the characteristics of a musical sound when we increase:
(i) its frequency,
(ii) its amplitude

Ans. (i) Pitch of the sound increases
(ii) Loudness increases

53. Complete the following sentences:
(i) The .... of sound increases if its frequency increases.
(ii) If the amplitude of a sound wave is doubled, its loudness becomes ....

Ans. (i) pitch
(ii) four times.

54. What is a musical sound? Give an example.

Ans. It is a pleasant, continuous and uniform sound which is produced by a vibrating body making periodic and regular vibrations, *e.g.*, sound produced by flute, violin etc.

55. Name the factors which determine: (i) the loudness, (ii) pitch, and (iii) the quality of a note.

Ans. (i) The loudness of a note is determined by its amplitude.
(ii) The pitch of a note is determined by its frequency.
(iii) The quality of a note is determined by its wave form.
(*i.e.*, the presence of secondary notes along with the principal note)

56. How can one distinguish the sounds of two musical instruments even if they are of same pitch and same loudness?

Ans. The sounds of two musical instruments of the same pitch and same loudness can be distinguished by their wave forms.

57. If the amplitude of a wave is doubled, how is its loudness affected?

Ans. The loudness will become four times.

58. How do you account for the fact that two strings can be used to give notes of the same pitch and loudness but of different quality?

Ans. If the two strings are plucked at different places in the same way and their lengths be adjusted such as they give notes of the same pitch and loudness, their wave forms will differ. Hence, they will be of different quality.

59. The same musical note when played on a piano and when played on a flute sounds different.

Ans. The quality of sound produced by the two instruments is different. The quality of the sound depends upon the resultant wave form which depends upon the number of overtones and their relative intensities.

60. What is difference between loudness and intensity of sound?

Ans. For a sound wave, loudness is a subjective quantity, while intensity can be measured and is an objective quantity.

61. How is it possible to recognize a person by his voice without seeing him?

Ans. We can recognize a person just by listening to his voice because the three characteristics of sound, *i.e.*, loudness, pitch and quality are not the same for each person.

62. Two identical guitars are played by two persons to give notes of the same pitch. Will they differ in quality? Give reason.

Ans. Even if the notes played have same pitch and loudness, their wave forms will differ. Hence, the notes will be of different quality.

63. Comment on the statement, "Loudness of sound is a subjective quantity, while intensity is an objective quantity."

Ans. Loudness is a subjective quantity because it depends not only on the energy reaching the eardrum of the listener but also on the sensitivity of the ears of the listener. Intensity does not depend on the sensitivity of the ears of listener, hence it is an objective quantity.

64. How do you account for the fact that two strings can be used to give notes of the same pitch and loudness but of a different quality?

Ans. The 'quality' of a given note is determined not by its frequency and loudness but by the overall effect of the harmonics present in it. The harmonics are multiples of the fundamental or basic frequency of the 'note'. Depending on the conditions under which vibrations are taking place, sometimes we get one set of harmonics and sometimes another set. The quality of the two notes would then be different even though their fundamental frequencies may be the same.

65. What is noise pollution?

Ans. The disturbance produced in the environment due to undesirable loud and harsh sound of level above 120 dB, from various sources is called noise pollution.

66. Name the type of waves which are used for sound ranging. Why are radio waves not audible to us?

Short Answers

Ans. Radio waves. These waves are not audible to us because their frequency range is 3×10^4 Hz to 3×10^9 Hz, which is more than audible range of us.

67. What is a noise? Give an example.

Ans. Noise is a harsh and unpleasant sound produced by a body making a succession of irregular and discontinuous disturbances, *e.g.*, a sound produced by a stone thrown on a thin sheet, thunder of lightening, etc.

68. How does frequency affect a musical sound?

Ans. Frequency affects the pitch of a musical sound. Greater the frequency, higher is the pitch.

69. Name one main factor on which each characteristic such as intensity, pitch and quality of musical sound depends.

Ans. Intensity depends on amplitude, pitch on frequency and quality on wave form.

70. Three musical instruments give out notes at the frequencies listed below. Flute : 400 Hz; Guitar: 200 Hz; Trumpet: 500 Hz. Which one of these has the highest pitch?

Ans. Trumpet (500 Hz) will have the highest pitch.

71. State two differences between music and noise.

Ans.

Music	Noise
1. The wave form is regular.	1. The wave form is irregular.
2. It is regular, smooth and pleasant to the ears.	2. It is harsh, discordant and unpleasant to the ear.

72. Explain why are stringed musical instruments provided with large sound boxes?

Ans. The large sound box contains large amount of trapped air. When the vibrations of the vibrating string are impressed upon the air, it starts vibrating with forced vibrations. As large volume of enclosed air vibrates, a loud sound is produced.

73. (i) Name the system which enables us to locate underwater objects by transmitting ultrasonic waves and detecting the reflecting impulse.

(ii) What are acoustically measurable quantities related to pitch and loudness?

Ans. (i) SONAR or sound navigation and ranging.

(ii) The acoustically measurable quantities related to pitch is frequency or wavelength and for loudness is amplitude.

74. It is observed that during march-past we hear a base drum distinctly from a distance compared to the side drums.

(i) Name the characteristics of sound associated with the above observation.

(ii) Give a reason for the above observation.

Ans. (i) Pitch and Timbre.

(ii) The sound produced from the base drum is louder than the sound produced by the side drums, hence it can be heard distinctly from a distance as compared to the side drums.

75. A man playing a flute is able to produce notes of different frequencies. If he closes the holes near his mouth, will the pitch of the note produced, increase or decrease? Give a reason.

Ans. If the man closes the holes in a flute near his mouth a sound of lower frequency note will be produced because the length of vibrating air column increases and the frequency of vibrating air column is inversely proportional to the length of vibrating air column.

76. (i) Define resonant vibrations.

(ii) Which characteristic of sound, makes it possible to recognize a person by his voice without seeing him?

Ans. (i) When the frequency of the externally applied periodic force on a body is equal to its natural frequency, the body readily begins, to vibrate with an increased amplitude. Such large amplitude vibrations are called resonant vibrations.

(ii) Quality of sound or Timbre.

Chapter 7. Electricity

1. Define the term potential and state its S.I. unit.

Ans. Potential at a point is defined as the work done in bringing a unit positive charge from infinity to that point. Its S.I. unit is volt, where $1 \text{ V} = \dfrac{1 \text{ joule}}{1 \text{ coulomb}}$.

2. Explain why the potential difference across the terminals of a cell is more when the cell is not in use than it is when the cell is being used.

Ans. The reason is that when the cell is being used, there is a drop in potential across the internal resistance of the cell.

Short Answers

3. What determines the direction of flow of electrons between two charged conductors kept in contact?

Ans. The potentials of the two conductors determine the direction of flow of electrons. The electrons flow from the conductor at low potential to the conductor at high potential.

4. Define the term current and state its S.I. unit.

Ans. Current is defined as the amount of charge flowing per second.

i.e., $I = \dfrac{Q}{t}$

Its S.I. unit is ampere (A), where 1 A

$= \dfrac{1 \text{ coulomb}}{1 \text{ second}}.$

5. How is direction of current related to the direction of flow of electrons?

Ans. The direction of current is opposite to the direction of flow of electrons.

6. What is the cause of resistance?

Ans. The cause of electric resistance is the collision between the free electrons with each other as well as with the fixed oppositely charged ions.

7. State the order of resistivity of (i) a metal, (ii) a semiconductor, and (iii) an insulator.

Ans. (i) 10^{-8} ohm × m

(ii) 10^{-3} ohm × m

(iii) 10^{10} ohm × m.

8. Name the material of wire used for making standard resistances. Give a reason.

Ans. Constantan is used for making standard resistances. The reason is that its resistance remains almost unaffected with change in temperature.

9. What material is used for making wire to prepare heating coils? Give a reason.

Ans. Nichrome is used for making wire to prepare heating coils. The reason is that its melting point is high and its resistivity is also high.

10. Write an expression for the resistance of a conducting wire in terms of its length and area of cross-section.

Ans. Let R be the resistance of a conducting wire of length l and area of cross-section A. We then have:

$$R = \rho l/A$$

Here ρ is a constant characteristic of the material of the wire and is called its resistivity. Thus,

$$\text{Resistance} = \frac{\text{(Resistivity of the material of the wire)} \times \text{(Length of the wire)}}{\text{(Area of cross-section of the wire)}}$$

11. What do you understand by the resistance of a wire?

Ans. The obstruction offered in the flow of electrons (or current) by the material of the wire, is called its resistance.

12. State the S.I. unit of electrical resistance and define it.

Ans. The S.I. unit of electrical resistance is ohm (Ω). The resistance of a conductor is said to be 1 ohm if 1 ampere current flows through it when the potential difference across its end is 1 volt.

13. How does the resistivity of a metallic wire depend on temperature?

Ans. The resistivity of a metallic wire increases with increase in temperature.

14. How does the resistivity of an alloy such as constantan depends on temperature.

Ans. The resistivity of an alloy almost remains unaffected with change in temperature.

15. Define the term resistivity and state its S.I. unit. Which will have higher resistivity, a conductor or an insulator?

Ans. The resistivity of a material is the resistance of a wire of that material of unit length and unit area of cross-section. Its S. I. unit is ohm × m. An insulator will have higher resistivity.

16. The voltage applied to a given rheostat is made four times its initial value. By what factor will the resistance of the rheostat change?

Ans. The resistance of the rheostat will not change at all as it does not depend on the voltage applied across it.

17. Why is the filament in an electric bulb coiled?

Ans. To have a high resistance, the filament wire is long and thin and also to fit in a small space, the filament is coiled.

18. State the expression for current (I), if 'n' electrons pass through the cross-section of a conductor in 't' seconds.

Ans. Charge (Q) = ne

Time = t second

$$\therefore \quad I = \frac{Q}{t} = \frac{ne}{t}$$

(e = charge on an electron = $-1{\cdot}6 \times 10^{-19}$C)

19. Define conductance. Name its SI unit.

Ans. The reciprocal of resistance is called conductance. Its S.I. unit is $(\text{ohm})^{-1}$ or Ω^{-1}.

20. State the conditions for the flow of charge in a circuit from one point to the other.

Ans. The main condition for the flow of charge between two points is the difference in their potentials. If the potential difference is greater, a strong current flows and if both points are at the same potential, no current will flow.

21. Why is it not advisable to use a piece of copper wire as fuse wire in an electric circuit?

Ans. Copper wire should not be used as a fuse wire because it has high melting point and its specific resistance is much less than a normal fuse wire.

22. State and explain Ohm's law.

Ans. Ohm's law states that at a constant temperature, the current flowing in a conductor is directly proportional to the potential difference across its ends, *i.e.*, if the potential difference is halved, the current is also halved and if p.d. is doubled, the current is also doubled. *i.e.*,

V (potential difference) µ I (current)

or V = IR, where R is constant, called the resistance of the conductor

So V = IR represents Ohm's law.

23. Define the unit ohm.

Ans. One ohm is the resistance of a conductor in which 1 ampere current flows when the potential difference across its ends is 1 volt.

24. Define ampere and volt with respect to Ohm's law.

Ans. **Ampere:** It is the S.I. unit of current. If a current flows in a conductor of resistance of 1Ω, when the potential difference across its ends is 1 volt, the current is said to be 1 ampere, *i.e.*,

$$1 \text{ ampere} = \frac{1 \text{ volt}}{1 \text{ ohm}}$$

Volt: It is the S. I. unit of potential difference. If a current of 1 ampere flows in a conductor of resistance 1 ohm, then potential difference across its ends is said to be 1 volt, *i.e.*,

$$1 \text{ volt} = 1 \text{ ampere} \times 1 \text{ ohm}.$$

25. State the limitations of Ohm's law.

Ans. There are two main limitations :

(i) Temperature,

(ii) Physical condition of the conductor.

It is found that the ratio V/I is no longer constant when the temperature is not kept constant. Generally, resistance increases with the rise in the temperature of the conductor.

Physical conditions of any conductor mainly include:

(i) Its length, (ii) Its cross-sectional area, (iii) The kind of the material.

If there is no change in any of the above three conditions and also other condition like temperature remains constant, then Ohm's law holds good, *i.e.*, the ratio V/I remains constant.

26. What is a non-ohmic resistor?

Ans. This is a resistor which does not obey Ohm's laws.

V – I graph for non-ohmic resistor is not a straight line.

27. Two copper wires are of the same length, but one is thicker than the other.

(i) Which wire will have more resistance?

(ii) Which wire will have more specific resistance?

Ans. (i) Thinner wire will have more resistance because resistance is inversely proportional to the area of cross-section.

(ii) Both wires will have same specific resistance, since, specific resistance is independent of shape and size of a conductor and both wires are made of same material (*i.e.*, copper).

28. What does $\frac{\Delta V}{\Delta I}$ in the V-I graph for non-ohmic conductor represent and why?

Ans. The slope $\left(\frac{\Delta V}{\Delta I}\right)$ in the V-I graph for a non-ohmic conductor represents dynamic resistance, since its value is different for different values of V and I.

29. State the effect of temperature on the specific resistance of semiconductors?

Ans. Specific resistance of semiconductors decreases with the increase in temperature.

30. What are the characteristic properties of fuse wire?

Ans. Characteristic properties of a fuse wire:

(i) It is made up of an alloy of lead and tin.

(ii) It has high resistivity and low melting point.

31. What is an ohmic resistor?

Ans. An ohmic resistor is a resistor which obeys Ohm's law. The V-I graph is a straight line for an Ohmic resistor.

32. State four factors on which resistance of a conductor depends.

Ans. (i) Length of conductor

(ii) Area of cross-section of conductor

(iii) Material of conductor
(iv) Temperature of conductor

33. How does the resistance of a metallic wire depend on the following:
(i) the length of wire,
(ii) the thickness of wire,
(iii) the temperature of wire?

Ans. (i) The resistance of a wire increases with increases in the length of wire ($R \propto l$).

(ii) The resistance of a wire decreases with increase in its thickness ($R \propto \frac{l}{a}$, where $a = \pi r^2$, the area of cross-section of the wire).

(iii) The resistance of a metallic wire increases with increase in its temperature.

34. A given wire is stretched to double its length. How will its resistance change?

Ans. When a given wire is stretched to double its length, its area of cross-section will be halved, so the resistance of wire will become four times.

35. A given metallic wire of resistance R is doubled on itself. What will be its new resistance?

Ans. When a given metallic wire is doubled on itself, its length is reduced to half, but its area of cross-section gets doubled. So, the resistance of the wire will become one-fourth *i.e.*, the new resistance of wire will be $\frac{R}{4}$ (or 0.25 R).

36. How does the resistance of (i) a semiconductor, and (ii) an alloy, vary with temperature?

Ans. (i) The resistance of a semiconductor decreases with increase in temperature.
(ii) The resistance of an alloy remains nearly unaffected by the change in temperature.

37. Bends in a rubber pipe reduce the flow of water through it. How would the bends in a wire affect its electric resistance?

Ans. Bends in the wire will not affect its resistance. This is because the electrons are extremely small in size as compared to the bends. So, the electrons can easily change their direction of motion.

38. What is meant by internal resistance of a cell?

Ans. The resistance offered by the electrolyte inside the cell to the flow of current is called internal resistance of cell.

39. What do you understand by voltage drop in a cell?

Ans. The work done in carrying a unit charge through the electrolyte is called the voltage drop in the cell.

40. State the factors on which the internal resistance of a cell depends.

Ans. The factors on which the internal resistance of cell depends are:

(i) The surface area of the electrodes: The larger the surface area, lesser is the internal resistance.

(ii) The distance between the electrodes: As the distance between the electrodes increases, the internal resistance of the cell also increases. It is also affected by the nature, concentration and temperature of the electrolyte.

41. State expression for internal resistance of a cell in terms of e.m.f. (E), terminal voltage (V) and current (I).

Ans. $$r = \frac{E - V}{I}$$

42. A substance has nearly zero resistance at a temperature of 1 K. What is such a substance called?

Ans. Superconductor.

43. What is the e.m.f. of a cell?

Ans. When no current is drawn from a cell, *i.e.*, when the cell is not connected to any external circuit (or is in open circuit), then the potential difference between its electrodes (or plates) is called its electromotive force, *i.e.*, e.m.f. It is denoted by E.

44. State the factors on which the e.m.f. of a cell depends.

Ans. The e.m.f. of a cell depends on the following factors:

(i) The material of the electrodes.
(ii) The electrolyte used in the cell.

It is to be noted that the e.m.f. of a cell does not depend on its shape or size.

45. A cell of e.m.f. E and internal resistance r is used to send current I in an external resistance R. Write expressions for (i) the current drawn from the cell, (ii) the terminal voltage V of the cell, and (iii) the voltage drop across the internal resistance. How are E and V related?

Ans. Total resistance of circuit = $R + r$

(i) The current drawn from the cell, $I = \frac{E}{R + r}$

(ii) The terminal voltage of the cell, V = IR

(iii) The voltage drop across the internal resistance = Ir

E and V are related as $E = V + Ir$.

46. Point out two differences between e.m.f. and terminal voltage as applied to electric circuits.

Ans. Electromotive force or e.m.f. of a cell is the terminal voltage between its two electrodes when it is in open circuit, *i.e.,* not sending current in the external circuit, while terminal voltage is the difference in potential between the terminals when the cell is in closed circuit or it is sending current in the given circuits. E.m.f. is greater than terminal voltage of a cell on account of energy being spent in driving charge through the circuit.

47. State expression for resistance connected in series.

Ans. Let n resistors are connected in series then its equivalent resistance is given by

$$R = r_1 + r_2 + r_3 + \ldots\ldots + r_n.$$

48. State expression for resistance connected in parallel.

Ans. Let n resistance are connected in parallel then equivalent resistance is given by

$$\frac{1}{R} = \frac{1}{r_1} + \frac{1}{r_2} + \frac{1}{r_3} + \ldots + \frac{1}{r_n}$$

49. How should the electric lamps in a building be connected?

Ans. Electric lamps are connected in parallel.

50. When should resistors be connected in series?

Ans. Resistors should be connected in series, when total resistance in a current is to be increased in order to limit the current in the circuit.

51. State two facts about series combination of resistors?

Ans. (i) Same current passes through each resistor.

(ii) The potential difference across the entire circuit is equal to the sum of potential difference across the individual resistors. (*i.e.,* $V = V_1 + V_2 + V_3$).

52. State two facts about parallel combination of resistors?

Ans. (i) Potential difference across each resistor is same, which is equal to the potential difference across the terminals of the battery.

(ii) The main current (I) divides itself in the different arms having the resistors such that:

$$I = I_1 + I_2 + I_3.$$

53. Why are thick copper wires used as connecting wires?

Ans. The resistance of connecting wires should be almost negligible. For this, thick wires are used since a thicker wire has a small resistance and the wires are made of copper because the resistivity of copper is low.

54. State the effect of rise in temperature on the resistance of semiconductors.

Ans. In the case of semiconductors like carbon, silicon, germanium, etc., the resistance decreases with increase in temperature.

55. When should resistors the connected in parallel?

Ans. Resistors should be connected in parallel, when total resistance in a circuit is to be decreased in order to pass a heavy current in the circuit.

56. (i) Write one advantage of connecting electrical appliances in parallel combination.

(ii) What characteristics should a fuse wire have?

(iii) Which wire in a power circuit is connected to the metallic body of the appliance?

Ans. (i) Each appliance gets connected to 220 V supply for its normal working.

(ii) A fuse wire must have low melting point and its specific resistance must be more than that of copper or aluminium.

(iii) Earth wire.

Chapter 8. Electric Power and Household Circuits

1. What does the unit kilowatt hour measure? Define it.

Ans. The unit kilowatt hour measures the electrical energy. 1 kWh is the electrical energy consumed by an electric appliance of power 1 kW when it is used for 1 hour.

2. Define watt hour.

Ans. Watt hour is a smaller commercial unit of electric energy, which is spend by an electric appliance of 1 watt power in 1 hour, *i.e.*

$$1 \text{ watt hour} = 1 \text{ watt} \times \text{hour} = 1 \text{ watt} \times (60 \times 60)\text{s}$$
$$= 3600 \text{ joule}$$

3. Express watt hour and kilowatt hour in joule.

Ans. 1 watt hour = 1 watt × 1 hour

$= 1\ Js^{-1} \times 60 \times 60s = 3600\ J$

1 kilowatt hour

= 1000 watt × 1 hour

$= 1000\ Js^{-1} \times 60 \times 60s$

$= 3{\cdot}6 \times 10^6\ J$

4. What is the energy conversion when an electric current passes through a metallic wire ?

Ans. When an electric current passes through a metallic wire, the wire becomes hot and the whole of electric energy is converted into heat energy.

5. An electric current is passed through a metallic wire. The wire gets heated up. Give the reason. Is it possible to melt the wire by passing a heavy current?

Ans. The metallic wire gets heated up on passing an electric current through it due to collisions of free electrons and energy is transferred to the positive ions which begin to vibrate more violently, causing to produce heat. It is possible to melt the wire by passing a heavy current. The wire will melt when its temperature due to heating reaches its melting point.

6. Why is heating coil wound in the form of a helix in electric stove?

Ans. This reduces the area over which element of heater is spread. As heat is produced over smaller area, therefore, temperature of its element rises to about 1000°C.

7. Why is the heating coil placed in a circular porcelain plate in an electric stove?

Ans. The porcelain plate (i) is bad conductor of electricity (ii) does not melt till 1000°C.

8. How does the heat produced by the passage of current in a metallic wire depend on: (i) the current in the wire (ii) the resistance of the wire, and (iii) the time of passage of current in the wire?

Ans. (i) The heat produced is proportional to the square of current passed in the wire.

(ii) The heat produced is proportional to the resistance of the wire.

(iii) The heat produced is proportional to the time of passage of current in the wire.

9. An electric filament lamp is connected to a supply of voltage higher than the recommended value. Give reasons, why compared with a normal performance, the lamp emits a brighter light and its life is shortened.

Ans. The lamp will have the filament of a particular resistance. If it is connected to a supply of higher voltage, the heat energy produced in it will be more $\left(\text{since } H = \frac{V^2 t}{R}\right)$, *i.e.*, the filament will get more heated so it will glow more and the lamp will emit a brighter light. On the other hand, due to excess heating, the filament may vaporize and so its life may shorten.

10. "The current rating of a fuse is 5A". Comment on this statement.

Ans. This statement means that the current in the circuit should not exceed 5A. As the current in circuit will exceed 5A, the fuse will melt and the circuit will break.

11. Which cable, one rated 15A and the other rated 5A will be thicker? Give a reason.

Ans. The cable rated 15A will be thicker. The reason is that thicker the wire, lesser is its resistance and hence higher is its current carrying capacity.

12. What is electric power? State and define its S.I. unit.

Ans. Electric power is the rate of doing work or it is the work done in 1 second.

The S.I. unit of power is watt (W).

If a current of 1 ampere flows through a metallic conductor at a potential difference of 1 volt, the power is said to be 1 watt (W).

or 1 watt = 1 volt × 1 ampere.

13. State three factors on which the heat produced in a metallic wire due to passage of current in it depends.

Ans. (i) On resistance of wire

(ii) On strength of current through the wire

(iii) On time for which the current flows in the wire

14. Explain the meaning of the statement, "The power of an appliance is 100 W".

Ans. This means that if the appliance is used for 1h, it will consume 100W of electrical energy.

15. Where is a fuse placed in an electrical circuit? Name a material suitable for making a fuse wire.

Ans. The fuse is placed in the live wire just after the 'electricity meter' of the consumer. Fuse wires are generally made of an alloy of tin and lead and have a relatively low melting point.

16. Why is the electric power from the generating station transmitted at high voltage?

Ans. The electric power from the generating station is transmitted at high voltage to minimize the loss of energy in the live wires.

17. Name three main characteristics of a heating element.

Ans. The three main characteristics of a heating element are:

(i) It should have a high resistance so that a major amount of electric energy is converted into heat energy.

(ii) It should not get oxidised easily up to a temperature of 1000°C.

(iii) It should have a high melting point and a low thermal expansion so that it does not get loosened on getting heated.

18. Name two common material used for heating elements.

Ans. Nichrome and manganin are the common alloys that are mostly used for making heating elements.

Nichrome is: 60% nickel + 25% iron + 15% chromium.

Manganin is: 83% manganese + 13% copper + 4% nickel.

19. In an electric radiator or room-heater, state with reason, why the porcelain tube wound with a nichrome wire is placed between the pole and principal focus of the parabolic concave reflector.

Ans. In this position, when current passes through the nichrome wire, it becomes red hot and the concave reflector helps in spreading the heat radiation in the room because it forms a divergent beam on account of the heating element being placed between the pole and the principal focus of the reflector.

20. Why is that the same current flowing through the bulb heats up the filament, while the leads are not heated up?

Ans. The leads are much thicker than the filament wire. Thus, the resistance of the leads is much smaller than that of filament wire. So, with the same current, the filament gets heated while the leads are not.

21. Why is the filament in an electric bulb coiled?

Ans. To have a high resistance, the filament wire is long and thin. To fit in a small space it is coiled.

22. Why is the heating element of an electric oven wound on a helix? State the reason.

Ans. In an electric oven, the china clay plate is provided either with lateral grooves or grooves in the form of a spiral helix. This type of spiral winding reduces the area of spread of nichrome wire and helps to concentrate heat in a small region.

23. The electric bulbs used these days have a coil-coiled filament. State its reason.

Ans. On account of having a coil-coiled filament, the area of spread of tungsten is not much and so the heat now produced in a very small area, raises the temperatures to a high degree *i.e.*, 2300°C, producing more bright light.

24. What is 'Rating' of an electric appliance?

Ans. Rating of an electric appliance indicates energy consumed per second or power when connected to a supply of given voltage.

25. If the rating of an electric bulb is 100 W – 230 V, explain its meaning.

Ans. If an electric bulb is rated as 100 W – 230 V, it means that the bulb is lighted on a 230 V supply and it consumes 100 W electric power (*i.e.*, 100 J of energy per sec).

26. What changes in energy occur in an (i) electric bulb and (ii) electric heater?

Ans. (i) In an electric bulb, electrical energy changes to heat and light energy.

(ii) In an electric heater, electrical energy changes to heat energy.

27. Why mica is preferred to other insulators in an electric iron?

Ans. Mica is preferred to other insulators in an electric iron because mica is a bad conductor of electricity, but it easily conducts away heat. So when current flows through nichrome wire, the heating element is fully protected against any leakage of electric current, whereas heat is easily conducted.

28. State one advantage of a.c. over d.c.

Ans. Advantage of a.c. over d.c. is that it is able to travel long distance without losing power.

29. What is meant by earthing of an electrical appliance? Why is it essential?

Ans. Earthing of an electrical appliance means that outer metallic case of appliance is connected with earth wire, which is welded to the end of a copper rod and it is buried in the ground.

It saves the appliance during short circuiting by passing excessive current to earth.

30. Where are the cartridge type fuse used?

Ans. Cartridge type fuse are used to provide additional safety to an electric appliance. The fuse wire used must have a lower rating than the maximum current which the appliance or the circuit can withstand. Hence, the fuse blows off just as the current nearly approaches the maximum value and thus, it protects the appliance from any damage done due to heating.

31. Why does the heating coil not produce any visible light in an electric stove?

Ans. It is because, the temperature of heating element is around 1000°C, whereas visible light is produced if the temperature rises above 1500°C.

32. At what voltage and frequency is the a.c. supplied to our houses?

Ans. The voltage is 220 V and the frequency at which a.c. is supplied to our houses is 50 Hz.

33. At what voltage and frequency is the electric power generated at the power generating station?

Ans. The voltage is 11,000 volt or 11 kV and the frequency is 50 Hz.

34. What are the two main systems of wiring commonly used in houses? Which of these systems mentioned by you is better and why?

Ans. The two main systems of wiring are: (i) the tree system, and (ii) the ring system.
Out of these systems, the ring system is better. The reason is that this system of wiring is cheaper because a low rating wire can be used for connection of appliances.

35. State two advantages of 'the ring system of house wiring'.

Ans. The two advantages of ring system are:

(i) Length of wiring used is much less and so less costly, since a single ring feeds the whole house.

(ii) In case of short-circuit, only the fuse of that particular appliance blows off, disconnecting it from the electric circuit; while the circuits of other appliances remain unaffected.

36. Household wiring for lamp connections can either be done in parallel or in series. Which one would you prefer? Give a reason for your answer.

Ans. We would prefer the parallel wiring. This is done for the following reasons :

(i) In parallel wiring, each lamp can have its independent switch and the disconnection (putting off) of one of them will not affect the other lamps.

(ii) All lamps have the same potential difference (equal to that of the mains) across the terminals and are thus able to work as per their specified rating.

37. What is a switch? State its function in an electric circuit.

Ans. A switch is an on-off device for current flowing in a circuit or in an electric appliance. The switch should always be connected in the live wire. Its main function is either to connect or disconnect an electric appliance in an electric circuit.

38. What are 'dual switches'? Mention their one main use.

Ans. Dual switches are double pole type switches. With such switches, an appliance, *e.g.*, an electric bulb can be switched off or on from two different places. They are generally used in staircases, long corridors etc.

39. Two electric lamps each rating 100 W, 110 V are connected in series to a 220 V power supply and two other electric lamps each of marking 100 W, 220 V are connected in parallel to the same power supply. Will any one of the two combinations give more light than the other ? Give a reason for your answer.

Ans. Both combinations will give same light.

Reason: Each combination has same power namely 200 W.

40. Two sets A and B of four bulbs each, are glowing in two separate rooms. When one of the bulbs in set A is fused, the other three bulbs also cease to glow. But in set B, when one bulb fuses, the remaining ones continue to glow. Explain the difference.

Ans. In set A, the bulbs are joined in series so there is only one path of current for each bulb. But in set B, the bulbs are joined in parallel so there is separate path of current for each bulb.

41. How should the electric lamps in a house be connected so that the switching ON or OFF in a room has no effect on other lamps in the same building?

Ans. In parallel.

42. What is meant by 'MCB' ? Which part of the domestic electric circuit does it replace ?

Ans. 'MCB' stands for miniature circuit breaker'. This a device that automatically switches off the main supply if the current drawn exceeds the prescribed upper safe limit. It can be 'reset' with the help of a flip-over switch. It protects the entire circuit and appliances against short circuiting, etc. MCB has replaced the fuse in electric circuits.

43. What is the rating of fuse used in (i) light and fan circuits, (ii) power circuits, in domestic use?

Ans. (i) The rating of fuse used in light and fan circuits is 5A

(ii) The rating of fuse used in power circuits is 15A.

44. Two lamps are rated 220V, 50W and the other rated 220V, 100W are connected in series with the mains of 220 V. Explain, why the 50W lamp glows more?

Ans. From relation $R = V^2/P$, the resistance of filament of 50W lamp will be more than that of 100W lamp. If the two lamps are connected in series with the mains, same current will flow through each lamp. From relation $H = I^2Rt$, since resistance of 50W lamp is more than that of 100W lamp, so the 50 W lamp will consume more electrical energy than that of 100W lamp, in the same time. Hence, the 50W lamp will glow more.

45. What would happen and why, to an electric bulb when it is connected across a supply of voltage (i) lower (ii) higher than its proper rating?

Ans. We know, Current = Voltage/Resistance:

For a given bulb, the filament would have a definite characteristic resistance. Hence,

(i) When the bulb is connected to a lower voltage supply, it would draw less current than its 'proper current'. Hence, it would not glow to its full capacity.

(ii) When the bulb is connected to a higher voltage supply, it would draw more current than its 'proper current'. Hence, its filament would get heated to a temperature higher than its proper permitted limit and is, therefore, likely to melt or fuse.

46. What is the order of current drawn by appliances connected to the 'light circuit' of a household?

Ans. The appliances, connected to the light circuit, are supposed to draw current of the order of 5A to 10A.

47. What is the order of current drawn by appliances connected to the 'power circuit' of a household?

Ans. The appliances connected to the 'power circuit' are supposed to draw current of the order of 10A to 15A.

48. State two advantages of filling an inert gas in an evacuated electric filament lamp.

Ans. (i) The inert gas is filled at very low pressure to prevent oxidation as well as evaporation and thus increases its life.

(ii) While bulb is working, a convection current is set-up within it which carries the spurting particles of tungsten upward towards the cap and hence the blackening of the lower part (transparent portion) of the bulb is avoided.

49. What is the advantage of using a double pole switch as the main switch in the house wiring?

Ans. A double pole switch disconnects the live as well as the neutral wires simultaneously from the main supply and thus, ensures greater safety when we want to switch off the main supply.

50. In a three-pin plug top, one of the pins is thicker and longer than the other two. Give reason. Name this pin.

Ans. The thicker and longer pin is the earth pin. It is longer so that the earth connection is made first of all and it is thicker so that even by mistake it cannot be inserted into the live socket.

51. State the purpose served by the terminals of a three pin plug.

Ans. The earth pin, which is thicker and longer than the other two pins first connects the appliances to the earth and ensures safety from an electric shock to the user. The live pin on the right and the neutral pin on the left connects the appliance to the live wire and the neutral wire of the circuit with the supply voltage.

52. A lady working in her kitchen received a shock from an electric hot plate. What could be the reason for the shock?

Ans. The reason can be that the metallic case of the electric hot plate is not earthed and the live wire of hot plate have come in contact with its case.

53. Name the markings of a three-pin plug. Give the colour codes of the connecting leads.

Ans. The three pins of a three-pin plug are:

(i) 'Live' pin (ii) 'Neutral' pin (iii) 'Earth' pin

The colour of the corresponding connecting leads are (i) 'Live' pin—Brown (or Red), (ii) 'Neutral' pin—Blue (or Black), (iii) 'Earth' pin—Yellow (or Green).

54. Why fuse wire must always be connected in 'live' wire? Explain.

Ans. Under any circumstances, fuse wire always be connected in a live wire and never in neutral wire. The reason is that in the case of over-loading, if fuse wire is connected in neutral wire, it will certainly melt and break, but the electric appliance will still be in contact with the live wire, so if any person happens to touch the appliance, he will get a shock which may prove fatal.

55. Of the three connecting wires in a household circuit : Which two of the three wires are at the same potential ? In which of the three wires should the switch be connected?

Ans. Neutral wire and earth wire are at same potential and the switch should be connected to live wire.

56. What will happen if switch is connected in 'neutral wire'?

Ans. If switch is connected in the neutral wire, an electric appliance, *e.g.*, an electric iron will remain connected to the live wire even when the switch is off and so if any one happens to touch the appliances; he or she will get a shock which is sometimes fatal.

57. What precaution do you observe while connecting a plug top into a socket?

Ans. The plug pins should fit in the socket tightly. Any loose coonection will give rise to sparking and burning of the socket or the plug.

58. Why a switch should not be touched with wet hands?

Ans. A switch should not be touched with wet hands because the water forms a conducting layer between the live wire of the switch and the hand and so we may get a fatal shock when the current passes from the 'live' wire to the hand.

59. Name two precautions to be observed while using an immersion heater.

Ans. The two main precautions to be observed are:

(i) Never try to find the temperature of water by dipping your fingers in water because some times when water comes in contact with live or neutral wire, there is a possibility of getting electric shock.

(ii) The heater must be surrounded by water to the desired level before it is switched on. If not so, the heat produced in the nichrome wire will not dissipate easily and so the heating element will melt.

60. State two precautions for a plug and socket arrangement.

Ans. (i) The plug pins should fit in the socket tightly.

(ii) The hands must be completely dry while using them.

61. Give two precautions that you would take while putting off an electric switch.

Ans. We will make sure that no naked wire is touching the button of the switch and neither the hand nor the switch is wet.

62. State the colour coding of wires in the flexible lead used for connecting the appliances to the mains in a house circuiting.

Ans. The flexible lead has three wires with insulation of different colours. The red (or brown) is for live, the black (or light blue) is for neutral and the green (or yellow) is for earth connections.

Chapter 9. Magnetic Effect of Current

1. State whether a magnetic field is associated or not around (i) a static charge, (ii) a moving charge, (iii) current carrying conductor.

Ans. (i) A static charge is not associated with a magnetic field around it.

(ii) A moving charge is associated with a magnetic field around it.

(iii) A current carrying conductor is associated with a magnetic field around it.

2. What is the nature of the lines of force of the magnetic field due to a straight current carrying wire?

Ans. The lines of force, in this case, are concentric circles centered on the wire.

3. What will happen to a compass needle when the compass is placed below a wire and a current is made to flow through the wire? Give a reason to justify your answer.

Ans. The compass needle will show deflection. This happens because when a current is passed through a conductor (here, wire), a magnetic field is produced around the conductor due to which the compass needle gets deflected.

4. State two factors on which the strength of induced e.m.f. depends.

Ans. (i) The time of change of magnetic flux.

(ii) The strength of magnetic field.

5. Describe briefly one way of producing an induced current in a coil. State one factor that determines the magnitude of induced e.m.f.

Ans. To produce an induced current in a coil, a magnet is moved along the axis of coil.

The magnitude of induced e.m.f. depends on the strength of magnet.

6. State the two properties of the magnetic lines of force around a circular coil.

Ans. The important properties are:

(i) The magnetic lines of force near the wire are nearly circular, especially at the points where the current enters or leaves the coil.

(ii) At the centre of the coil, the lines of force are almost parallel to each other, showing that the magnetic field at the centre of the coil may be considered uniform.

7. A straight wire is held vertical and current is passed in it in the upward direction. A compass needle is kept near the wire in north of it. How is the needle deflected? What is the change in the deflection of the needle if the current in wire is reversed?

Ans. When the current in the wire is upwards, the needle will deflect with its north pole towards west.

If the current in the wire is reversed, the north pole of the needle will deflect towards east.

8. Describe a simple experiment to demonstrate that there exists a magnetic field around a current carrying conductor.

Ans. The conductor is kept in N-S direction and its ends are connected to a battery, rheostat and a tapping key. When the key is open *i.e.*, no current flows in the conductor, the needle points in N-S direction. As the key is pressed and a current passes in the conductor, the needle deflects. This shows that a magnetic field exists around a current carrying conductor.

9. State the law which determines the direction of magnetic field around a current carrying conductor.

Ans. The direction of magnetic field produced due to flow of current in a conductor can be determined by the right hand thumb rule. It states that if we hold the current carrying conductor in our right hand such that the thumb points in the direction of current then the fingers encircling the wire indicate the direction of the magnetic lines of force.

10. A straight current carrying conductor is kept in a magnetic field. It experiences a force. State three factors on which the magnitude of force experienced by the conductor depends. Under what condition will the conductor experience no force?

Ans. The three conditions are: (i) Current in the conductor, (ii) Length of conductor in the magnetic field, (iii) Strength of magnetic field. The conductor will experience no force if the direction of current in the conductor is parallel to the direction of magnetic field.

11. State under what conditions force acting on a current carrying conductor which is freely suspended in a magnetic field can be (i) Maximum (ii) Zero.

Ans. (i) The force F is maximum when the magnetic field due to flow of current in the conductor and the given magnetic field B are perpendicular to each other.

(ii) The force F is zero when the magnetic field due to flow of current in the conductor is parallel to the given magnetic field B.

12. How does a magnetic field set up by a solenoid change when number of turns are increased?

Ans. By having more number of turns, the intensity of magnetic field increases.

13. How does a magnetic field set up by a solenoid change when cross-sectional area of solenoid is increased?

Ans. By having a solenoid of large diameter or more cross-sectional area, the intensity of magnetic field increases.

14. How does a magnetic field set up by a solenoid change when a soft iron core is inserted inside solenoid?

Ans. By placing a soft iron core inside the solenoid, the magnetic field becomes more intense.

15. How does a magnetic field set up by a solenoid change when strength of current in the solenoid is decreased?

Ans. By decreasing the current flowing in the solenoid, the intensity of magnetic field decreases.

16. State two ways by which the magnetic field of a solenoid can be made stronger.

Ans. (i) By increasing current.

(ii) By increasing number of turns.

17. Name two factors on which the magnitude of an induced e.m.f. in the secondary coil depends.

Ans. Magnitude of induced e.m.f. depends on

(i) Magnitude of current.

(ii) Number of turns in the coil.

18. State the condition when an electric charge can give rise to a magnetic field.

Ans. When an electric charge flows in a conductor, a magnetic field is produced around it.

19. How will the direction of force be changed if the current is reversed in the conductor placed in a magnetic field?

Ans. The direction of force will be reversed.

20. State the unit of magnetic field in terms of force experienced by a current carrying conductor placed in a magnetic field.

Ans. Newton/(ampere-metre) or $NA^{-1} m^{-1}$.

21. State Lenz's law.

Ans. The Lenz's law states that the direction of induced current is such that it opposes the cause which produces it.

22. When a current carrying conductor is placed in a magnetic field, it experiences a mechanical force.

What should be the angle between the magnetic field and the length of the conductor so that the force experienced is:

(i) Zero

(ii) Maximum?

Ans. (i) 0°

(ii) 90°

23. State the purpose of the iron core.

Ans. The purpose of the iron core is to intensify the strength of the magnetic field inside the solenoid and thereby increase the strength of the electromagnet.

24. Why is soft iron generally used as the core of the electromagnet?

Ans. (i) Soft iron has less retentivity so it acquires the magnetic properties only when the current flows through the coil wound on it and loses the magnetic properties as soon as the current is switched off.

(ii) The soft iron intensifies the magnetic field of the electromagnet because of its high permeability.

25. What is an electromagnet?

Ans. An electromagnet is a solenoid containing a soft iron core within it. It is a temporary strong magnet. It acquires the magnetic properties only when an electric current flows through the solenoid and loses the magnetic properties as soon as the current is switched off.

26. List some of the practical applications of an electromagnet.

Ans. Electromagnets are used to lift heavy loads of scrap iron etc., to separate magnetic substances from non-magnetic ones and in devices like the electric bell, telegraphs, telephones, loud-speakers, etc.

27. State two advantages of an electromagnet over a permanent magnet or a bar magnet.

Ans. The two advantages are:

(i) An electromagnet can produce a strong magnetic field and its strength can be controlled by changing the strength of the current or the number of turns in the coil.

(ii) The polarity of an electromagnet can be changed by reversing the direction of current that flows in the solenoid.

28. Name a common device that uses electromagnets.

Ans. Electromagnets are used in electric bell, door alarm, motor, dynamo etc.

29. What material is used for making the armature of an electric bell? Give a reason for using this material.

Ans. A soft iron metal piece is used for making the armature of an electric bell.

Reason: Soft iron can be easily magnetised.

30. What is a permanent magnet? Give one use of it.

Ans. A permanent magnet is a magnet made from steel such that once magnetised, it does not lose its magnetism easily.

Use: It is used in galvanometer.

31. State two ways to increase the speed of rotation of a D.C. motor.

Ans. (i) By increasing the strength of current in the coil.

(ii) By reducing the number of turns in the coil.

32. What is an electric motor ? State its principle.

Ans. An electric motor is a device which converts the electrical energy into the mechanical energy.

The working of an electric motor is based on the principle that a current carrying coil placed in a magnetic field with its plane parallel to the field, experiences a deflecting couple.

33. State three factors which govern the speed of rotation of an electric motor.

Ans. (i) The strength of current in the coil. (ii) The number of turns in the coil. (iii) The strength of the magnetic field.

34. State the factors that can increase the speed of rotation of the coil in a D.C. motor.

Ans. The speed of rotation of the coil in a D.C. motor can be increased by:

(i) Increasing the strength of the current,

(ii) Having more turns in the coil,

(iii) Increasing the area of the coil,

(iv) By increasing the strength of the magnetic field.

35. State the principle of a D.C. motor.

Ans. It is based on the principle that when an electric current is passed through a conductor placed normally in a magnetic field, a force acts on the conductor as a result of which the conductor begins to move. The direction of the force is obtained with the help of Fleming's left hand rule.

36. State the position of armature coil relative to the direction of magnetic field, in a D.C. motor when the deflecting couple acting on the coil is (i) maximum, (ii) minimum.

Ans. (i) The deflecting couple acting on the coil is maximum when the plane of the coil is parallel to the direction of magnetic field.

(ii) The deflecting couple acting on the coil is minimum when the plane of the coil is normal to the direction of magnetic field.

37. What do you understand by electromagnetic induction?

Ans. Whenever there is a change in number of magnetic lines of force associated with a coil, the e.m.f. induced in the coil lasts as long as the change takes place. This phenomenon is called electromagnetic induction.

38. What will be the effect on the working of an electric bell if instead of a direct current an alternating current is used?

Ans. If electric bell is connected to a.c. instead of d.c. the bell will still ring on pressing the switch because, the attraction of armature does not depend on the polarity of the electromagnet.

39. What energy conversion takes place during the working of a D.C. motor?

Ans. Electrical energy to mechanical energy.

40. What is dynamo (a.c. generator)? State the principle on which it works.

Ans. Dynamo (or a.c. generator) is a device which converts mechanical energy into electrical energy.

It works on the principle of electromagnetic induction.

41. Suggest two ways in an a.c. generator to produce higher e.m.f.

Ans. (i) The speed of rotation of the coil is increased.

(ii) More powerful magnet is used to increase the strength of magnetic field.

42. What is a transformer? On what principle does it work?

Ans. A transformer is device by which the amplitude of an alternating e.m.f. can be increased or decreased. It works on the principle of electromagnetic induction.

43. How are the e.m.f. in the primary and secondary coils of a transformer related with the number of turns in these coils ?

Ans. $$\frac{\text{E.m.f. in secondary coil}}{\text{E.m.f. in primary coil}} = \frac{\text{Number of turns in primary coil}}{\text{Number of turns in secondary coil}}$$

44. How do the input and output powers in a transformer compare?

Ans. In an ideal transformer, the output power is equal to input power. But in an actual transformer, the output power is less than the input power.

45. State the principle of a transformer.

Ans. It is based on the principle of 'electromagnetic induction'. The magnitude of induced e.m.f. is given by the formula.

$$\frac{\text{No. of turns in primary coil}}{\text{No. of turns in secondary coil}} = \frac{\text{E.m.f. in primary}}{\text{E.m.f. in secondary}}$$

Neglecting power losses,

Electric power fed in the primary (input)

= Electric power produced in secondary (output)

or $$E_p \times i_p = E_s \times i_s$$

or $$\frac{E_p}{E_s} = \frac{i_s}{i_p}.$$

46. Can a transformer be used with direct current source? Give reason.

Ans. No, a transformer cannot be used with direct current source. The reason is that with direct current source, there will not be any change in the magnetic flux linked with the primary coil, so no e.m.f. will be induced in the secondary coil.

47. State the factor on which the frequency of the alternating e.m.f. depends.

Ans. The frequency of the alternating e.m.f. depends on the frequency of rotation of the coil.

48. What is the function of a step-up transformer?

Ans. We use a step-up transformer to push up or step up the voltage of a given a.c. source.

49. Why is the core of a transformer made of soft iron?

Ans. The core of a transformer is made of soft iron because it has high permeability so it provides complete linkage of magnetic flux of the primary coil to the secondary coil. Secondly, the hysteresis or loss of energy in soft iron is less.

50. Why is the core of a transformer laminated?

Ans. The core of a transformer is laminated to reduce the energy loss due to eddy currents.

51. What energy conversion does take place in a generator?

Ans. The mechanical energy changes to electrical energy.

52. The primary windings of a step-up transformer are usually made of thicker wire than the secondary. Explain.

Ans. In case of a step-up transformer, the e.m.f. across the secondary coil is greater than across the primary coil. So the current in the secondary coil is less than in the primary coil. Therefore, to reduce heat loss in the windings, the primary coil is made of thicker wire than the secondary coil.

Chapter 10. Specific Heat Capacity and Latent Heat

1. What is heat? What is the S.I. unit of heat?

Ans. Heat is a form of energy, which when absorbed by a body makes it hot and when extracted from a body makes it cold. It is also defined as the total kinetic energy possessed by all the molecules of a body. The S.I. unit of heat is joule (J).

2. Define the term calorie. How is it related with joule (the S.I. unit of heat)?

Ans. One calorie is the heat required to raise the temperature of 1 g of water from 14·5°C to 15·5°C

1 calorie = 4·2 joule.

3. What are other units of heat? Name and define them.

Ans. **(i) Calorie:** It is the quantity of heat required to raise the temperature of 1gm of water by 1°C (more accurately the rise in temperature is taken from 14.5°C to 15.5°C).

(ii) Kilo calorie: It is the quantity of heat required to raise the temperature of 1 kg of water by 1°C. It is equivalent to 1000 calories.

4. What are the factors on which the quantity of heat given to a body depends.

Ans. The quantity of heat given to body depends on :

(i) The mass of the body,

(ii) The rise (or fall) in the temperature of the body, and

(iii) Nature of the material of the body or its specific heat capacity.

5. Ice-cream at 0°C feels colder than water at 0°C. Give reason for this observation.

Ans. We know that we need to supply 336 joule of heat per g, to convert ice at 0°C, into water at 0°C. It follows that ice-cream at 0°C, will draw more heat (336 joule more per gram) from our body than water at 0°C. It, therefore, feels colder than water at 0°C.

6. Bottled drinks are cooled more effectively, when surrounded by lumps of ice than iced water.

Ans. Every 1 kg of ice at 0°C absorbs 336,000 J of heat energy to form water at 0°C. As ice can extract 336,000 J of heat energy, more than water at 0°C, therefore, it cools the bottled drinks more effectively.

7. Write an expression for the heat energy liberated by a hot body.

Ans. Expression for the heat energy liberated by hot body = $mc\Delta T$

i.e., $Q = mc\Delta T$

where m is mass, c is specific heat capacity, ΔT is the change in temperature.

8. 1 kg of water freezes to form ice at 0°C. What amount of heat is withdrawn ?

Ans. 3,36,000 J.

9. What energy change would you expect to take place in the molecules of a substance when it undergoes:

(i) a change in its temperature?

(ii) a change in its state without any change in its temperature?

Ans. (i) Average kinetic energy of molecules increases.

(ii) Potential energy of molecules increases.

10. Name (i) S.I. unit (ii) most commonly used unit of heat energy and write the relation between them.

Ans. (i) joule (J) (ii) calorie, 1 calorie = 4·18 joule (or 4·2 J)

11. Differentiate between heat and temperature.

Ans.

Heat	Temperature
(i) It is a form of energy.	It is the sensation of hotness and coldness.
(ii) Unit of heat is joule.	Unit of temperature is °C or kelvin.

12. State the condition for the flow of heat energy from one body to another.

Ans. Heat energy always flows from a body at a higher temperature to a body at a lower temperature.

13. Explain, why are big tubs of water kept in underground cellars for storing fresh fruit and vegetables, in cold countries?

Ans. During sub zero temperatures, the water within the cells of fruits and vegetables freezes and hence damages them. To avoid such a damage, the huge water tubs are kept. The water gives off 4200 J of heat energy for every one kg for 1°C fall in temperature. Thus, it does not allow the temperature of cell to fall below 0°C and hence there is no damage to fresh fruits and vegetables.

14. Why does the heat supplied to a substance during its change of state not cause any rise in its temperature ?

Ans. During the state change heat supplied increases potential energy of molecules. As distance between molecules increases, work is done by heat supplied against attractive force.

15. How is the heat capacity of a body related to the specific heat capacity of its substance?

Ans. Heat capacity of a body = Mass of the body × Specific heat capacity of its substance.

16. An equal quantity of heat is supplied to two substances A and B. The substance A shows a greater rise in temperature. What can you say about the heat capacity of A as compared to that of B?

Ans. Heat capacity of substance B is more than substance A.

17. Is it possible to freeze the water formed, back to ice by adding ice at 0°C? Explain with reason.

Ans. No, it is not possible to freeze the water formed, back to ice, by adding ice at 0°C because ice has to loose heat in order to freeze and the ice added at 0°C is unable to take the heat loss, otherwise it will also melt.

18. Why does stone lying in the sun get heated up much more than water lying for the same duration of time?

Ans. The specific heat capacity of stone is much less than the specific heat capacity of water, hence for the same heat supplied the temperature of stone rises much more than water for the same duration and gets heated up much more.

19. The specific heat capacity of a substance A is 3,800 $Jkg^{-1}K^{-1}$ and that of a substance B is 400 $Jkg^{-1}K^{-1}$. Which of the two substances is a good conductor of heat? Give a reason for your answer.

Ans. Substance B is a good conductor of heat because specific heat capacity of B is less than that of A and specific heat capacity is the heat energy required to raise the temperature of 1 kg of a substance by 1°C, so substance B gets heated faster.

20. How does an increase in the temperature affect the specific resistance of a:
(i) Metal and
(ii) Semiconductor?

Ans. (i) Specific resistance of a metal increases with the increase in temperature.
(ii) Specific resistance of a semiconductor decreases with the increase in temperature.

21. (i) State whether the specific heat capacity of a substance remains the same when its state changes from solid to liquid.
(ii) Give one example to support your answer.

Ans. (i) No, specific heat capacity of a substance is different in its different phases (states).
(ii) Specific heat capacity of water is 4200 $Jkg^{-1}K^{-1}$ and that of ice is 2100 $Jkg^{-1}K^{-1}$.

22. m kg of a substance of specific heat capacity s J/kg °C is heated so that its temperature rises from θ_1°C to θ_2 °C. Write down the expression for the heat Q supplied.

Ans. Heat supplied Q
= mass × Specific heat capacity × Rise in temperature
= $m \times s \times (\theta_2 - \theta_1)$ joule

23. Name the substance which has maximum specific heat capacity.

Ans. Water has the maximum specific heat capacity.

24. Write two advantages of high specific heat capacity of water.

Ans. (i) Water is used as an effective coolant in radiator of a car.
(ii) Hot water bottles are used for fomentation.

25. Does the specific heat capacity of a substance depend upon its mass and rise in temperature only?

Ans. No, for the same mass and for the same rise in temperature for different substance, the specific heat capacity is found to be different. Thus, specific heat capacity depends upon the nature of the substance also and so it is different for different substances.

26. The farmers fill their fields with water in winter. Give reason.

Ans. The farmers fill their fields with water in winter to protect the crops from frost because water due to its high specific heat capacity does not allow the temperature in the surrounding area of plants to fall below 0°C.

27. Explain, why water is sprayed on roads in evening in hot summer?

Ans. Water has the highest specific heat capacity of 4.2 $Jg^{-1}°C^{-1}$. Thus, water absorbs large amount of heat from the roads and surroundings but its own temperature does not rise much. Thus, on the whole, the roads and surrounding get cooled.

28. Explain, why temperature in hot summer, falls sharply after a light shower?

Ans. Water has the highest specific heat capacity of 4.2 Jg^{-1} C^{-1}. Thus, when there is a light shower,

Short Answers

the water evaporates. In doing so, it absorbs large amount of heat from surroundings, hence temperature falls sharply.

29. Explain, why do sandy soils get heated up quickly as compared to wet soils?

Ans. The specific heat capacity of sand is about five times less than water. Thus, when sun shines equally on sandy soil and wet soil, the sandy soil gets heated up rapidly as compared to wet soil. It is because water absorbs large amount of heat energy, but its temperature does not rise sufficiently.

30. Explain, why water is considered as best liquid for quenching thirst?

Ans. Water has the highest specific heat capacity of 4.2 $Jg^{-1}\,°C^{-1}$ and hence it can absorb large amount of heat energy, without rising in temperature sufficiently. Thirst is the natural signal, when body produces more heat energy than required. Thus, water is ideal for quenching thirst, because it can absorb large amount of heat energy.

31. Explain, why is it advisable to pour cold water over burns, caused on human body, by hot solids?

Ans. Water has the highest specific heat capacity of 4.2 $Jg^{-1}\,°C^{-1}$. Thus, it can extract large amount of heat energy rapidly from the site of burn and hence gives a lot of relief.

32. Explain, why is water used as a coolant in motor car radiators?

Ans. Water has the highest specific heat capacity of 4.2 $Jg^{-1}°C^{-1}$ and its temperature does not rise beyond 100°C. Thus, it can absorb large amount of heat from the working engine and at low temperature the engine works efficiently.

33. Name two factors on which the heat absorbed or given out by a body depends.

Ans. The mass and specific heat capacity are two factors on which the heat absorbed or given out by a body depends.

34. Explain why water is used in hot water bottles for fomentation and also as a universal coolant.

Ans. Hot water bottles are used for fomentation since water does not cool quickly due to its large specific heat capacity.

Water is used as an effective coolant because of its large specific heat capacity due to which it can extract more heat.

35. Specific heat capacity of substance A is 3.8 $Jg^{-1}K^{-1}$ whereas the specific heat capacity of substance B is 0.4 $Jg^{-1}K^{-1}$. Which of the two is a good conductor of heat ? How is one led to this conclusion ?

Ans. 'B' is a good conductor of heat, because for the same heat energy and same mass, the rise in temperature of B will be more. Hence, 'B' is a good conductor of heat.

36. If substances A and B are liquids then which one would be more useful in car radiators ? Given: Specific heat capacity 'A' 3.8 J/g/K. Specific heat capacity 'B' 0.4 J/g/K

Ans. 'A' which has more specific heat will be preferable for car radiators as it will act as coolant.

37. How is the heat capacity of the body related to its specific heat capacity?

Ans. $$\text{Heat capacity} = \frac{Q}{\Delta T} = \frac{mc\Delta T}{\Delta T} = mc$$

i.e., Heat capacity = Mass × Specific heat capacity

38. A certain amount of heat Q will warm 1 g of material X by 3°C and 1 g of material Y by 4°C. Which material has a higher specific heat capacity?

Ans. X has higher specific heat capacity.

39. Give one example where high specific heat capacity of water is used as a heat reservoir.

Ans. For fomentation purpose in hot water bottles.

40. Give one example where high specific heat capacity of water is used for cooling purposes.

Ans. In car radiators.

41. State the value of specific heat capacity of (i) Copper (ii) Ice, in S.I. unit.

Ans. (i) 390 $J\,kg^{-1}\,K^{-1}$.

(ii) 2100 $J\,kg^{-1}\,K^{-1}$.

42. What do you mean by the statement, specific heat capacity of water is 4200 $Jkg^{-1}\,°C^{-1}$?

Ans. The above statement means that, the temperature of 1kg water rises by 1°C when 4200J of heat energy is supplied to it. It also means that, if the temperature of 1kg water falls by 1°C, it releases 4200 J of heat energy.

43. Does the specific heat capacity of a substance remain same in its different phases? Explain with an example.

Ans. No, specific heat capacity of a substance does not remain same in different phases (or states). Example: Specific heat capacities of ice and water are 2100 $Jkg^{-1}K^{-1}$ and 4200 $Jkg\,K^{-1}$ respectively.

44. Why is the base of a cooking pan made thick?

Short Answers

Ans. It is made thick to increase its thermal capacity so that it gets heated slowly and it imports heat energy at a slow rate to the food for its proper cooking.

45. Explain, why no tracks are left on the ice during ice skating?

Ans. The weight of the body acts on a very small area of skate, in comparison to that of ice. Under this high pressure, ice melts to form water below 0°C. The water so formed makes the surface of skating more slippery. But as soon as the skater moves further, the water does not experience any pressure and re-freezes into ice leaving no track on ice surface.

46. Why is the base of an electric iron made thick and heavy.

Ans. It is made thick and heavy to increase the thermal capacity of base of electric iron so that it remains hot for a long duration even after switching off the current.

47. Why is specific heat capacity taken as a measure of thermal inertia?

Ans. We have, $Q = mc\,\Delta T$

or $\Delta T = \dfrac{Q}{mc}$

It follows that if c is more, ΔT will be small (for given values of Q and m). Thus, for a given body, its specific heat capacity determines the change in temperature produced by a given quantity of heat. It is thus like mass in mechanics which determines the change in velocity (or the acceleration) produced by a given force. It is quite appropriate, therefore, to regard specific heat capacity as a measure of thermal inertia.

48. Differentiate between heat capacity and specific heat capacity.

Ans. **Heat capacity:** It is the amount of heat energy required to raise the temperature by 1°C.

Specific heat capacity: It is defined as the heat capacity per unit mass of that body.

49. What is 'Calorimetry'?

Ans. The measurement of the quantity of heat is termed as calorimetry.

50. State the principle of calorimetry.

Ans. According to the principle of calorimetry, if two bodies at different temperatures are mixed together (or kept in contact), heat flows from a body at higher temperature to the body at lower temperature till both the bodies attain the same temperature. If there is no heat lost to the surroundings, then

The heat lost by the hot body
= The heat gained by the cold body.

51. Give two reasons as to why copper is preferred over other metals for making calorimeters.

Ans. Copper is preferred over other metals for making calorimeters because :

(i) It has a low specific heat capacity.

(ii) It is a good conductor of heat.

52. What is a calorimeter?

Ans. A calorimeter is a cylindrical vessel made of a thin sheet of copper, which is used to measure the amount of heat gained or lost by a body.

53. What is a calorimeter made of? Why?

Ans. A calorimeter is made of a thin sheet of copper because copper has a low specific heat capacity and also it is a good conductor of heat. By making it of a thin sheet, its mass and hence thermal capacity also becomes low.

54. Why is a calorimeter placed inside a wooden jacket and the space between them is filled with some poor conductor?

Ans. To avoid heat loss by conduction.

55. Why is a calorimeter covered with a wooden lid?

Ans. To avoid heat loss by convection.

56. Why are the inner and the outer surfaces of a calorimeter polished?

Ans. To reduce the heat loss due to radiation.

57. Two metallic blocks P and Q having their mass in the ratio 1 : 2 are given same amount of heat. If their temperature rises by same amount, compare their specific heat capacities.

Ans. Ratio of specific heat capacities of P and Q = 2 : 1.

58. What do you mean by change of phase?

Ans. The process of change from one state to another at a constant temperature is called the change of phase.

59. Some heat is provided to a body to raise its temperature by 25°C. What will be the corresponding rise in temperature of the body as shown on the Kelvin scale?

Ans. Increase in Kelvin will be 25 K.

60. If, in a central heating system, steam enters a radiation pipe at 100°C and water leaves the radiation pipe at 100°C, can this radiation pipe heat a room ? Give an explanation for your answer.

Ans. Yes, because steam at 100°C gives out its latent heat of vaporisation to condense to water at the same temperature of 100°C.

61. Ice is more effective in cooling than the ice-water. Explain.

Ans. Each 1g of ice absorbs nearly 336 J of heat when melts to water at 0°C. Hence, it is more effective in cooling than the ice-water.

62. Why does atmospheric temperature fall after hail storm?

Ans. Ice has the highest specific latent heat of fusion of 336,000 Jkg^{-1}. Thus, when hails melt, they absorb large amount of heat energy from surroundings. Thus, atmospheric temperature drops.

63. What happens to the average kinetic energy of the molecules as ice melts at 0°C?

Ans. The average kinetic energy of molecules (or temperature) remains same.

64. State effect of pressure on the melting point of substances which expand on melting (*e.g.* wax, lead etc.).

Ans. The melting point of substances which expand on melting increases by the increase in pressure.

65. State the effect of impurities on the melting point.

Ans. The melting point of a substance decreases by the presence of impurities in it.

66. Is there any change in temperature during change of phase?

Ans. No, temperature remains constant during change of phase.

67. Does a substance absorb or liberate any heat energy during change of phase ?

Ans. Yes, heat is either absorbed or released during change of phase.

68. Explain the meaning of the term latent heat. State its S.I. unit.

Ans. The heat absorbed or liberated by a substance during its change of state at a constant temperature is called latent heat. Its S.I. unit is joule (J).

69. What happens to the heat supplied to a substance when the heat supplied causes no change in the temperature of the substance?

Ans. The heat supplied is used in the change of state. This heat is known as latent heat.

70. When 1 g of ice at 0 °C melts to form 1 g of water at 0 °C then, is the latent heat absorbed by the ice or given out by it?

Ans. Latent heat is absorbed by the melting ice.

71. Which contains more heat: 1g water at 0°C or 1g ice at 0°C? Give reason.

Ans. 1g water at 0°C contains more heat than 1g ice at 0°C. The reason is that 1g water at 0°C, when converts to 1g ice at 0°C, liberates 336 J heat.

72. Define the term 'specific latent heat of fusion' of a substance.

Ans. **Specific latent heat of fusion:** The specific latent heat of fusion of a substance is the heat energy released when a unit mass of substance converts from liquid to solid state without the change in temperature.

73. Define melting.

Ans. The change from solid phase to liquid phase at a constant temperature is called melting.

74. Define melting point.

Ans. The constant temperature at which a solid changes to liquid is called melting point of the solid.

75. Define freezing or fusion.

Ans. The change from liquid to solid phase at a constant temperature is called freezing or fusion.

76. Define freezing point.

Ans. The constant temperature at which a liquid changes to solid is called its freezing point.

77. Write the approximate values of: (i) the specific latent heat of fusion of ice, (ii) the specific heat capacity of water.

Ans. (i) The specific latent heat of fusion of ice is 3,36,000 J/kg.

(ii) The specific heat capacity of water is 4200 J kg^{-1} K^{-1}.

78. The specific latent heat of fusion of ice is 336 J/g. Comment on this.

Ans. The heat required to convert 1 g of ice at 0°C to 1 g water at 0°C is 336J.

79. Give one consequence of the high specific latent heat of fusion of ice.

Ans. Due to high specific latent heat of fusion of ice, snow on the mountains does not melt all at once but changes into water slowly as it gets heat from the sun.

80. Why does weather become pleasant when it starts freezing in cold countries?

Ans. Ice has the highest specific latent heat of fusion of 336,000 Jkg^{-1}. Thus, every 1 kg of water at 0°C, on freezing releases 336,000 J of heat energy. As enormous amount of heat energy is released in atmosphere, therefore, weather becomes pleasant.

81. Is it possible to melt ice below 0°C? Explain.

Ans. Yes, it is possible by increasing the pressure.

82. State the effect of pressure on the melting of substances which contract on melting (*e.g.* ice).

Ans. The melting point of substances which contract on melting (*e.g.* ice) decreases by the increase in pressure.

Chapter 11. Radioactivity and Nuclear Energy

1. Define atomic number of an element.

Ans. The number of protons in the nucleus is known as the atomic number of the element and is denoted by Z.

2. Define mass number of an element.

Ans. The total number of neutrons and protons in the nucleus is called the mass number of the element and is denoted by A.

3. What are isotopes? Give examples.

Ans. The atoms of same element which have same atomic number but different mass number, are called isotopes.

Examples: Protium (^1_1H), Deuterium (^2_1H), Tritium (^3_1H).

4. What are isobars? Give examples.

Ans. The atoms of different elements which have the same mass number but differ in their atomic numbers are called isobars.

Examples : Sodium ($^{23}_{11}N$), Magnesium ($^{23}_{12}Mg$).

5. What are isotones? Given examples.

Ans. The atoms of different elements having same number of neutrons are called isotones.

Example: Carbon and Boron.

6. What are radio isotopes? Give examples.

Ans. The isotopes of some elements with atomic number less than 82 are also found to be radioactive and are called radio isotopes.

Examples: Carbon ($^{14}_6C$) and Cobalt ($^{60}_{27}Co$).

7. Can radio isotopes be prepared artificially? If yes, name the process.

Ans. Yes, radio isotopes can be prepared artificially by nuclear transmutation.

8. What are tracers? How are they useful?

Ans. The salt of weak radioactive isotopes such as radio-sodium chloride are used for diagnosis. Such radio isotopes are called tracers. Tracers are used to detect suspected brain tumor and brain clots.

9. How is the age of materials of archaeological importance determined?

Ans. The age of excavated materials of archaeological importance is estimated by the study of decay of carbon-14 in the remains. The process is called carbon dating.

10. What do you understand by the following terms?

(i) Nucleus,

(ii) Nucleons.

Ans. (i) **Nucleus:** The central core of an atom is called nucleus. The nucleus contains protons and neutrons.

(ii) **Nucleons:** Neutrons and protons are collectively called nucleons.

11. State two industrial uses of radio isotopes.

Ans. (i) The consisting effect of radiations from radio isotopes are used in making luminescent signs.

(ii) Knowing the penetrating power of β-radiations from radio isotopes, they are used to control thickness of paper, plastic etc. during their manufacture.

12. A certain nucleus has a mass number 20 and atomic number 9. Find the number of neutrons and protons present in it.

Ans. (i) Number of neutrons

= Mass number – Atomic number

= 20 – 9 = 11

(ii) Number of protons

= Atomic number = 9.

13. What is Einstein's mass-energy equivalence relation?

Ans. Einstein's mass-energy equivalence relation is represented by :

$E = mc^2$; where E is the energy produced, 'm' is the loss in mass (mass defect) during nuclear fission or fusion and c is the speed of light (3×10^8) ms^{-1}).

14. What do you understand by the term radioactivity?

Ans. The process of self, spontaneous and random emission of α, β or γ radiations from the nucleus of elements of atomic number higher than 82, is called radioactivity.

15. What are 'Becquerel rays'?

Ans. Becquerel rays are 'radioactive radiations' emitted from radioactive substances like uranium, thorium etc. Since these radiations were discovered by Henry Becquerel, so they are named as 'Becquerel rays'.

16. 'Radioactivity is a nuclear phenomenon'. Comment on this statement.

Ans. Any physical change (such as change in pressure and temperature) or chemical change (such as chemical treatment, oxidation etc.) does not alter

Short Answers

the rate of decay of the radioactive substance, which suggests that the phenomenon of radioactivity cannot be due to orbital electrons. Hence, it is a nuclear phenomenon.

17. The isotope of $_{92}U^{238}$ decays by alpha emission to an isotope of Thorium (Th). The Thorium isotope decays by beta emission to an isotope of Protactinium (Pa). Write down the equations to represent these two nuclear changes.

Ans. $_{92}U^{238} \longrightarrow {}_{90}Th^{234} + {}_{2}He^{4}$ (α-particle)

$_{90}Th^{234} \longrightarrow {}_{91}Pa^{234} + {}_{-1}e^{0}$ (β-particle)

18. Which part of the atom undergoes a change in the process of radioactive decay ?

Ans. The nucleus of the atom undergoes a change in the process of radioactive decay.

19. A nucleus $_{84}X^{202}$ of an element emits an alpha particle followed by a beta particle. The final nucleus is $_{a}Y^{b}$. Find a and b.

Ans. $_{84}X^{202} \xrightarrow{\alpha} {}_{82}X_1^{198} \xrightarrow{\beta} {}_{83}Y^{198}$

$\therefore a = 83, b = 198$

20. Is it possible for a hydrogen (^1_1H) nucleus to emit an alpha particle? Give a reason for your answer.

Ans. No, it is not possible because an alpha particle (^4_2He) consists of two protons and two neutrons.

21. Name the different radiations which are emitted by the radioactive substances.

Ans. The following three radiations are emitted by the radioactive substances :

(i) Alpha (α), (ii) Beta (β), and (iii) Gamma (γ).

22. Are all the radiations *viz.* α, β and γ emitted in a single radioactive decay?

Ans. All the above radiations are not emitted in a single decay. There will be either an α emission or a β emission, which may be accompanied by γ-emission.

23. Compare the penetrating powers of α, β and γ-radiations.

Ans. γ-radiations are nearly 100 times more penetrating than β-radiations and 10,000 times more penetrating than α-radiations.

24. Compare the ionizing powers of α, β and γ radiations.

Ans. The ionizing power of α-radiation is nearly 100 times that of β radiations and nearly 10,000 times that of γ-radiations.

25. A mixture of radioactive substances gives off three types of radiations :

(i) Name the radiation which travels with the speed of light.

(ii) Name the radiation which has the highest ionizing power.

Ans. (i) γ-rays.

(ii) α-particles.

26. A mixture of radioactive substances gives off three types of radiations.

(i) Name the radiations which are charged.

(ii) Name the radiation similar in nature to cathode rays.

(iii) Name the radiation similar in nature to X-rays.

(iv) Name the radiation which is deviated most in a magnetic field.

(v) Name the radiation which is not affected by an electric field.

(vi) Name the radiation which has the maximum penetrating power.

(vii) Name the radiation which has the highest ionizing power.

(viii) Name the radiation which has the highest speed.

(ix) Name the radiation which when becomes neutral, become the atom of a rare gas.

Ans. (i) Alpha and beta radiations are charged.

(ii) Beta radiation is similar in nature to cathode rays.

(iii) Gamma radiation is similar in nature to X-rays.

(iv) Beta radiation is deviated most in a magnetic field.

(v) Gamma radiation is not affected by an electric field.

(vi) Gamma radiation has the maximum penetrating power.

(vii) Alpha radiation has the highest ionizing power.

(viii) Gamma radiation has the highest speed.

(ix) Alpha radiation when become neutral become the atom of rare (helium) gas.

27. How does the position of an element change in the periodic table when it emits out an alpha particle?

Ans. When an element emits out an alpha particle, a new element is formed with properties similar to those of an element two places earlier in the periodic table.

28. What changes occur in the nucleus of a radioactive element when it emits an alpha particle? Show by an example?

Ans. Example : $^{238}_{92}U \rightarrow {}^{234}_{90}Th + {}^{4}_{2}He$

	92 protons	90 protons	2 protons
	146 neutrons	144 neutrons	2 neutrons

29. What will an alpha particle change into when it absorbs:

(i) One electron;

(ii) Two electrons?

Ans. (i) Single ionized Helium (He^+),

(ii) Helium atom.

30. State three properties of α-rays.

Ans. (i) Alpha rays are composed of helium nuclei.

(ii) Alpha rays ionize the gas through which they pass.

(iii) Alpha rays are deflected by the electric and magnetic fields.

31. What is the effect on the atomic mass and the atomic number of a radioactive element, after the emission of an α-particle?

Ans. During the emission of an α-particle from the nucleus of a radioactive element, its atomic mass decreases by 4 and the atomic number also decreases by 2. If $_ZX^A$ represents a radioactive element having atomic mass A and atomic number Z, then,

$$_ZX^A \longrightarrow {}_{Z-2}Y^{A-4} + {}_2He^4 \text{ (α-particle)}$$

32. An atomic nucleus A is composed of 84 protons and 128 neutrons. It emits an α-particle and is transformed into nucleus B. Write down the composition of B.

Ans. A emits an α-particle, so in the nucleus of B, neutrons will be 128 – 2 = 126 and protons will be 84 – 2 = 82.

33. State charge and mass of (i) α-particle, (ii) β-particle.

Ans. (i) The charge on α-particle is 3.2×10^{-19} coulomb and its mass is 6.67×10^{-27} kg (i.e., nearly 4 times the mass of proton).

(ii) The charge on β-particle is -1.6×10^{-19} coulomb and its mass is 9.1×10^{-31} kg.

34. What are α, β and γ-radiations?

Ans. α radiations are positively charged particles with charge + 2*e* and mass 4mp where e is electronic charge (i.e., charge on a proton) and mp is the mass of a proton. β radiations are negatively charged particles (i.e., electrons) with charge – *e* and mass times the mass of a proton.

γ-radiation are electromagnetic waves of wavelength in the range 0.001 Å.

35. The nucleus of an atom does not contain electrons. Explain how is it possible for the nucleus to emit a beta particle which is fast moving electron.

Ans. The nucleus emits a beta particle when a neutron inside it changes into a proton. The reaction is

$$\underset{\text{neutron}}{{}^1_0n} \rightarrow \underset{\text{proton}}{{}^1_1p} + \underset{\text{β-particle}}{{}^{0}_{-1}e} + \underset{\text{anti-neutrion}}{v}$$

36. What happens to the atomic number of an element when it emits:

(i) An alpha particle

(ii) A beta particle.

Ans. (i) $^A_ZX \xrightarrow{\alpha} {}^{A-4}_{Z-2}Y + {}^4_2He$

(ii) $^A_ZX \xrightarrow{\beta} {}_{Z+1}^{A}P + {}_{-1}\beta^0$

37. How does the position of an element change in the periodic table when it emits out a beta particle?

Ans. When an element emits out a beta particle, a new element is formed with properties similar to those of an element one place later in the periodic table.

38. State three common properties of beta rays and cathode rays.

Ans. (i) Beta rays and cathode rays both are composed of electrons.

(ii) Beta rays and cathode rays both are deflected by the electric and magnetic fields in the same direction.

(iii) Beta rays and cathode rays both ionize the gas through which they pass.

39. Explain briefly what changes take place within the nuclei when beta particles are emitted by a radioactive substance.

Ans. The mass number remains unchanged but atomic number increases by 1 of the nuclei undergoing such a change.

40. What changes will take place in the mass number and atomic number of the nucleus Y if it emits gamma radiations ?

Ans. No change in mass number and atomic number when γ radiations are emitted out.

41. A radioactive carbon nucleus emits a β-particle to form a nitrogen nucleus. Represent the change by an equation showing the atomic and mass number of each elements.

Ans. When the nucleus of an element emits a beta particle, the number of protons increases by 1 and number of neutrons decreases by 1

Example: $^{14}_{6}C \rightarrow\ ^{14}_{7}N +\ ^{0}_{-1}e$

6 protons 8 neutrons; 7 protons 7 neutrons; β-particle

42. Two radioactive nuclei are represented by and when p and q are mass numbers and x, y are the atomic number. How can the products be represented, *i.e.*, what are the new values of p, q, x and y after the emission of an α-particle and a β-particle from A and B respectively.

Ans. (i) $^{p}_{x}A \xrightarrow{\alpha}\ ^{p-4}_{a-2}A_1 +\ ^{2}_{4}He$

Its mass number decreases by 4 and atomic number decreases by 2.

(ii) $^{q}_{y}B \xrightarrow{\beta}\ ^{q}_{y+1}B_1 +\ ^{0}_{-1}e$

Its mass number remains the same but the atomic number increases by 1.

43. A nucleus of an element has the symbol $^{202}_{84}P$ and emits an α-particle and then a β-particle. The final nucleus is $^{a}_{b}Q$. Find a and b.

Ans. (i) On the emission of an alpha ($^{4}_{2}He$) particle:

$$^{202}_{84}P \xrightarrow{\alpha}\ ^{198}_{82}X \text{ (new nucleus)}$$

(ii) On the emission of a beta particle :

$$^{198}_{82}X \xrightarrow{\alpha\text{-particle}}\ ^{198}_{83}Q$$

$^{198}_{83}Q$ is the same as a $^{a}_{b}Q$. Hence $a = 198$, $b = 83$.

44. How do infrared and γ-rays differ in (i) wavelength, (ii) penetrating power?

Ans. (i) Infrared rays are of much longer wavelength than that of γ rays. The wavelength of infrared rays lies between 8000 Å to 107 Å while that of γ rays are less than 0.1 Å.

(ii) γ-rays are most penetrating while infrared rays have almost negligible penetrating power.

45. From α, β and γ-rays, name the one which travels with the speed of light?

Ans. γ-rays travel with the speed of light.

46. From α, β and γ-rays, name the one which is not affected by electric or magnetic field?

Ans. γ–rays are not affected by electric or magnetic field.

47. Arrange α , β and γ-rays in ascending order with respect to their (i) Penetrating power, (ii) Ionizing power and (iii) Biological effect.

Ans. Given below:

		Minimum	More	Maximum
(i)	Penetrating power	α	β	γ
(ii)	Ionizing power	γ	β	α
(iii)	Biological effect	α	β	γ

48. What changes occur in the nucleus of radioactive element when it emits gamma radiation? Give an example.

Ans. When the nucleus of a radioactive element emits gamma radiation, there is no change inside the nucleus except that its excess energy is released in the form of γ-radiation.

Example: $^{A}_{Z}Z^{*} \rightarrow\ ^{A}_{Z}Z + \gamma$

The * indicates the excited state of nucleus.

49. What is the value of the speed of gamma radiations in air or vacuum?

Ans. Speed of γ radiation = 3×10^8 m/s in air or vacuum.

50. What are beta particles and gamma rays ?

Ans. β-particles are a stream of electrons emitted by the nucleus of an atom and γ-rays are electromagnetic waves of high frequency given off by nuclei.

51. Explain why alpha and beta-particles are deflected in an electric or a magnetic field but gamma rays are not deflected in such a field.

Ans. α and β are positive and negative charged particles respectively, therefore, these are deflected in electric or magnetic field whereas γ radiations are not charged particles so do not deflect.

52. State the medical use of radioactivity.

Ans. Radioactive radiations (gamma rays) are used to cure diseases such as cancer etc.

53. A radioactive substance is oxidized. What change would you expect to take place in the nature of radioactivity? Give reason for your answer.

Ans. No changes as the nucleus of the element is not involved.

54. Explain, why radium paint, consisting of zinc sulfide and a trace of radium salt, glows in the dark?

Ans. Radium is a radioactive substance. It gives out radioactive radiations which show fluorescence in zinc sulfide. Zinc sulfide is a fluorescent material, so it glows in dark.

55. State one use of radio-isotopes.

Ans. The most important use of radio-isotopes is in medicines which are used mainly to cure cancer.

56. State three safety precautions that you would take while using a radioactive substance.

Ans. (i) We should put on special lead lined aprons, lead gloves and special film badges.

(ii) We should not touch the radioactive substances but should handle them with long lead tongs.

(iii) The radioactive substance should be kept in a thick walled lead container.

57. It is advised not to touch a radioactive substance by hand. Give reason.

Ans. The reason is that the radiations (α, β and γ) emitted by a radioactive substance have large penetrating power.

58. Which of the radioactive radiations can cause severe genetical disorders ?

Ans. γ-radiations can cause severe genetical disorder.

59. Give any two important sources of background radiation.

Ans. Two sources of background radiations are :

(i) Potassium (K-40), Carbon (C-14) present in our body.

(ii) Cosmic rays coming from outer space.

60. What is mean by background radiation?

Ans. Background radiations are the radioactive radiations to which we all are exposed even in the absence of an actual visible radioactive source. Its total dose is not very large, so it does not cause any serious biological damage to us.

61. What are natural radioactive substances?

Ans. The isotopes of nearly all the elements of atomic number higher than 82 (*i.e.* after lead) in the periodic table are radioactive, and are called natural radioactive substances.

62. Name two main sources of harmful radiations.

Ans. The two main sources of harmful radiations are: (i) Nuclear waste and (ii) radioactive fall out from nuclear plants and other sources.

63. State one advantage of using nuclear energy?

Ans. A very small amount of nuclear fuel can produce a tremendous amount of energy.

64. What is a nuclear power plant?

Ans. Nuclear power plant is a setup where controlled chain reaction of nuclear fission of Uranium-235 or Plutonium-239 can be achieved.

65. What are radioactive substances? Give two examples of such substances.

Ans. Radioactive substances are those which have the property of self emission of α or β and γ radiations.

Examples: Uranium, Thorium.

66. State two disadvantages of using nuclear energy.

Ans. (i) The waste obtained from nuclear power plants causes environmental pollution.

(ii) It is not a clean source of energy because harmful radiations are produced.

67. What are the practical difficulties in controlling fusion reaction?

Ans. To carry out a controlled fusion reaction, it is necessary to maintain a temperature of the order of 10^8K at a very high pressure. No vessel of any substance can withstand such a high temperature and pressure.

68. What is nuclear energy? Name the process used for producing electricity using nuclear energy.

Ans. The energy released due to loss in mass during the processes of nuclear fission and fusion is called nuclear energy. The loss in mass (mass defect) gets converted into energy is accordance with Einstein's mass-energy relation ($E = mc^2$).

69. Why are radioactive substances stored in thick lead containers with a very narrow opening ?

Ans. To stop radiations coming out from other directions.

70. What is meant by nuclear waste ?

Ans. The radioactive material after its use is known as nuclear waste.

71. Suggest one effective way for the safe disposal of nuclear waste.

Ans. Nuclear waste must be stored in thick containers and sealed. These containers must be buried in specially constructed deep underground stores in remote areas.

72. State two applications of nuclear reactions?

Ans. (i) Energy produced in nuclear reactors is used to produce electricity.

(ii) Artificial radio isotopes of many elements are produced in the nuclear reactors.

(iii) Gamma radiations obtained from cobalt-60 are used to treat cancer.

73. What is used as fuel in a nuclear reactor?

Ans. Uranium-235, or plutonium-239.

74. What is used as a moderator in a nuclear reactor?

Ans. Heavy water, or graphite is used as a moderator to slow down the neutrons.

75. What are the control rods in a nuclear reactor made of?

Ans. Control rods are made of cadmium, or boron to absorb the extra neutrons produced in the chain reaction.

76. What is the difference between chemical reaction and nuclear reactions.

Ans. A chemical reaction involves only a rearrangement of electrons and does not involve the nucleus whereas a nuclear reaction involves a change in the nucleus producing a new element.

77. What is chain reaction?

Ans. During the fission of uranium, along with energy, two or three neutrons are also emitted. Each neutron emitted causes fission in a new uranium nucleus. This leads to a chain reaction.

78. Name two types of chain reaction.

Ans. (i) Controlled chain reaction, (ii) Uncontrolled chain reaction.

79. What is nuclear fission?

Ans. Nuclear fission is a process in which a heavy nucleus, after capturing a neutron, splits into two lighter nuclei of comparable mass with the release of a large amount of energy.

80. What is nuclear fusion?

Ans. When two light nuclei combine to form a heavy nucleus at a very high temperature and pressure, a tremendous amount of energy is released. This phenomenon is called nuclear fusion.

81. What is mass defect?

Ans. During nuclear reactions, there is some loss in mass, which gets converted into energy ($E = mc^2$). This loss in mass is called mass defect.

82. (i) Why is a nuclear fusion reaction called a thermo nuclear reaction?

(ii) Complete the reaction:

$$^3He_2 + {}^2H_1 \longrightarrow {}^4He_2 + \ldots + \text{Energy}$$

Ans. (i) Nuclear fusion takes place at extremely high temperature and pressure and the two nuclei due to their thermal agitations acquire sufficient kinetic energy so as to overcome the force of repulsion between them when they approach each other, so they get fused. Hence, it is called a thermo nuclear reactions.

(ii) $^3He_2 + {}^2H_1 \longrightarrow {}^4He_2 + \boxed{{}^1H_1} + \text{Energy}$

83. Justify with reasons, whether the following nuclear reactions are allowed or not.

(i) $^A{}_ZX \rightarrow {}^A{}_{Z+1}Y + {}^0{}_1b$

(ii) $^A{}_ZX \rightarrow {}^A{}_{Z-2}X' + {}^4{}_2He$

Ans. (i) This reaction is allowed because, the mass number as well as atomic number are conserved here.

(ii) This reaction is not allowed because even though charge (atomic number) is conserved in it, the mass number is not conserved.

84. A radioactive nucleus undergoes a series of decays according to the sequence

$$X \xrightarrow{\beta} X_1 \xrightarrow{\alpha} X_2 \xrightarrow{\alpha} X_3$$

If the mass number and atomic number of X_3 are 172 and 69 respectively, what is the mass number and atomic number of X?

Ans. ${}^b_aX_2 \xrightarrow{\alpha} {}^{172}_{69}X_3 \Rightarrow a = 69 + 2 = 71, b = 172 + 4 = 176$

${}^q_pX_1 \xrightarrow{\alpha} {}^{176}_{71}X_2 \Rightarrow p = 71 + 2 = 73, q = 176 + 4 = 180$

and ${}^y_xX \xrightarrow{\beta} {}^{180}_{73}X \Rightarrow x = 73 - 1 = 72, y = 180$

$\therefore$ Mass no. of X = 180

Atomic no. of X = 72

85. Complete the following nuclear changes:

(i) ${}^{24}_{11}Na \rightarrow \ldots Mg \ldots + {}^{0}_{-1}\beta$

(ii) ${}^{238}_{92}U \rightarrow {}^{234}_{90}Th + \ldots + \text{Energy}$

(iii) ${}_{92}P^{238} \xrightarrow{-\alpha} \ldots P_1 \ldots \xrightarrow{-\beta} \ldots P_2 \cdots \xrightarrow{-\beta} \ldots P_3 \cdots$

Ans. (i) ${}^{24}_{11}Na \longrightarrow {}^{24}_{12}Mg + {}^{0}_{-1}\beta$

(ii) ${}^{238}_{92}U \longrightarrow {}^{234}_{90}Th + {}^4_2He + \text{Energy}$

(iii) ${}_{92}P^{238} \xrightarrow{-\alpha} {}_{90}P_1^{234} \xrightarrow{-\beta} {}_{91}P_2^{234} \xrightarrow{-\beta} {}_{92}P_3^{234}$

86. An imaginary radioactive element ${}_{92}X^{235}$ decays to form X_1, X_2, X_3, X_4, X_5 and X6 nuclides by ejecting two β-particles, followed by α-particle and again two β-particles followed by α-particle. Represent the various nuclear changes in the form of an equation. State the mass number and atomic number of X_6. List the isotopes and isobars in the above nuclear equation.

Ans. ${}_{92}X^{235} \xrightarrow{-\beta} {}_{93}X_1^{235} \xrightarrow{-\beta} {}_{94}X_2^{235} \xrightarrow{-\alpha} {}_{92}X_3^{231} \xrightarrow{-\beta} {}_{93}X_4^{231} \xrightarrow{-\beta} {}_{94}X_5^{231} \xrightarrow{-\alpha} {}_{92}X_6^{227}$

The mass number and atomic number of X_6 in 227 and 92 respectively.

Isotopes: $\{_{92}X^{234},\ _{92}X_3^{231},\ _{92}X_6^{227};$

$_{93}X_1^{235},\ _{93}X_4^{241};\ _{94}X_2^{235},\ _{94}X_5^{231}\}$

Isobars: $\{_{92}X^{235},\ _{93}X_1^{235},\ _{94}X_2^{235};$

$_{92}X_3^{231},\ _{93}X_4^{231},\ _{94}X_5^{231}\}$

87. State, giving reasons, whether the following nuclear decays are allowed or not ?

(i) $^{A}_{Z}X \rightarrow ^{A}_{Z}Q + \gamma$

(ii) $^{A}_{Z}P \rightarrow ^{A}_{Z\text{-}2}Y + ^{4}_{2}He$

(iii) $^{A}_{Z}X \rightarrow ^{A\text{-}1}_{Z}Y + ^{0}_{-1}e$

Ans. (i) This nuclear decay is allowed since the atomic number and mass number are conserved.

(ii) This nuclear decay is not allowed since the mass number is not conserved.

(iii) This nuclear decay is not allowed since the atomic number and mass number both are not conserved.

88. A nucleus $_{11}Na^{24}$ emits a beta particle to change into Magnesium (Mg)

(i) Write the symbolic equation for the process.

(ii) What are numbers 24 and 11 called?

(iii) What is the general name $^{24}_{12}Mg$ with respect to $^{24}_{11}Na$?

Ans. (i) $_{11}Na^{24} \longrightarrow _{12}Mg^{24} + _{-1}e^{0}$ + energy

(ii) 24 —mass number (no. of protons and neutrons)

11 —atomic number (no. of protons)

(iii) $_{11}Na^{24}$ is called parent nucleus.

$_{12}Mg^{24}$ is called daughter product.

89. The nucleus $^{202}_{84}X$ emits an alpha particle and forms the nucleus Y. Represent this change in the form of an equation.

Ans. $_{84}X^{202} \rightarrow _{82}Y^{198} + _{2}He^{4}$.

90. State two advantages of nuclear fusion over nuclear fission.

Ans. (i) The fuel for fusion reaction is hydrogen which is found in abundance in sea water, whereas nuclear fission uses uranium.

(ii) No radioactive substance is formed in fusion reaction.

Short Answers

Figure Based Short Answers

Set 3

Chapter 1. Force, Work, Power and Energy

1. Is it possible to have an accelerated motion with a constant speed? Explain.

Ans. Yes, it is possible to have accelerated motion with constant speed. When a body is moving in a circular path, the body changes its in direction continuously but its speed remains the same and hence the velocity changes and there is a resultant uniform acceleration.

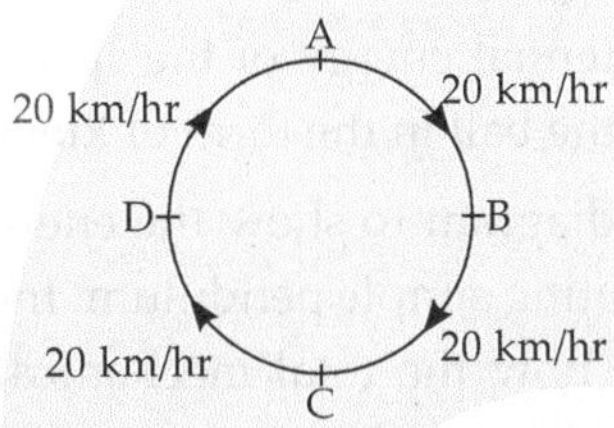

2. What is a couple?

Ans. Two equal, parallel and opposite forces acting on a rigid body at different points constitute a couple. The distance '*d*' is called arm of the couple.

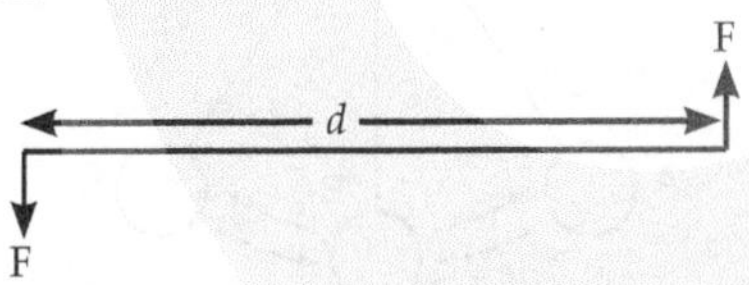

3. A boy of mass 30 kg is sitting at a distance of 2 m from the middle of a see-saw. Where should a boy of mass 40 kg sit so as to balance the see-saw?

Sol. By principle of moments,

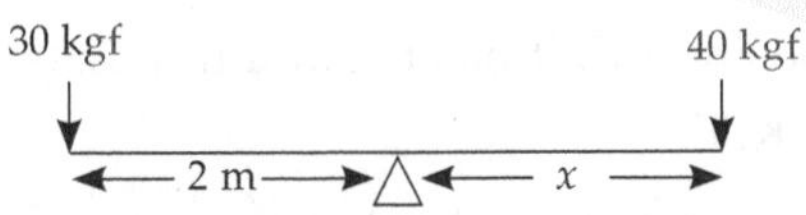

$$30 \times 2 = x \times 40$$

$\therefore$ Other boy should sit at a distance,

$$x = \frac{30 \times 2}{40}$$

$= 1{\cdot}5$ m from the middle of the see-saw. **Ans.**

4. Write the expression for calculating the moment of force about a given point.

Ans. Moment of force about the point O

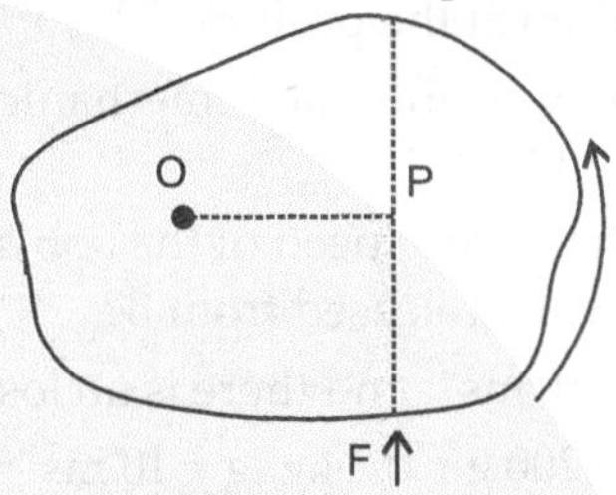

= Force × Perpendicular distance of force from the point O

= F × OP

5. The moment of a force of 10 N about a fixed point O is 5 Nm. Calculate the distance of the point O from the line of action of the force.

Sol. Given:

Moment of force = 5 Nm

Force = 10 N

According to the formula,

Moment of force = F × OA

$5 = 10 \times OA$

$$OA = \frac{5}{10}$$

$= 0{\cdot}5$ m **Ans.**

6. The centre of gravity of a boomerang lies outside the material of the body. Draw a diagram to show its position.

Ans.

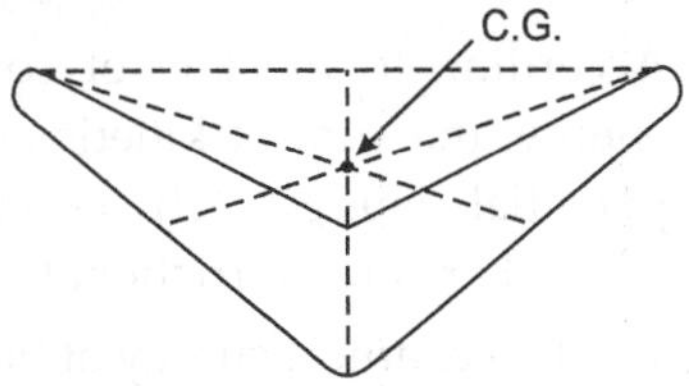

Boomerang

7. The figure below shows a simple pendulum of mass 200 g. It is displaced from the mean position A to the extreme position B. The potential energy at the position A is zero. At the position B the pendulum bob is raised by 5 m.

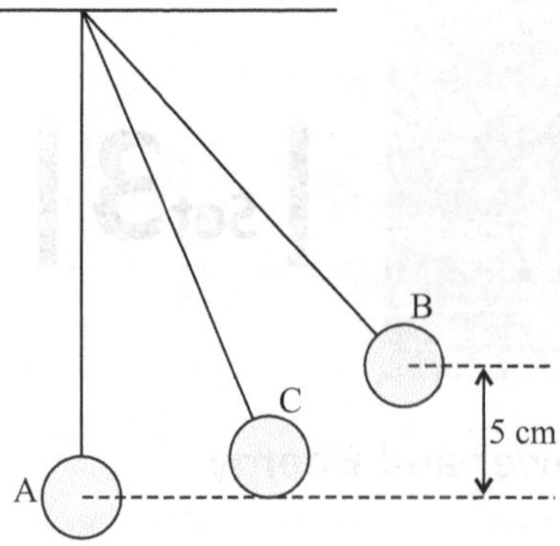

(i) What is the potential energy of the pendulum at the position B?

(ii) What is the total mechanical energy at point C?

(iii) What is the speed of the bob at the position A when released from B?

(Take $g = 10\ ms^{-2}$ and there is no loss of energy.)*

Ans. (i) $m = 200\ g = 0.2\ kg$, $g = 10\ ms^{-2}$, $h = 5\ m$

∴ Potential energy (U) = mgh

at position B = $0.2 \times 10 \times 5 = 10$ J

(ii) 10 J (∵ K + U = constant)

(iii) At position A, kinetic energy (K) = 10 J

$$\frac{1}{2}mv^2 = 10$$

$$\Rightarrow \frac{1}{2} \times 0.2 \times v^2 = 10$$

$$\Rightarrow v^2 = \frac{10 \times 2}{0.2} = 100$$

$$\Rightarrow v = \sqrt{100} = 10\ ms^{-1}$$

8. A body of mass 10 kg is kept at a height of 5 m. It is allowed to fall and reach the ground.

(i) What is the total mechanical energy possessed by the body at the height of 2 m assuming it is a frictionless medium?

(ii) What is the kinetic energy possessed by the body just before hitting the ground? (Take $g = 10\ m/s^2$.)*

Ans. Given: Mass (m) = 10 kg, height (h) = 5 m, g = 10 ms^{-2}, Potential energy (U) = mgh

= 10 × 10 × 5 J = 500 J

(i) According to the law of conservation of energy, the sum of kinetic energy (k) and potential energy (U) remains constant when there are no frictional forces.

∴ Total mechanical energy at height of 2 m
= Initial potential energy = 500 J

(ii) Similarly, according to the law of conservation of energy, the kinetic energy possessed by the body just before touching the ground = 500 J.

9. A boy sits and stands repeatedly. Draw a graph showing the variation of potential energy with time.

Ans.

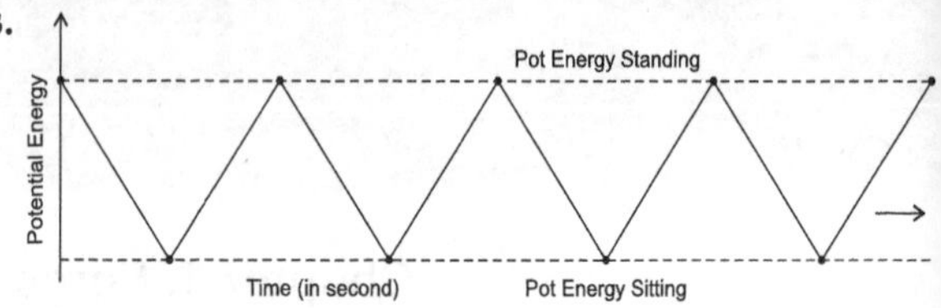

10. A ball is placed on a compressed spring. When the spring is released, the ball is observed to fly away.

(i) What form of energy does the compressed spring possess?

(ii) Why does the ball fly away?

Ans. (i) Compressed spring possesses potential energy.

(ii) Potential energy of the spring is imparted to the ball in the form of kinetic energy.

11. Draw a diagram to show the energy changes in an oscillating simple pendulum. Indicate in your diagram how the total mechanical energy in it remains constant during the oscillation.

Ans. Diagram shows a freely oscillating pendulum.

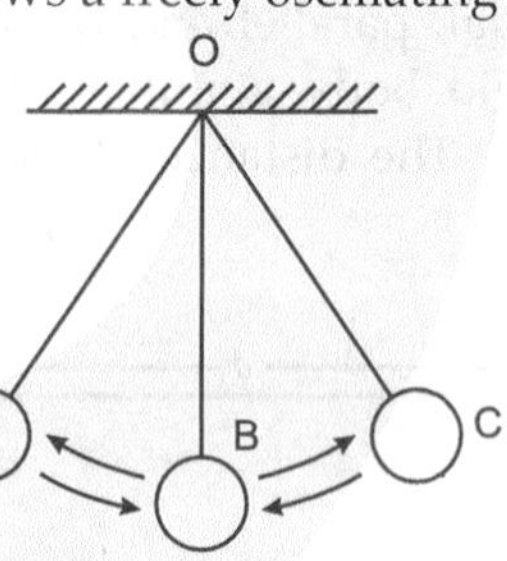

At position A, it has maximum P.E. (= mgh) and zero K.E.

At position B, it has zero P.E. and maximum K.E. (= mgh).

At position C, it has maximum P.E. (= mgh) and zero K. E.

Thus, experiment clearly proves that sum of total energy is conserved, but it can change its form.

12. What is the form of the graph between

(i) K.E. and velocity for a constant mass?

(ii) K.E. and mass for a constant velocity?

Ans. (i) For constant mass,

$$K.E. \propto v^2$$

Hence, the graph is a parabola.

* Frequently asked previous years Board Exam Questions.

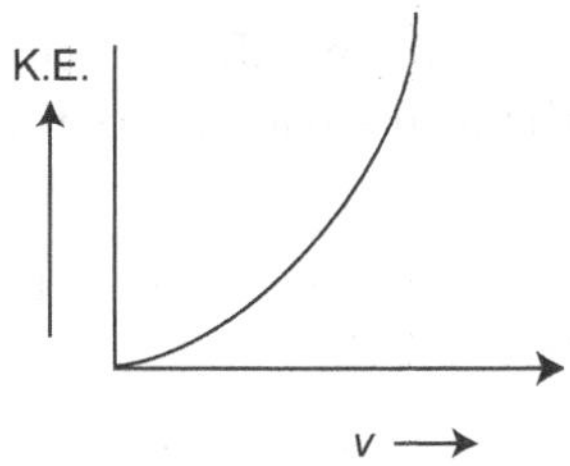

(ii) For constant velocity,

$$K.E. \propto m$$

Hence, the graph is a straight line.

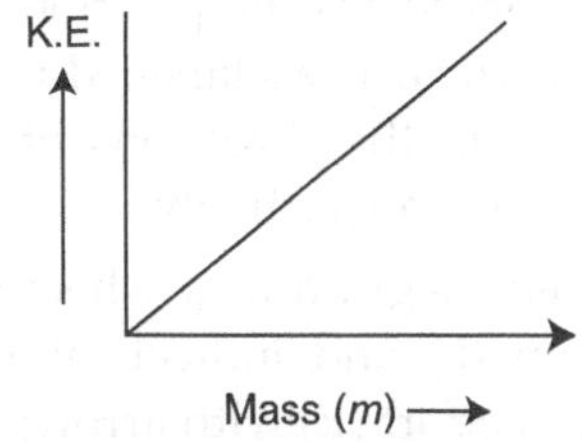

Chapter 2. Simple Machines

1. A crowbar of length 120 cm has its fulcrum situated at a distance of 20 cm from the load. Calculate the mechanical advantage of the crow bar.

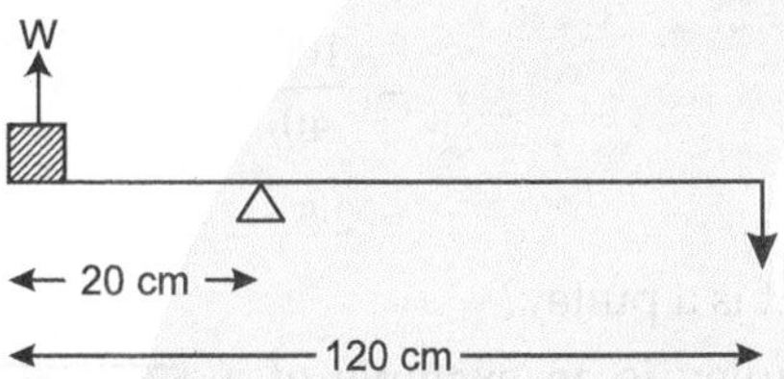

Ans. By the principle of moments,

Moments of load about the fulcrum

= Moment of efforts about the fulcrum

Load arm = 20 cm

Effort arm = (120 – 20) cm = 100 cm

$$\therefore \quad \text{Mechanical advantage} = \frac{\text{Effort arm}}{\text{Load arm}}$$

$$= \frac{100}{20} = 5.$$ **Ans.**

2. The diagram shows the combination of a movable pulley P_1 with a fixed pulley P_2 used for lifting up a load (L).

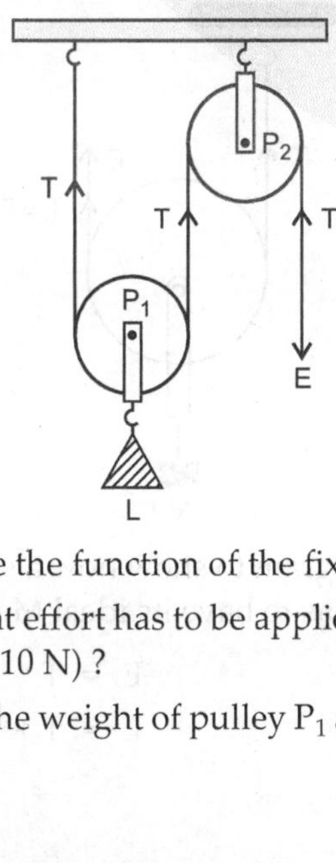

(i) State the function of the fixed pulley P_2.

(ii) What effort has to be applied to raise a load (L = 10 N) ?

(Neglect the weight of pulley P_1 and the friction).

Ans. (i) The fixed pulley P_2 is used to change the direction of effort (E) to be applied.

(ii) If we neglect the weight of pulley P_1 and the friction, then in equilibrium position:

$$E = T \text{ and } L = 2T$$

$$M.A. = \frac{L}{E} = \frac{2T}{T} = 2$$

$$\Rightarrow \quad \frac{L}{E} = 2$$

$$\Rightarrow \quad \frac{10}{E} = 2$$

$$\therefore \quad E = 5 \text{ N}.$$ **Ans.**

3. Draw a labelled sketch of a class II lever. Give one example of such a lever.

Ans. A labelled diagram of a class II lever is given below. Its example is a nut-cracker.

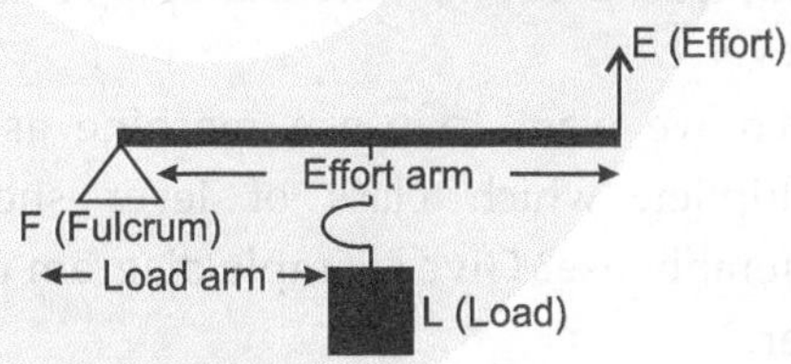

Second Class Lever

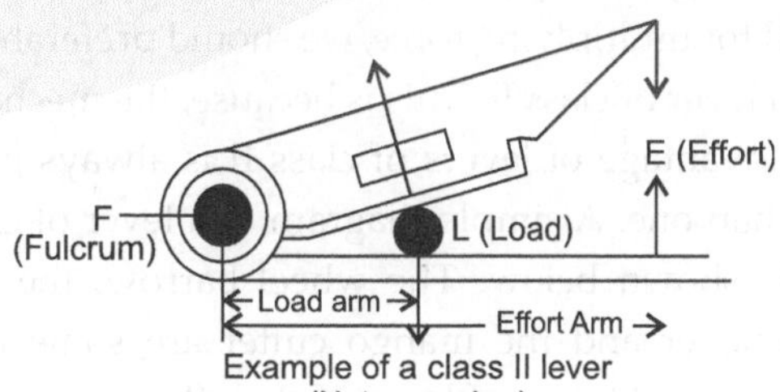

Example of a class II lever
(Nut – cracker)

4. In the diagram of a claw hammer, mark the fulcrum (F) and indicate the directions of load (L) and effort (E) with arrows. What class of lever is it ? Give one more example of this class of lever.

Ans. Claw-hammer is the lever of first order. One more example of this class of lever is see-saw. In the following diagram, F indicates the position of fulcrum. The arrows labelled L and E indicate the directions of load and effort, respectively.

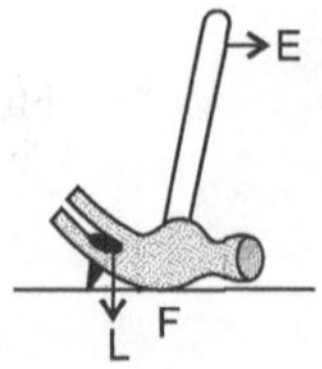

5. In the given diagram of a wheel barrow, mark the fulcrum (F) and indicate the directions of load (L) and effort (E) with arrows.

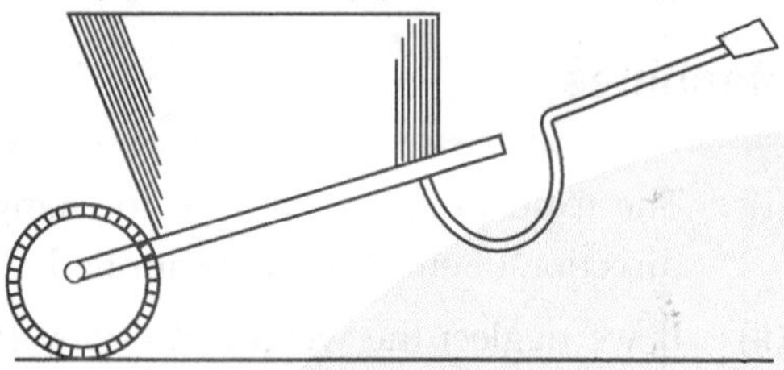

Wheel Barrow

What class of lever is it? Give one more example of the same class of lever.

Ans. In the diagram given below, F shows the position of fulcrum and arrows marked at L and E indicate the direction of load and effort respectively.

It belongs to class II lever.

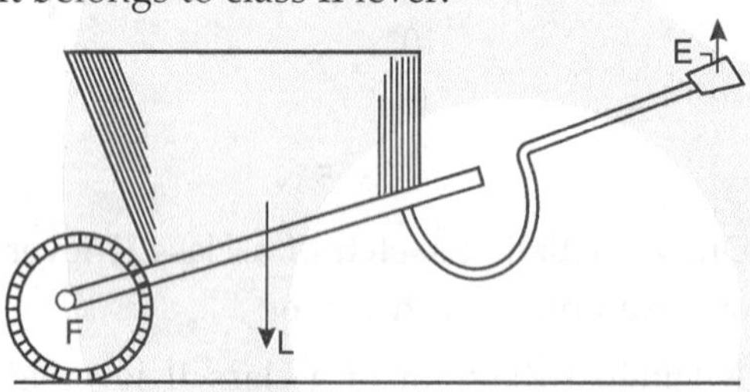

Wheel Barrow

A nut cracker is one more example of same class of lever.

6. When we want to use a machine as a force multiplier, which class of lever should we preferably use? Give a simple diagram of such a lever.

Ans. Though we can use either levers of class I or class II for multiplying force, we should preferably use a lever of class II. This is because, the mechanical advantage of levers of class II is always greater than one. A simple diagram of a lever of class II is shown below. The wheel barrow, the bottle opener and the mango cutter are some of the practical forms of lever of class II.

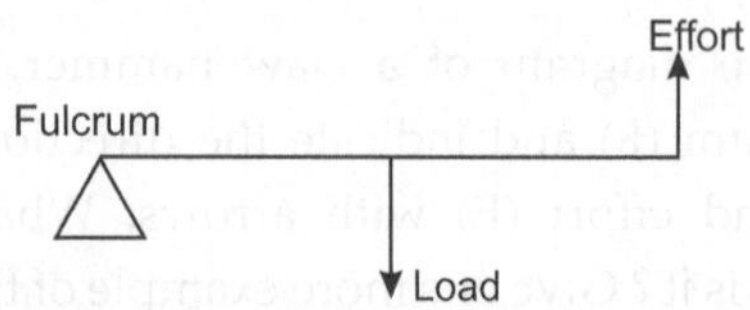

7. The diagram below shows a lever in use.

(i) To which class of lever does it belong?

(ii) If FA = 40 cm, AB = 60 cm, then find the mechanical advantage of the lever.

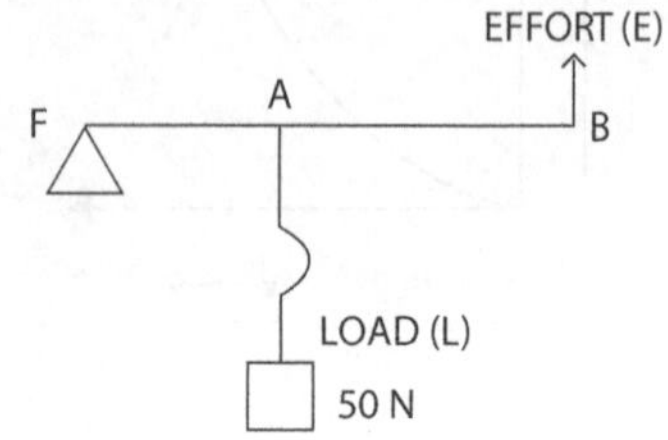

Sol. (i) Class 2 lever.

(ii) Given : FA = 40 cm, AB = 60 cm

∴ FB = (40 + 60) cm = 100 cm

We know, Load × Load arm = Effort × Effort arm

and, $$\text{M.A.} = \frac{\text{Load}}{\text{Effort}} = \frac{\text{Effort arm (FB)}}{\text{Load arm (FA)}} = \frac{100}{40} = 2{\cdot}5.$$ **Ans.**

8. What is a pulley?

Ans. A pulley is an example of a simple type of machine. It is either a wooden or a metallic disc which can rotate about a horizontal axis passing through its centre. It has a grooved rim along which a rope or chain can slide. It is generally mounted in a frame work called a block.

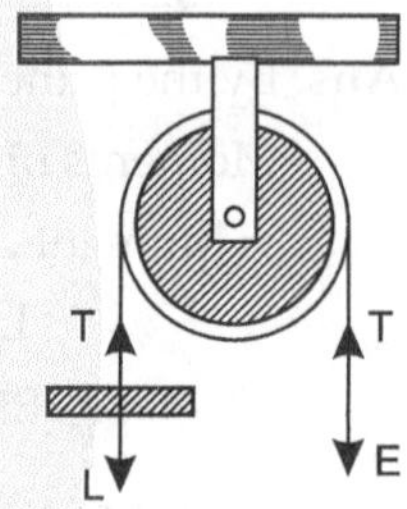

9. Name the type of single pulley that can act as a force multiplier. Draw a labelled diagram of the above named pulley.

Ans. Single Movable Pulley

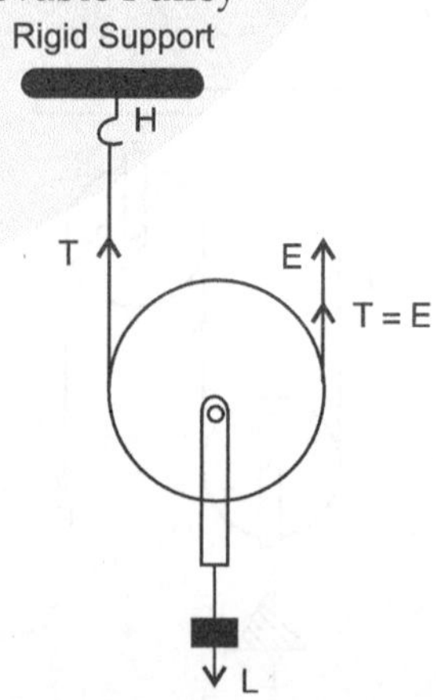

10. Draw a diagram to show how a single pulley can be used so as to have its ideal M.A. = 2.

Sol. E = T

L = T + T = 2T

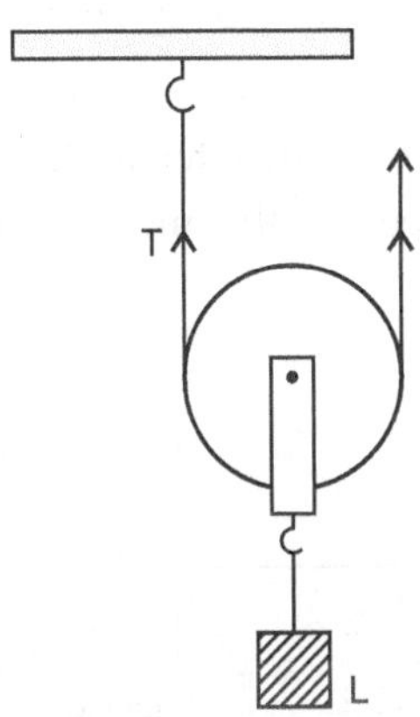

$$\text{M.A.} = \frac{L}{E} = \frac{2T}{T} = 2.$$

11. (i) Identify the class of the lever shown in the diagram below:

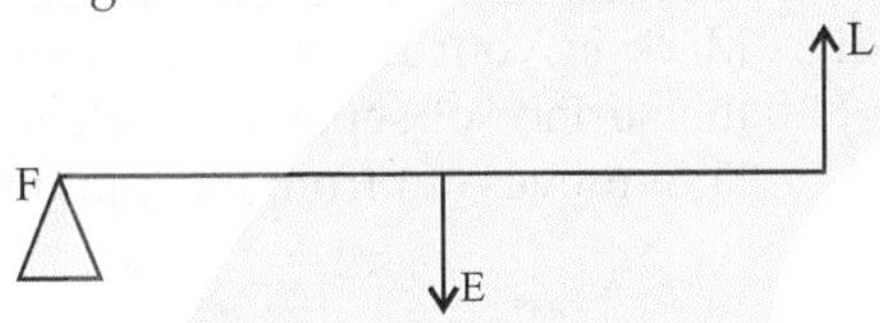

(ii) How is it possible to increase the M. A. of the above lever without increasing its length?*

Ans. (i) Class III lever.

(ii) Without increasing the length of the lever, its mechanical advantage can be increased by shifting the effort (E) towards the load (L), *i.e.*, by increasing the effort arm.

12. The diagram below shows a claw hammer used to remove a nail:

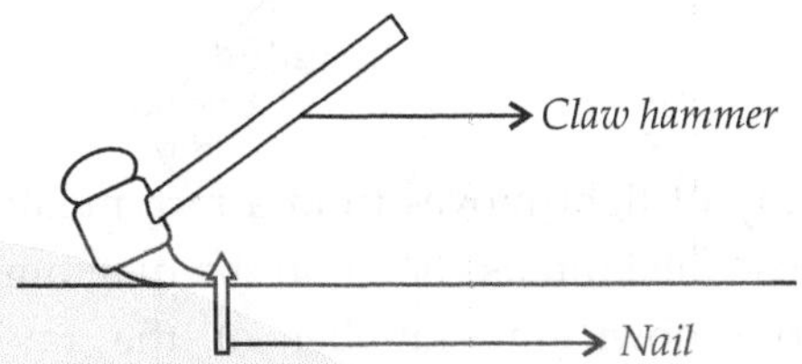

(i) To which class of lever does it belong?

(ii) Give one more example of the same class of lever mentioned by you in (i) for which the mechanical advantage is greater than one.*

Ans. (i) Class first lever.

(ii) Pliers.

Chapter 3. Refraction of Light

1. A ray of light falls normally on a rectangular glass slab.
Draw a ray diagram showing the path of the ray till it emerges out of the slab.*

Ans. AD → Normal
AB → Incident Ray
CD → Emergent Ray

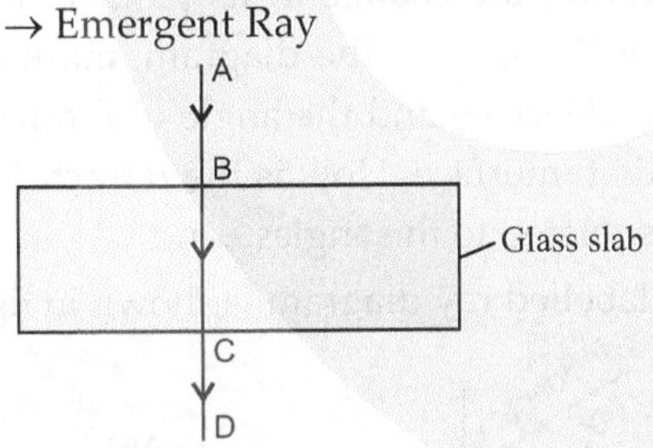

2. Complete the path of the monochromatic light ray AB incident on the surface PQ of the equilateral glass prism PQR till it emerges out of the prism due to refraction.

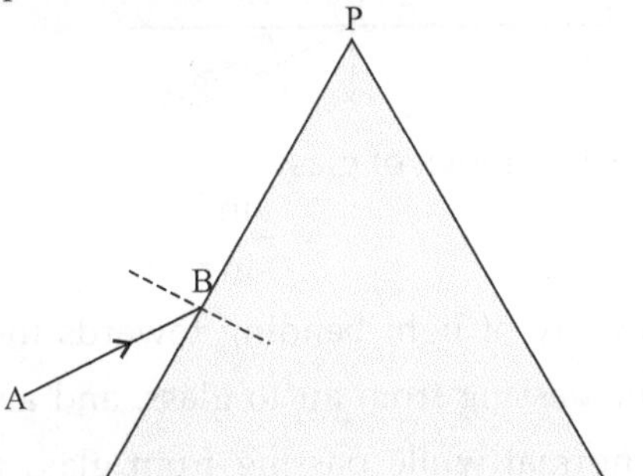

Ans. MM′ → Normal at B
NN′→ Normal at C
AB → Incident Ray
BC → Refracted Ray
CD → Emergent Ray.

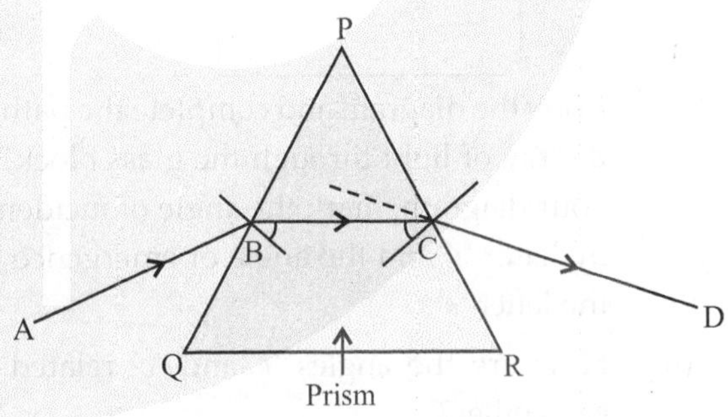

3. Complete the diagram given below and clearly show the path taken by the emergent ray.

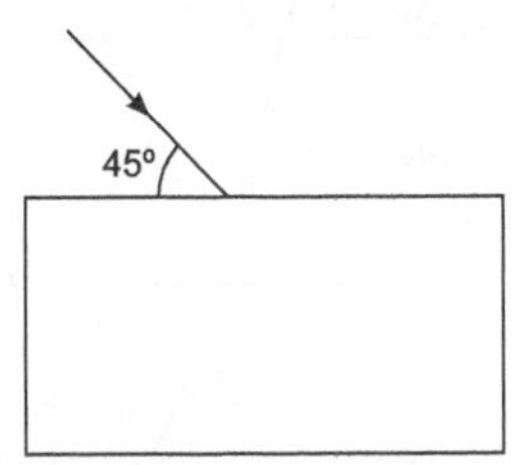

Ans. The ray undergoes refraction as it emerges out of the slab.

* Frequently asked previous years Board Exam Questions.

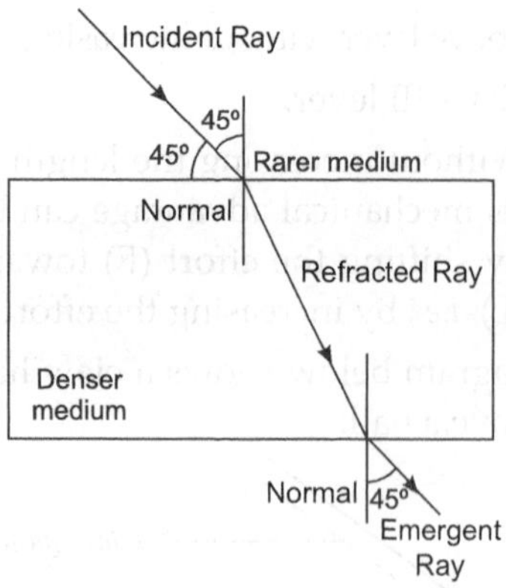

4. A ray of light moves from a rare medium to a dense medium as shown in the diagram below. Write down the number of the ray which represents the partially reflected ray.

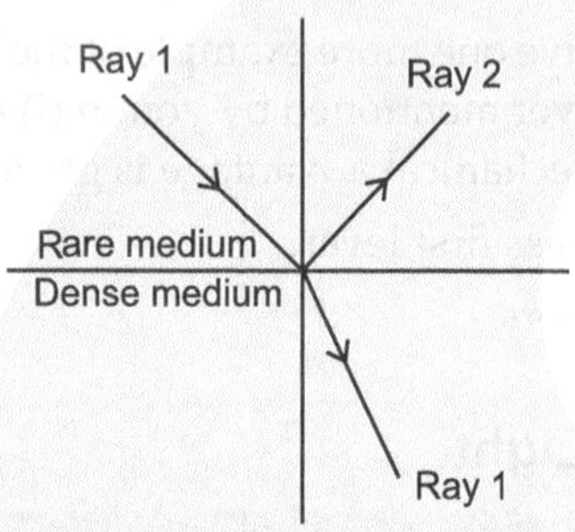

Ans. Ray 2 shows partially reflected ray.

5. In the diagram below, PQ is a ray of light incident on a rectangular glass block.

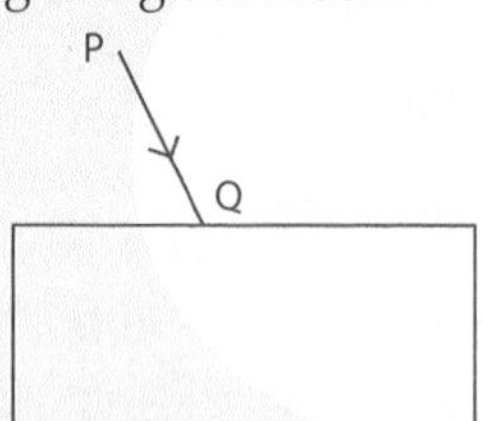

(i) Copy the diagram and complete the path of the ray of light through the glass block. In your diagram, mark the angle of incidence by letter '*i*' and the angle of emergence by the letter '*e*'.

(ii) How are the angles '*i*' and '*e*' related to each other?

Ans. (i)

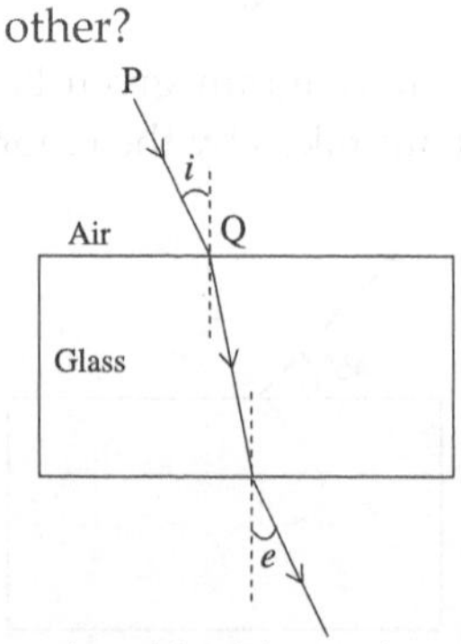

(ii) The angle of incidence is equal to angle of emergence, *i.e.*, $\angle i = \angle e$.

6. A ray of monochromatic light enters a liquid from air as shown in the diagram given below:

(i) Copy the diagram and show in the diagram the path of the ray of light after it strikes the mirror and re-enters the medium of air.

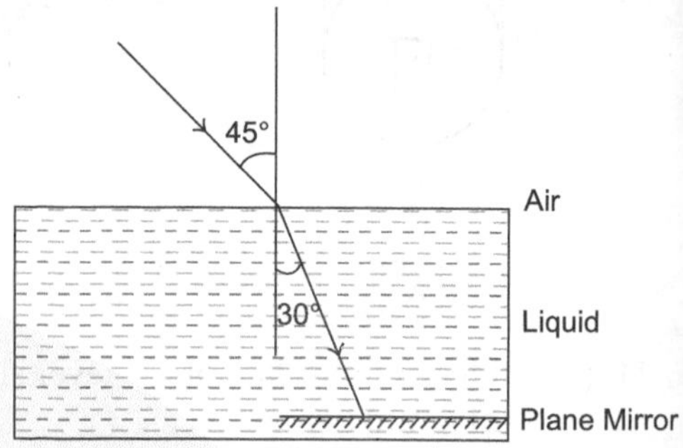

(ii) Mark in your diagram the two angles on the surface of separation when the ray of light moves out from the liquid to air.

Ans. (i)

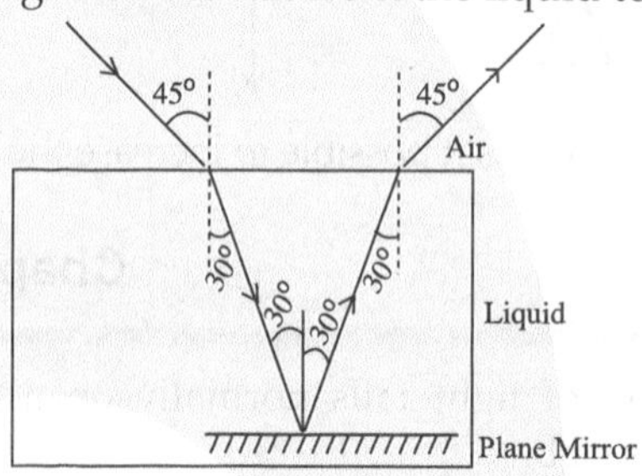

(ii) Angles are marked in the diagram.

7. A ray of monochromatic light is incident from air into a glass slab. Draw a labelled ray diagram indicating the change in its path till it emerges out of the slab. In the diagram, mark the angle of incidence (i) and the angle of refraction (r) at the first interface. How is the refractive index of glass related to the angles i and r ?

Ans. The labelled ray diagram is shown in figure:

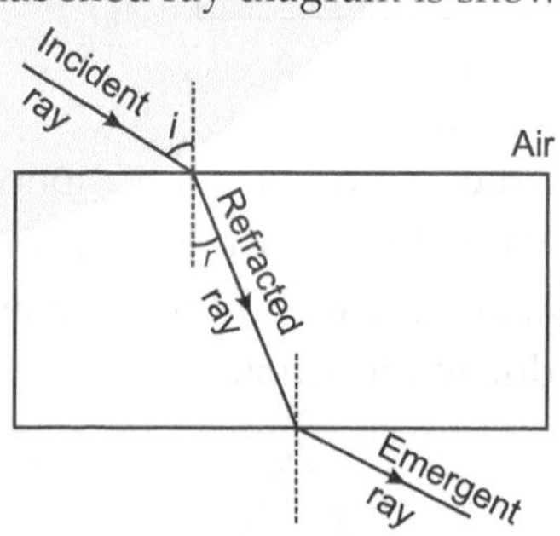

Refractive index of glass

$$= \frac{\sin i}{\sin r}$$

8. Draw ray of light bending towards the normal while passing from air to glass, and away from the normal while passing from glass to water. Label your diagrams.

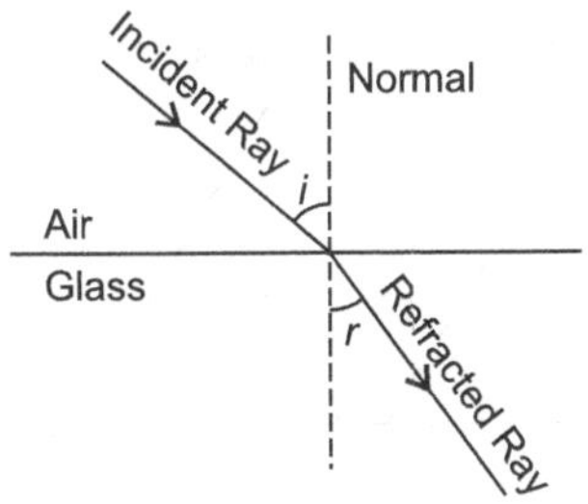

Ans.

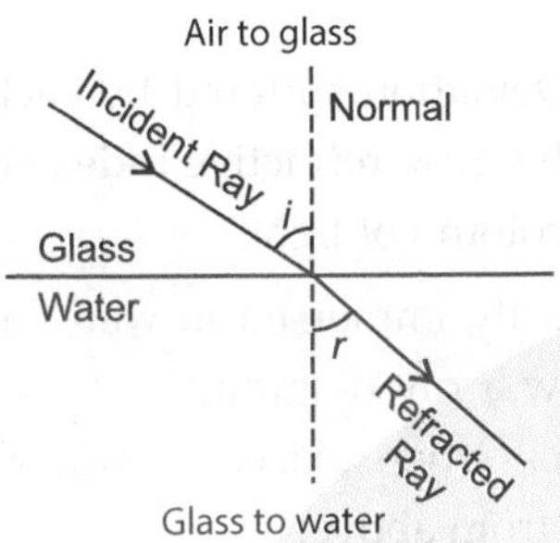

9. Show the path of a ray of light when it travels from air into water, the angle of incidence being 30°. Mark the angle of incidence and the corresponding angle of refraction.

Ans.

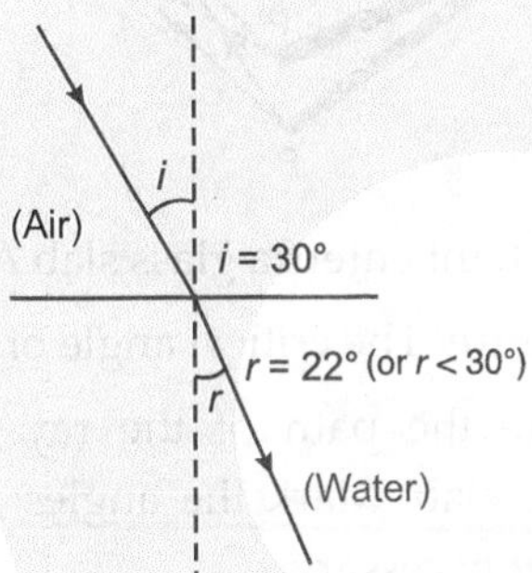

10. A ray of light is moving from a rarer medium to a denser medium and strikes a plane mirror placed at 90° to the direction of the ray as shown in the diagram.

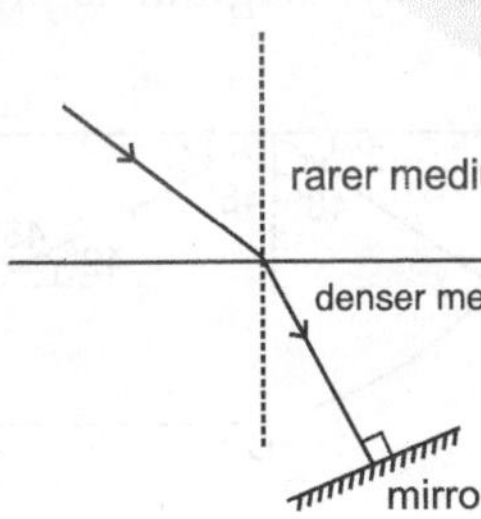

(i) Copy the diagram and mark arrows to show the path of the ray of light after it is reflected from the mirror.

(ii) Name the principle you have used to mark the arrows to show the direction of the ray.

Ans. (i)

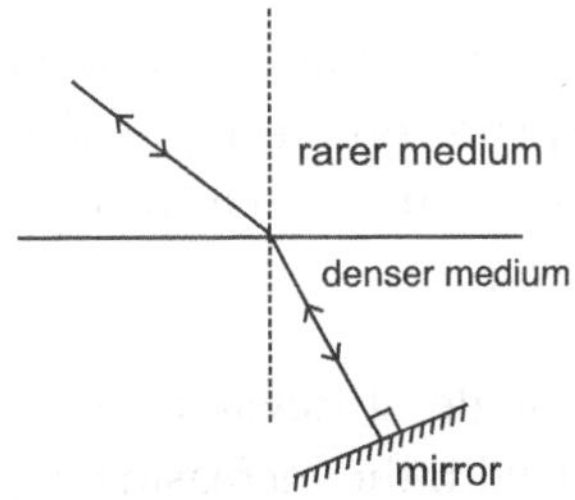

(ii) Principle of reversibility of light.

11. After a robbery, if a window has been broken, there will be tiny particles of glass. Some of these will be found at the scene of the crime and some may be caught in the thief's clothing. If the police can prove that these particles are identical, they have a strong case.

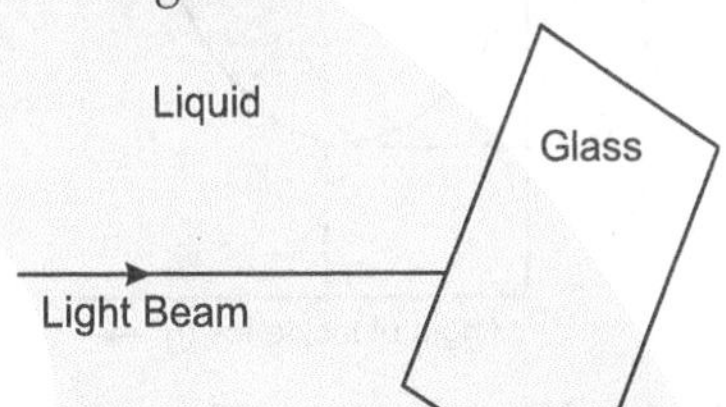

A method of doing this is to suspend the particles of glass in a special liquid. Light of a single colour is thrown through the liquid and the particles viewed through a microscope. The temperature of the liquid is then slowly altered. This alters the speed of light through the liquid (*i.e.,* it alters the refractive index). At one particular temperature, the particles of glass disappear. If this happens at the same temperature for both sets of glass particles, they probably came from the same broken pane of glass.

Complete and copy the diagram to show how light bends when it travels from the liquid to the glass and back to the liquid,

(i) If the light slows down in the glass, (ii) If the light speeds up in the glass.

Ans. (i)

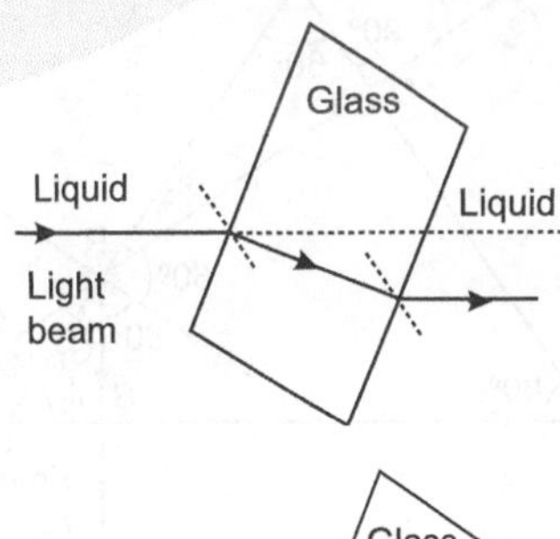

(ii)

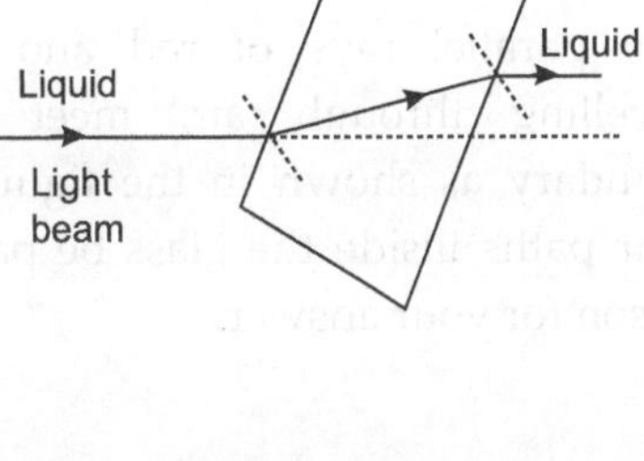

12. How does the angle of deviation produced by a prism depends on the angle of incidence of light at the prism surface? Draw a graph to illustrate your answer.

Ans. First the angle of deviation produced by prism decreases with the increasing angle of incidence but on further increasing the angle of incidence the angle of deviation increases. Figure given aside shows the graph for variation of angle of deviation with the angle of incidence.

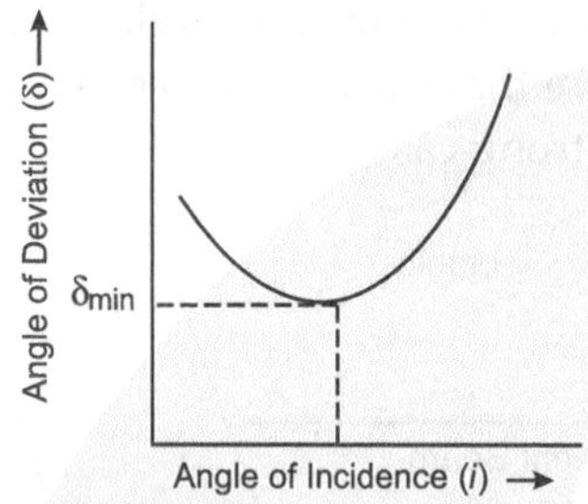

13. (i) In the diagram alongside, a ray of light PQ is incident normally on one face AB of an equilateral glass prism. What are the angles of incidence at the faces AB and AC?

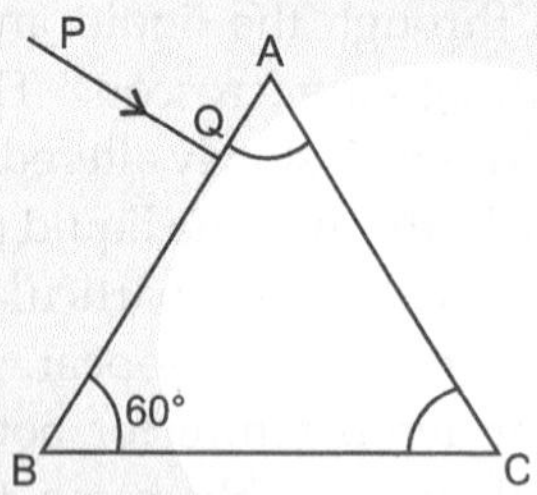

(ii) Complete the ray diagram showing its emergence into air after passing through the prism.

Ans. (i) Angle of incidence of face AB = 0°

Angle of incidence of face AC = 60°

(ii)

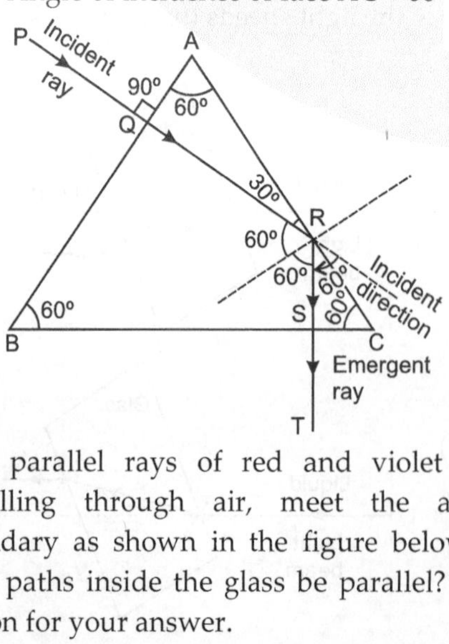

14. Two parallel rays of red and violet colour travelling through air, meet the air-glass boundary as shown in the figure below. Will their paths inside the glass be parallel? Give a reason for your answer.

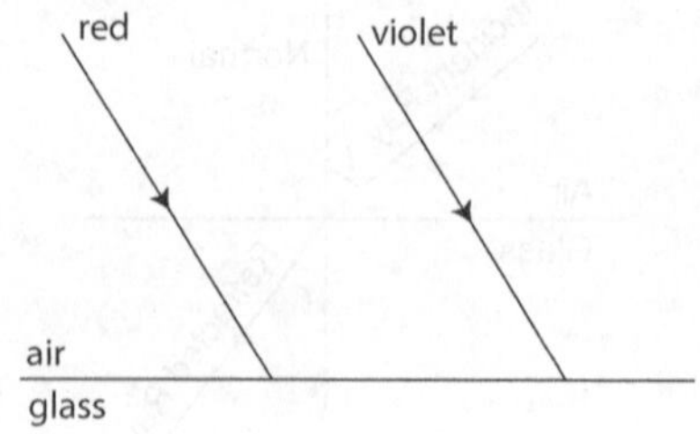

Ans. No.

Reason: Deviation suffered by each ray will be different because refractive index is different for different colours of light.

15. A stick partly immersed in water appears to be bent. Draw a ray diagram to show the bending of the stick when placed in water and viewed obliquely from above.

Ans.

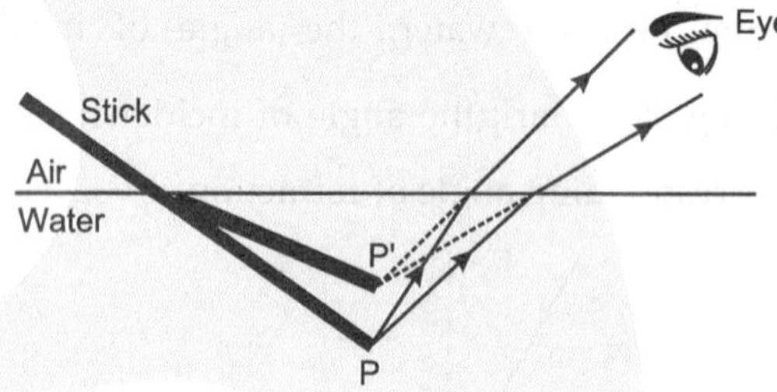

16. A ray of light enters a glass slab ABCD as shown in the figure. The critical angle of glass is 42°. Complete the path of the ray till it emerges from the slab. Mark the angles in the diagram wherever necessary.

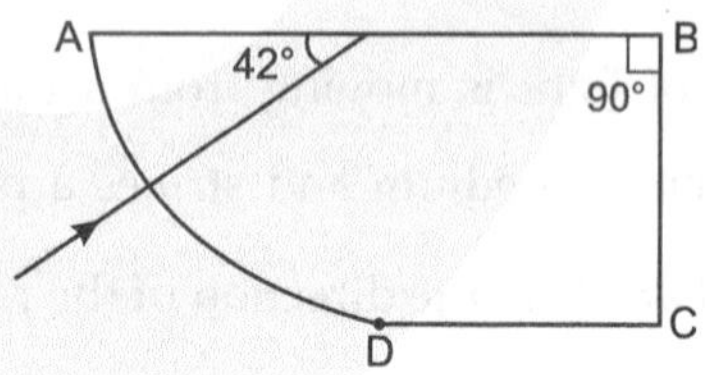

Ans. The complete ray diagram is shown in figure given above:

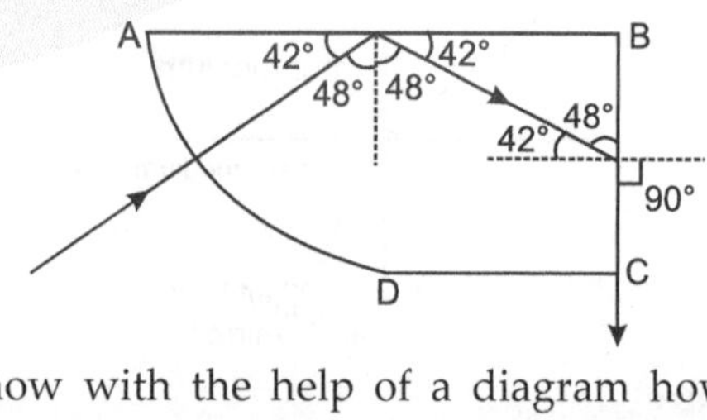

17. Show with the help of a diagram how a total reflecting prism can be used to turn a ray of light through 90°. Name an instrument in which this action of prism is used.

Ans. The diagram showing the use of a total reflecting prism to turn a ray of light through 90° is given in the figure.

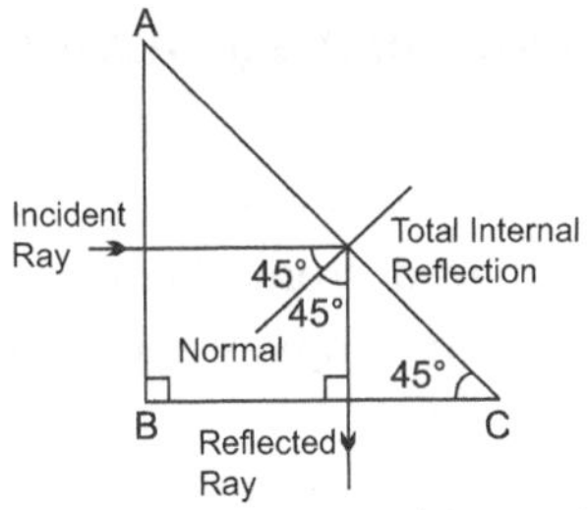

This action of prism is used in periscope.

18. Draw a diagram, properly labelled, to illustrate the use of a total reflecting prism (a right angled isosceles prism) to turn a ray of light through 180°. Name an instrument in which this device is used.

Ans. The diagram is shown below. This device is used in binoculars.

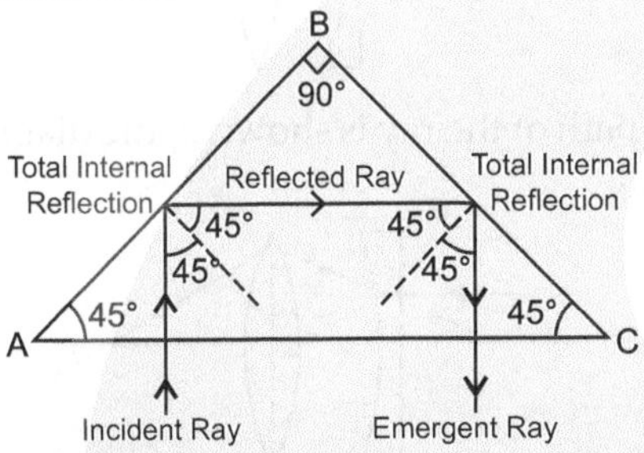

19. The figure shows two prisms A and B. A monochromatic ray of light PO is incident at the face of the prism A. Complete the diagram to show the path of the ray till it emerges out of the prism B.

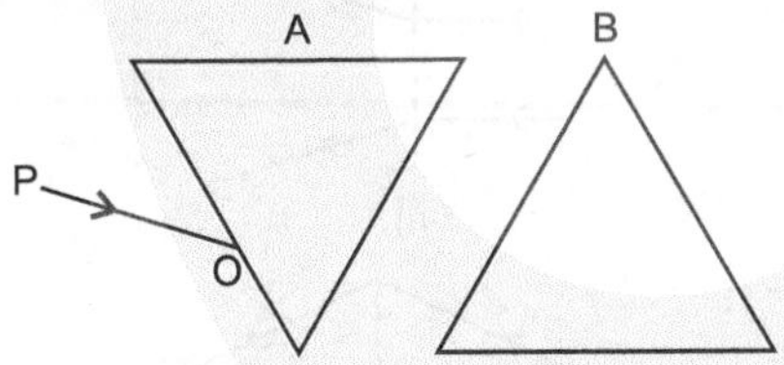

Ans. The completed diagram is shown in the figure below:

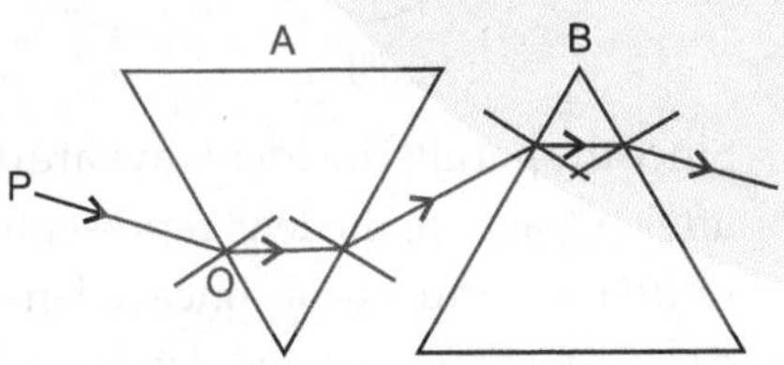

20. Draw a diagram to show the refraction of a monochromatic light ray through an equilateral prism. On the diagram, label the incident, refracted and emergent rays. Also indicate the angle of deviation by the letter δ.

Ans. The labelled diagram is given below:

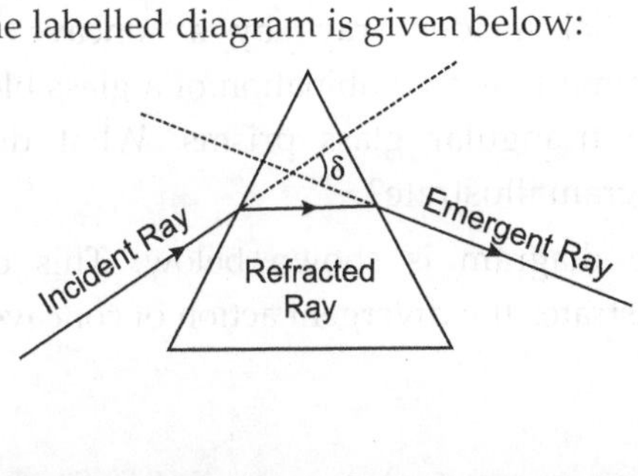

21. A ray of monochromatic light is incident from air on a glass slab:

(i) Draw a labelled ray diagram showing the change in the path of the ray till it emerges from the glass slab.

(ii) Name the two rays that are parallel to each other.

(iii) Mark the lateral displacement in your diagram.

Ans. (i)

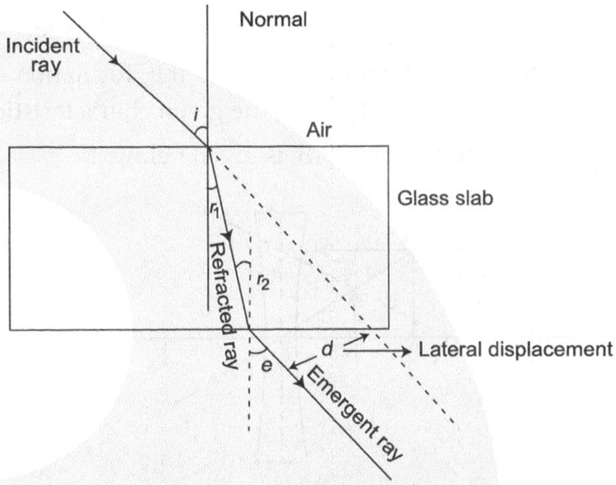

Ray diagram of monochromatic light

(ii) Incident ray and emergent rays are parallel to each other.

(iii) Lateral displacement is marked by d in the diagram.

22. (i) With the help of a well labelled diagram show that the apparent depth of an object, such as a coin, in water is less than its real depth.

(ii) How is the refractive index of water related to the real depth and the apparent depth of a column of water?

Ans. (i) AB = Apparent depth

AC = Real depth

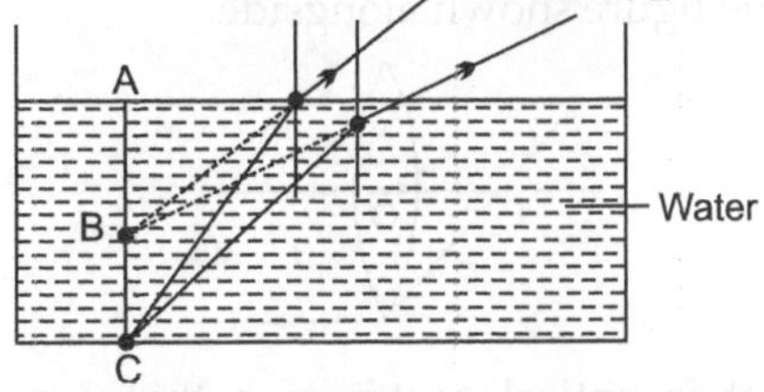

(ii) Relation between refractive index of water, real depth and apparent depth is

$$_{a}\mu_{w} = \frac{\text{Real depth}}{\text{Apparent depth}}$$

Chapter 4. Refraction through Lenses and Optical Instruments

1. A virtual, diminished image is formed when an object is placed between the optical centre and the principal focus of a lens.

(i) Name the type of lens which forms the above image.

(ii) Draw a ray diagram to show the formation of the image with the above stated characteristics.*

Ans. (i) A concave lens forms the given image.

(ii) Ray diagram to show the formation of image (A'B') with the given characteristics, of the object AB, is given below.

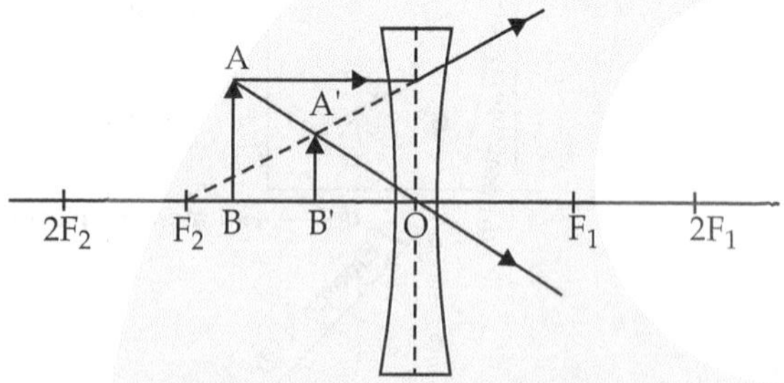

2. Draw a diagram to show the convergent action of a convex lens by treating it as a combination of glass block and two triangular glass prisms, with the aid of two parallel incident rays.

Ans. The diagram is shown adjacent:

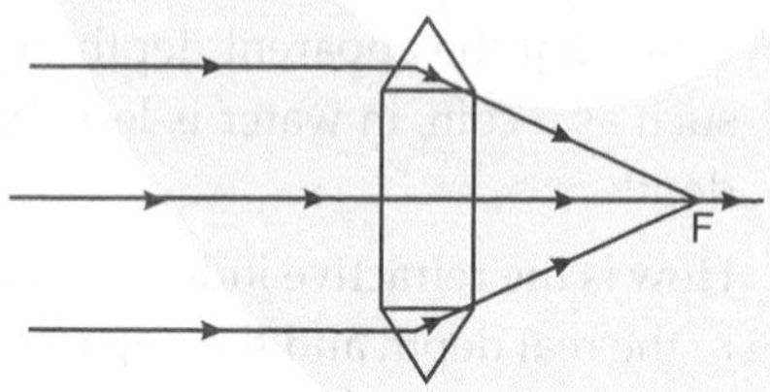

3. What is the path of ray after refraction, that passes through the first focus of the convex lens ?

Ans. The ray becomes parallel to the principal axis after refraction through the lens. This is shown in the figure shown alongside.

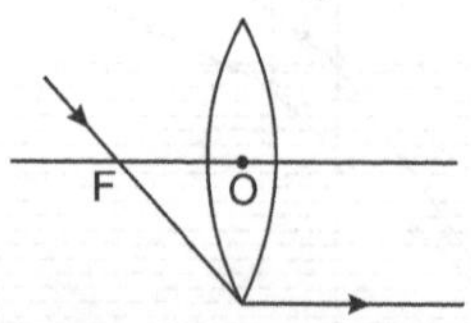

4. What is optical centre of a thin lens ? Draw diagram/ diagrams to illustrate it.

Ans. The optical centre of a thin lens is a point O on the principal axis of the lens such that a ray of light directed towards it, passes undeviated and undisplaced from the lens. The diagram illustrating it in a convex and a concave lens are shown alongside.

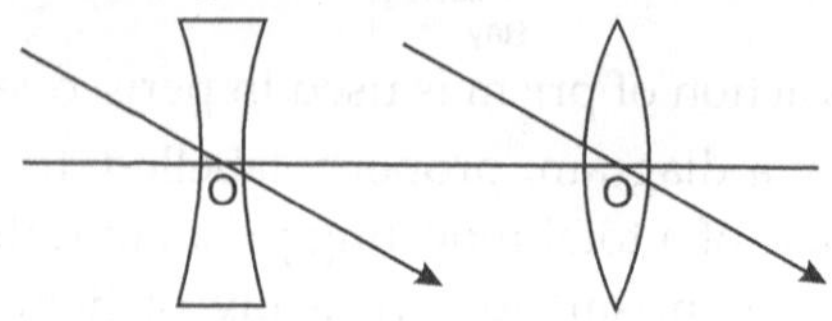

5. Complete the ray diagram to show how the rays pass into and out of the lens?

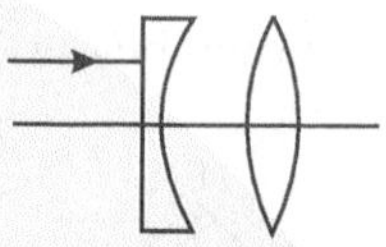

Ans. The path of the ray is shown in the diagram given below:

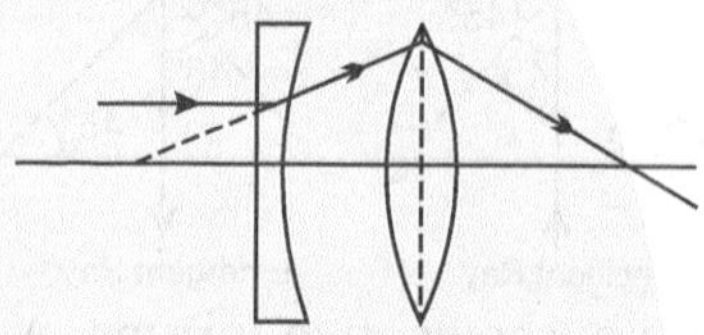

6. From the nature of the refracted rays, shown in the diagram [(i) and (ii)], interpret the nature of the lens represented by the dotted line.

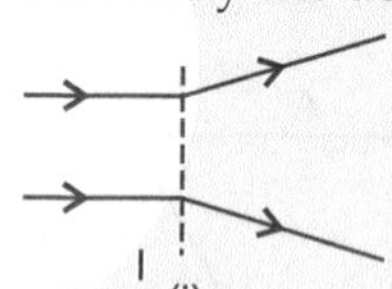

(i)

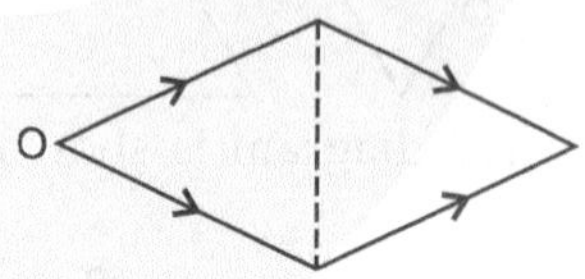

(ii)

Ans. (i) Since the parallel incident rays are diverging after refraction, the lens represented by the dotted line must be a concave lens.

(ii) Since the rays, starting from the object point 'O', are being made to converge to another point 'I', the lens used must be a convex lens.

7. Draw a diagram to show the refraction of two parallel incident rays by a concave lens by treating it as a combination of a glass block and two triangular glass prisms. What does this diagram illustrate?

Ans. The diagram is shown below. This diagram illustrates the divergent action of concave lens.

* **Frequently asked previous years Board Exam Questions.**

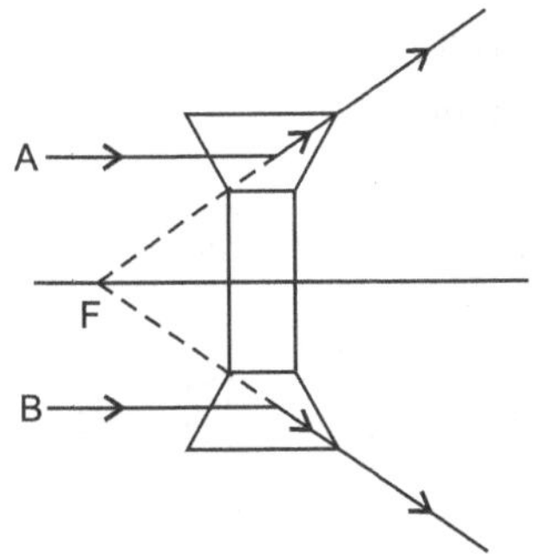

8. Where should an object be placed infront of a convex lens so as to form an inverted and enlarged image? Will the image be real or virtual ? Draw a ray diagram to illustrate your answer.

Ans. To form an inverted and enlarged image by a convex lens, the object should be placed between F_1 and $2F_1$ in front of the lens.

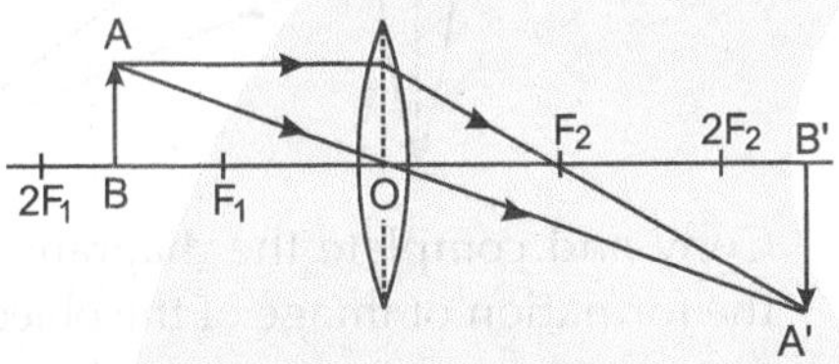

The image will be real. The ray diagram is shown in alongside figure. For the object AB, the image is A′B′.

9. Draw a ray diagram to show the formation of an inverted and diminished image of an object by a lens. Name the lens used. State the position of object with respect to the lens used. State one more characteristic of the image.

Ans. The ray diagram to show the formation of an inverted and diminished image of an object by a lens is shown alongside. In figure A′B′ is the image for the object AB.

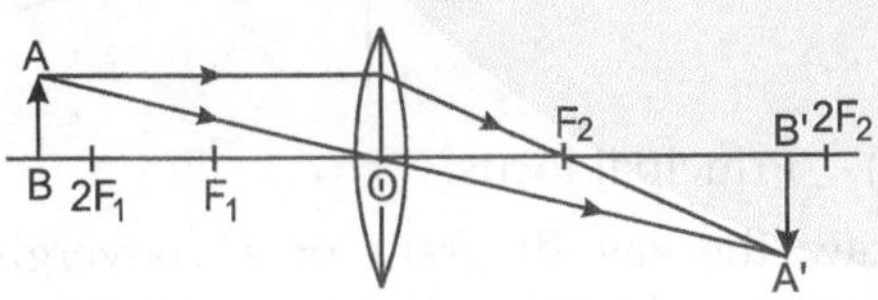

The lens used is convex (or converging). The object lies beyond $2F_1$ in front of the lens. The image is real.

10. A lens forms an upright and diminished image of an object, irrespective of its position. What kind of lens is this? Draw an outline ray diagram to show the formation of image. State the position and one more characteristic of the image.

Ans. A concave (or diverging) lens forms an upright and diminished image of an object irrespective of its position.

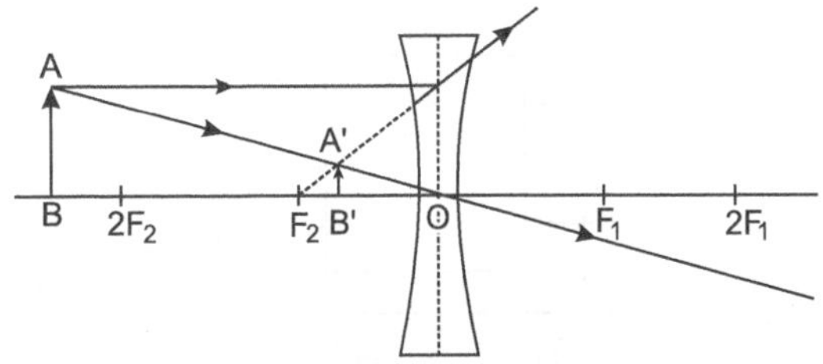

The outline ray diagram showing the formation of image is given in figure. For the object AB, the image is A′B′.

The image is formed between optical centre O and focus F_1, in front of the lens.

The image is virtual.

11. The following diagram shows the object O and the image I formed by a lens. Copy the diagram and on it mark the positions of the lens LL′ and focus (F). Name the lens.

Ans. The complete diagram is shown in the below figure in which the position of lens has been marked by the letter LL′. and the focus as F. The lens is convex (or converging).

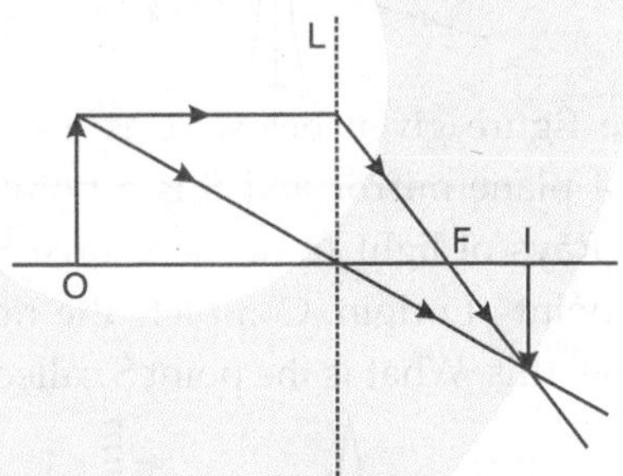

12. In figure given below of thin concave lens, F_1 and F_2 are its foci, complete the path of the given ray of light after it emerges out of the lens.

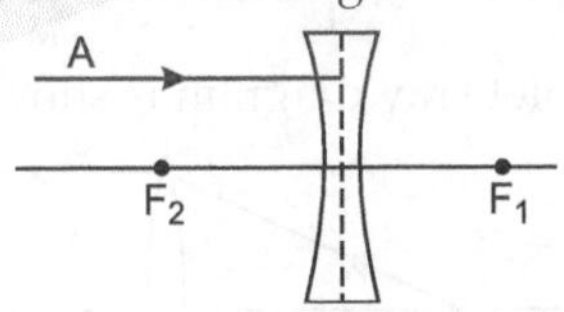

Ans. The required figure is given below:

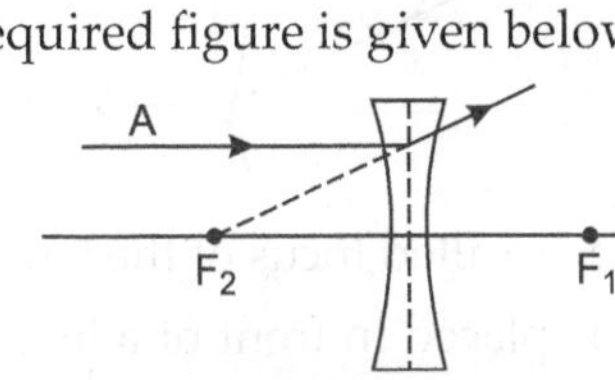

13. Figure shows the transmitted rays through lens kept in a box. Draw the lens and complete the path of rays.

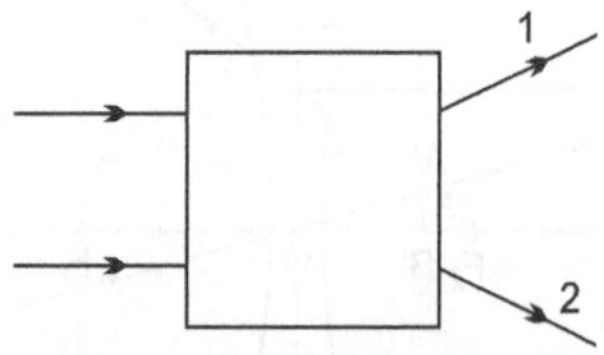

Ans.

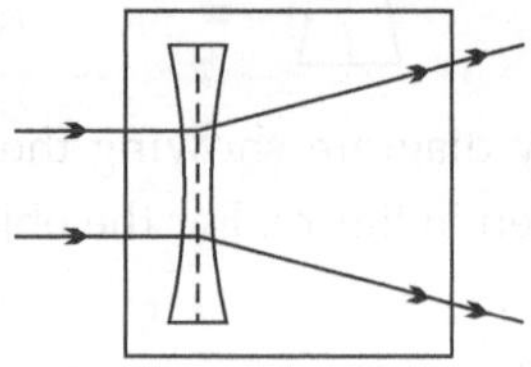

14. Copy the following figure and complete it to show the formation of the image of the object AB. Name the lens used in the figure.

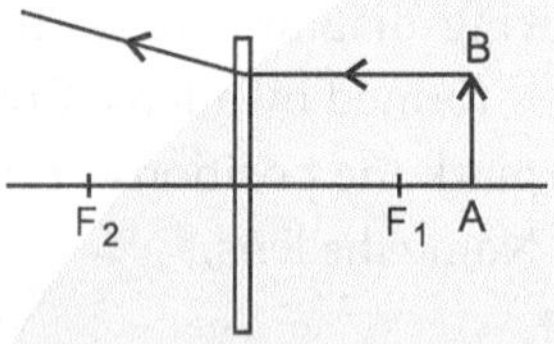

Ans. Since the ray is deviated away from the axis, the lens used is a concave lens. The complete figure showing the image formed, *i.e.*, B′A′, is given alongside. The image is virtual, erect and diminished.

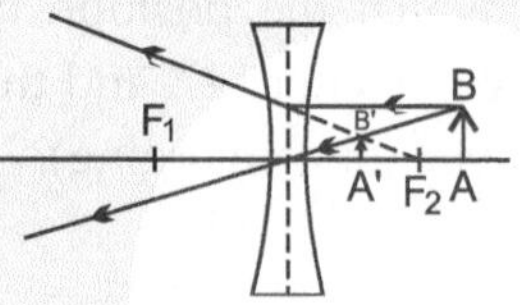

15. In the figure given below, L is a convex lens, M is a plane mirror and S is a point source of light. Rays of light from the source S return to their point of origin. Complete the ray diagram to show this. What is the point S called ?

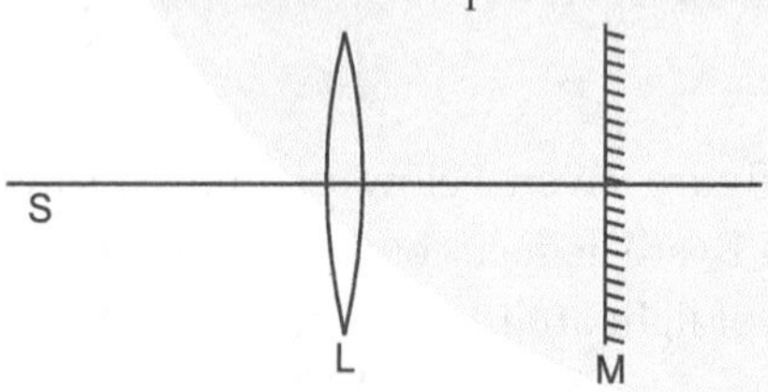

Ans. The complete ray diagram is shown below:

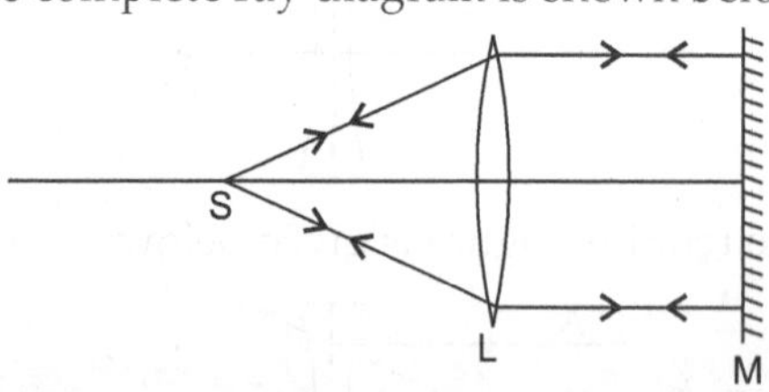

The point S is called focus of the lens.

16. An object is placed in front of a lens between its optical centre and the focus and forms a virtual, erect and diminished image. Draw a ray diagram to show the formation of the image with the above stated characteristics.

Ans. AB → Object

A′B′ → Image

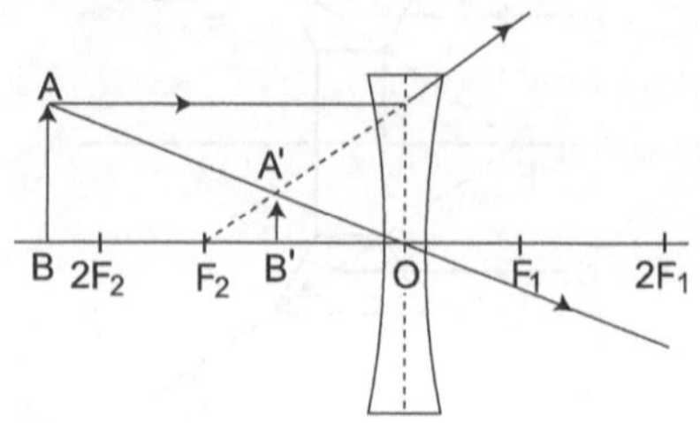

17. A candle is placed between f and 2f of a convex lens. Draw a ray diagram showing position of the image.

Ans. In the figure, $OF_1 = f$ and $OF_2 = 2f$

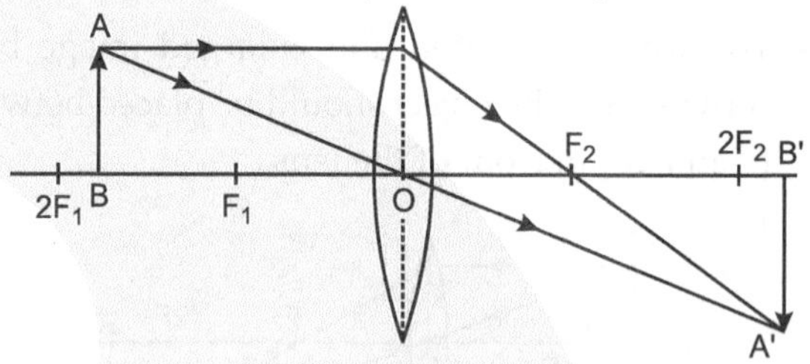

18. (i) Copy and complete the diagram, to show the formation of image of the object AB.

(ii) What is the name given to X?

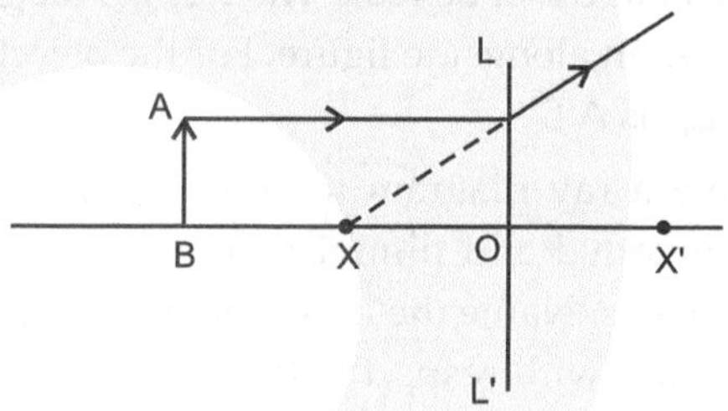

Ans. (i)

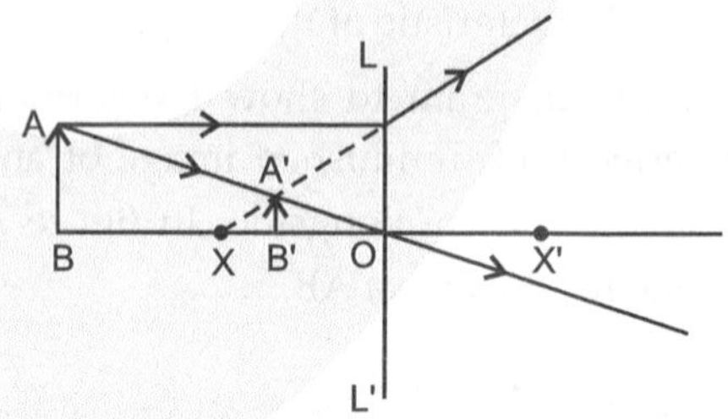

(ii) Principal focus.

19. Draw the ray diagram of a converging lens, when the object is placed at a distance greater than twice the focal length of the lens.

Ans. AB → Object

B′A′ → Image

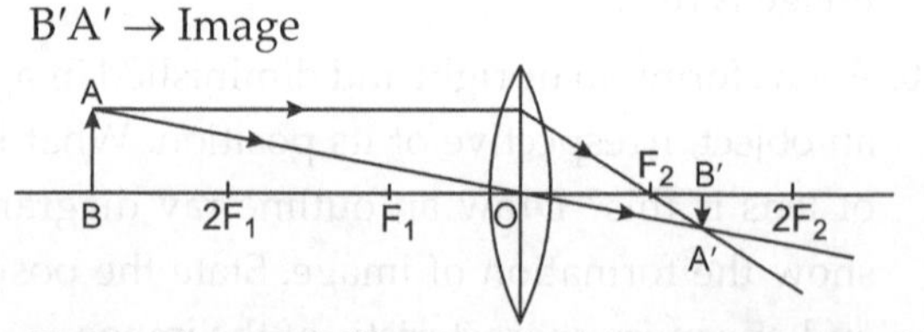

20. Draw the ray diagram of a converging lens, when the object is placed at $2F_1$ on the principal axis. Write down the nature, position and size of the image. Use only two rays in the diagram.

Figure Based Short Answers

Ans. Position of Object: Object is placed at $2F_1$.

Nature of Image: Real and inverted (B′A′).

Size of Image: Same (equal).

Position of Image: At $2F_2$.

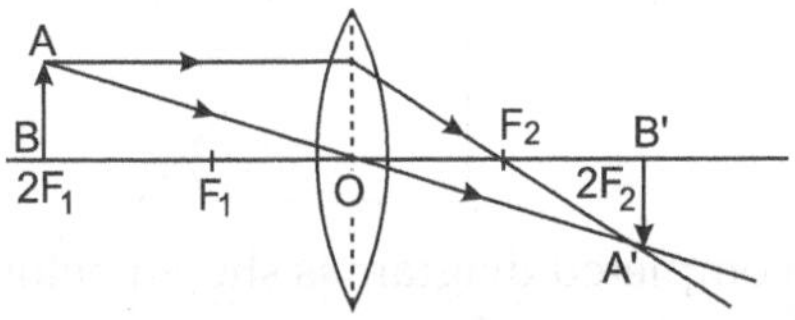

23. Show, using a ray diagram, how a converging lens can be used to form a virtual image.

Ans. The ray diagram is given below.

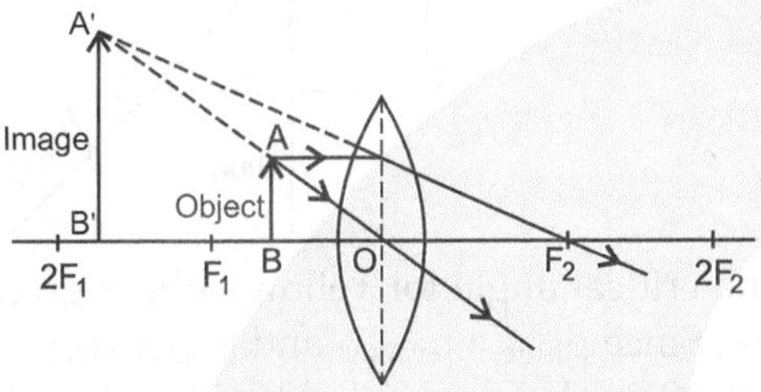

24. A linear object is placed on the axis of a lens. An image is formed by refraction in the lens. For all positions of the object on the axis of the lens, the positions of the image are always between the lens and the object.

(i) Name the lens.

(ii) Draw a ray diagram to show the formation of the image of an object placed in front of the lens at any position of your choice except infinity.

Ans. (i) Concave lens.

(ii) AB → Object

A′B′ → Image

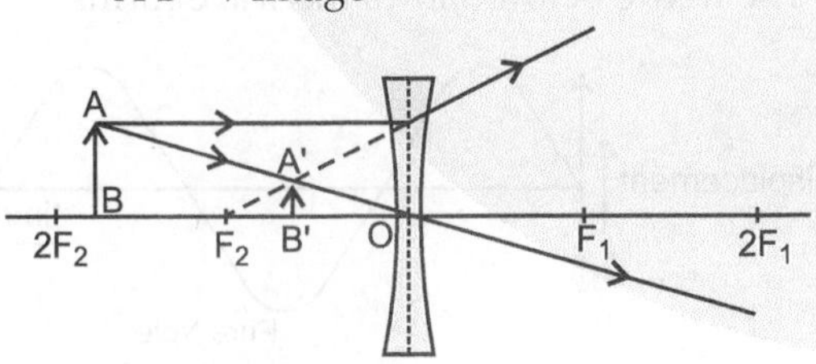

25. Draw a neat diagram to explain the action of a convergent (convex) lens as a reading glass (or magnifying glass). State the characteristics of the image formed.

Ans. For convex lens to act as a magnifying glass, the object to be seen must be placed at a distance less than its focal length.

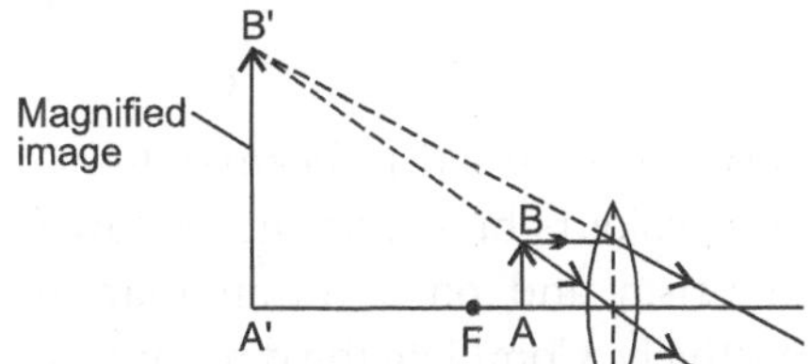

Characteristics of the image so formed are:

(i) It is virtual and upright.

(ii) It is magnified.

(iii) The image is formed on same side behind the object.

26. An object is placed in front of a convex lens such that the image formed has the same size as that of the object. Draw a ray diagram to illustrate this.

Ans.

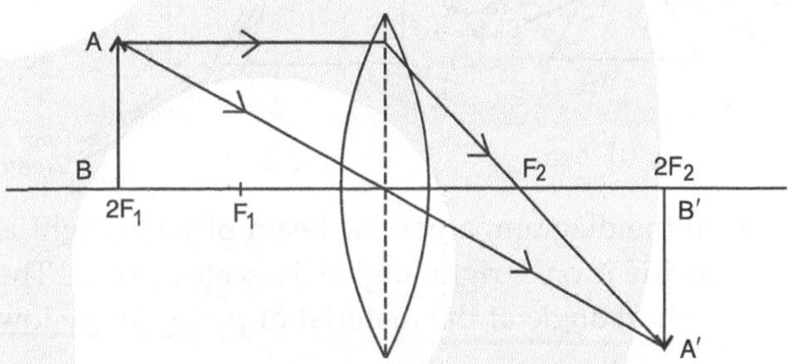

28. We can burn a piece of paper by focussing the sun rays by using a particular type of lens. Name the type of lens used for the above purpose. Draw a ray diagram to support your answer.

Ans. Convex Lens

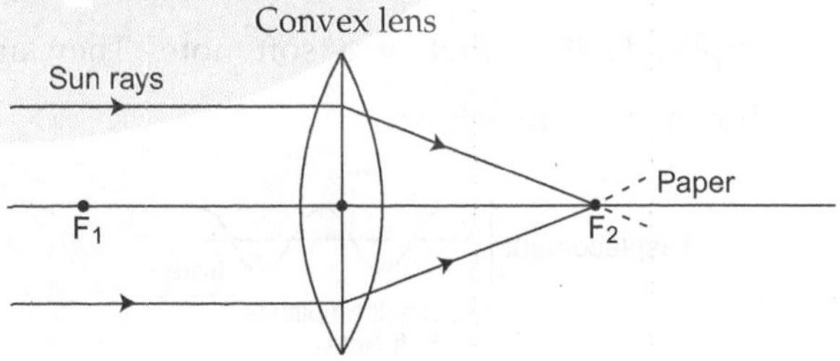

Chapter 5. Spectrum

1. Sunlight entering through a narrow aperture falls on a prism. Draw a neat labelled ray diagram to show the formation of spectrum on a screen. Name the colour obtained nearer the base of the prism.

Ans. The labelled ray diagram showing the formation of spectrum of sunlight on a screen is given below. The colour obtained nearer the base of the prism is violet.

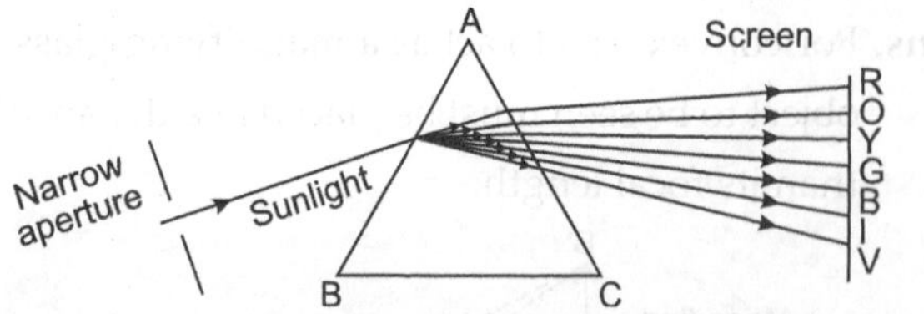

2. In the figure (i) and (ii) a beam of light consisting of three colours blue, red and yellow is incident on a prism and on a rectangular glass block respectively. Complete the diagrams by drawing the refracted and emergent rays.

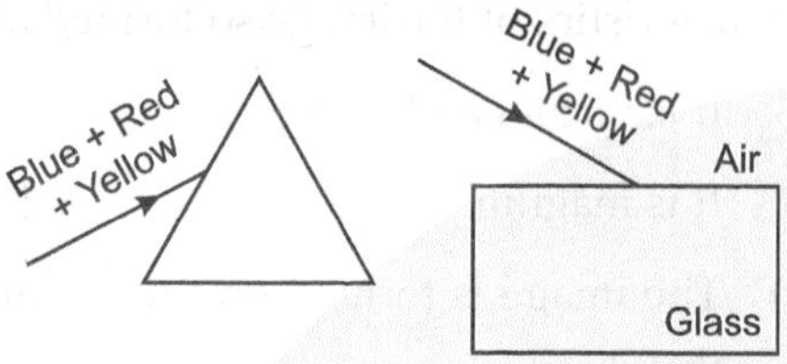

Ans. The complete diagrams are given below in figure (i) and (ii).

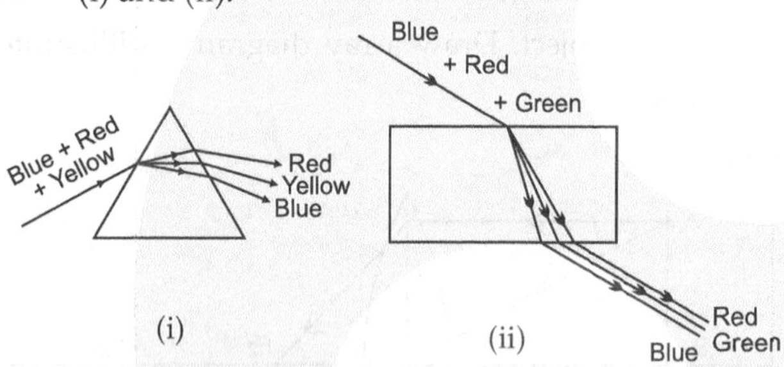

3. In the diagram, a narrow beam of white light is incident on a right-angled isosceles prism. The critical angle of the material of prism for yellow colour of white light is 45°. Complete the diagram to show the path of blue, yellow and red colours of white light till they emerge out of the prism.

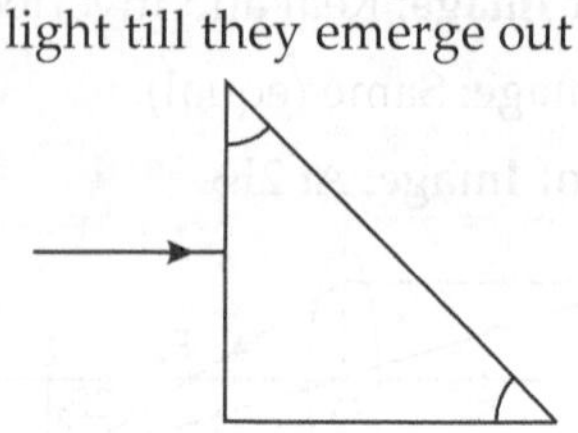

Ans. The completed diagram is shown below:

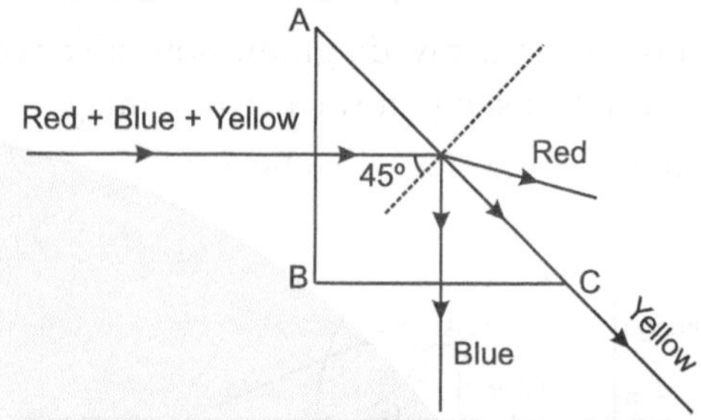

The critical angle for yellow light is given to be 45°. Since $\mu_{blue} > \mu_{yellow}$ and $\mu_{red} < \mu_{yellow}$ so the critical angle for blue light will be less than 45° and for red light will be more than 45°.

4. Draw a neat diagram to show the arrangement you would use to project a pure spectrum of white light on a screen.

Ans. The diagram is shown below:

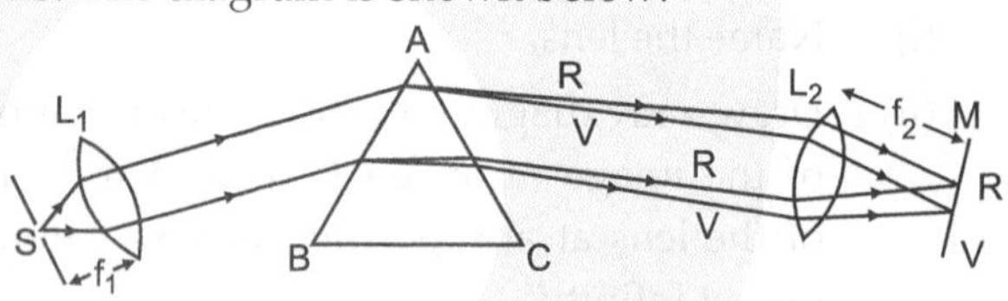

Chapter 6. Echoes and Vibrations of Sound

1. How does the wave forms of loud note differ from a soft note? Draw diagram.

Ans. The wave form of a loud note is of larger amplitude than that of a soft note. They are shown in figure below:

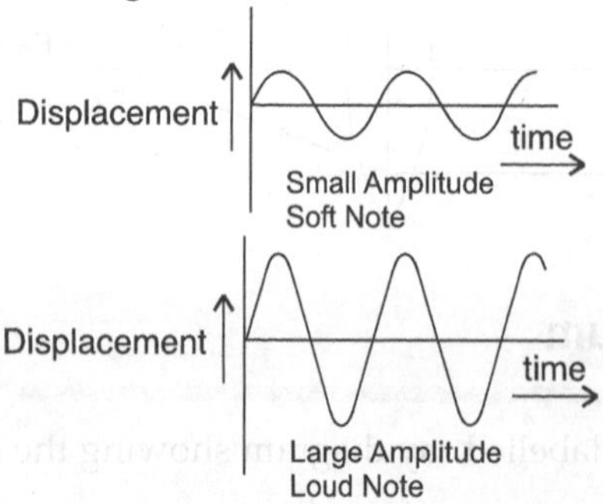

2. Draw a diagram to show the wave forms of a pure note and a musical note, of same pitch and same loudness.

Ans. The figure below shows the wave forms:

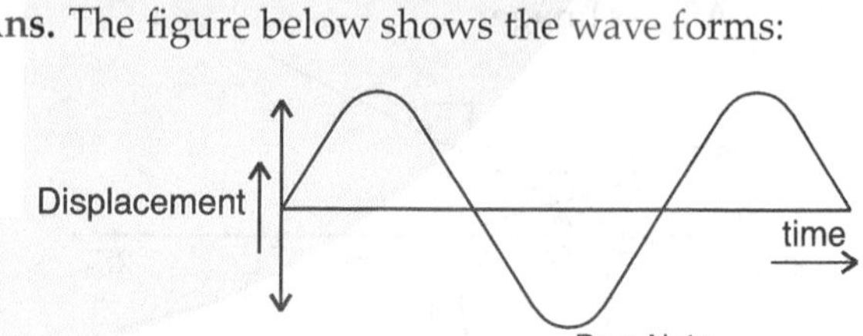

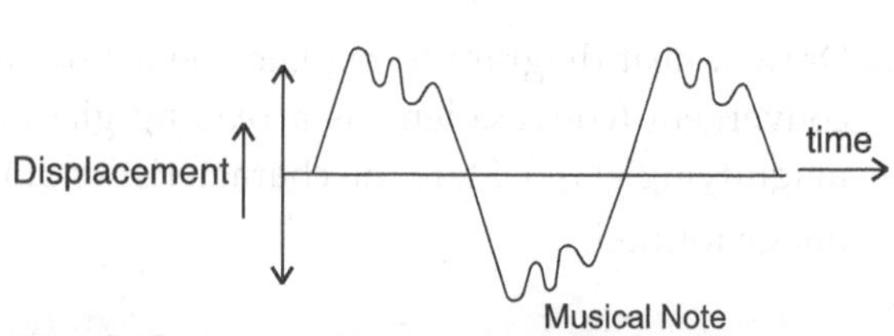

3. Draw a diagram representing a wave of amplitude 4 cm and wavelength 2 m. If frequency of wave is 150 hz, calculate its velocity.

Ans. The diagram below represents the wave:

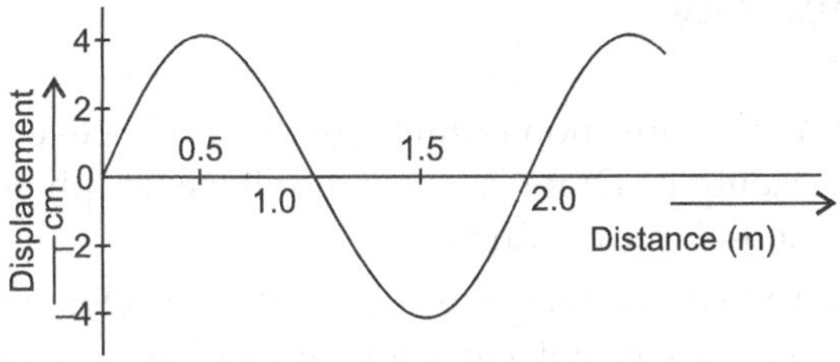

Given: $f = 150$ Hz, $\lambda = 2$ m

$\therefore \quad v = f\lambda = 150 \times 2 = 300 \text{ ms}^{-1}$.

4. What do you understand by free vibrations of a body ? Draw a displacement-time graph to represent them. Give one example.

Ans. The periodic vibrations of a body, of constant amplitude and constant frequency, are called free vibrations. The displacement-time graph to represent free vibrations of a body is given in figure.

Example: A tuning fork vibrating in vacuum.

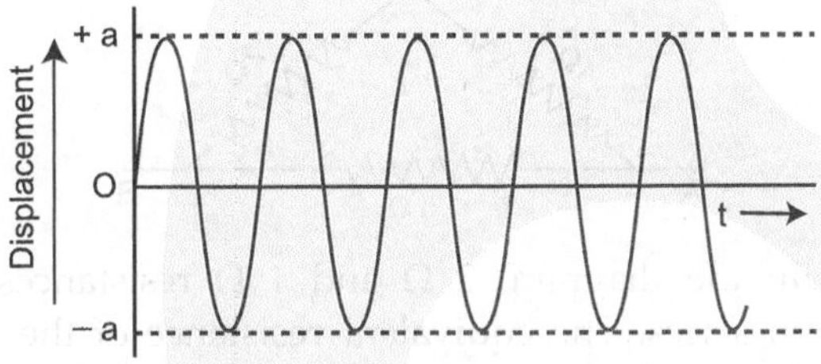

5. Draw a displacement-time graph to illustrate damped vibrations.

Ans. The displacement-time graph showing damped vibration is given in figure below:

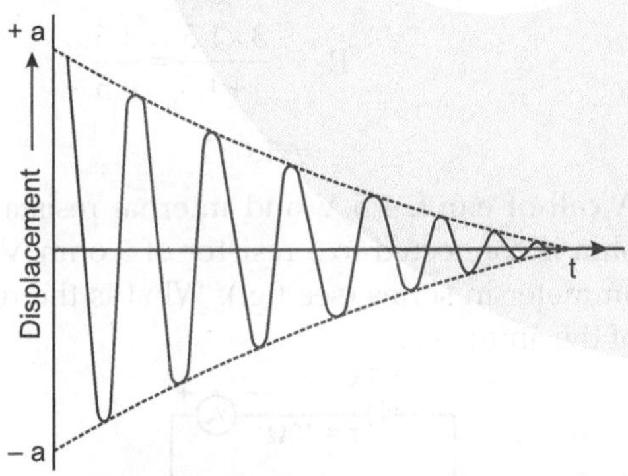

6. Two musical notes of the same pitch and loudness are played on a violin and a piano. Their wave forms are as shown in the figure below. Why are the two wave patterns different?

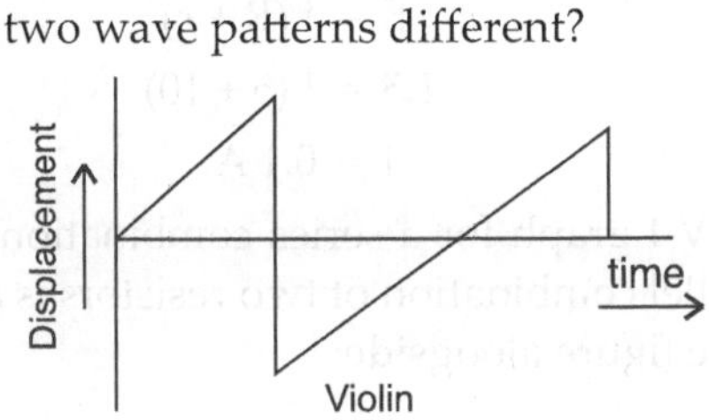

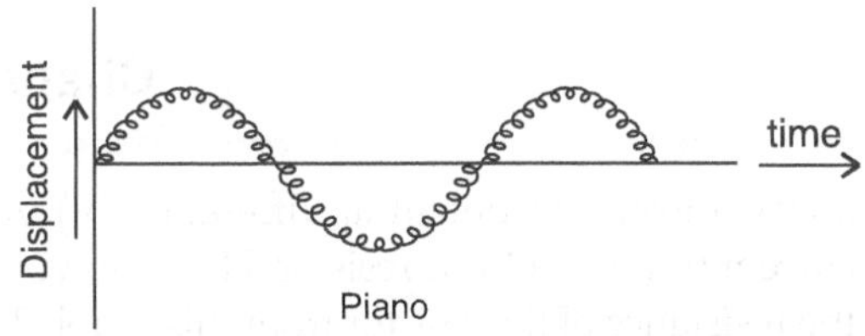

Ans. Though the musical notes of the same pitch and loudness are played on a violin and a piano but their wave patterns are different because the wave form depends upon the number of overtones and their relative intensities.

7. The frequency of a stretched string of fixed length is f when it is plucked in the middle. Where should it be plucked so that string vibrates with frequency (i) 2f, (ii) 3f ? Draw diagrams to show these vibrations and name them.

Ans. (i) The stretched string should be plucked at one-fourth part of the string to obtain frequency 2f .

(ii) The string should be plucked at one-sixth part of the string to obtain frequency 3f.

These vibration are shown in figure and are named as fundamental, first overtone and second overtone respectively.

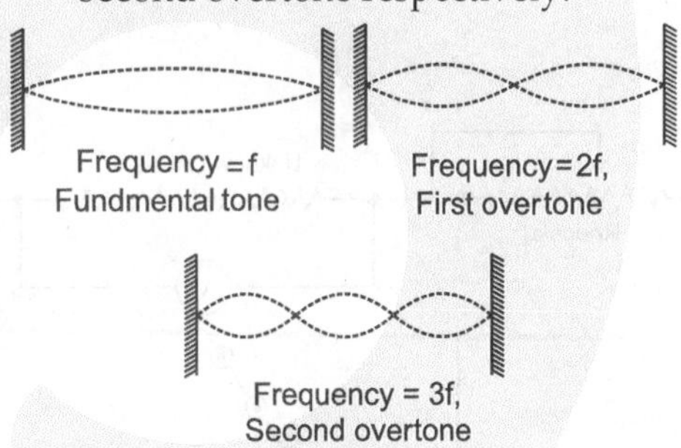

8. In what respect does wave forms of a noise and a musical note differ. Draw diagrams to illustrate your answer.

Ans. The wave form of a noise is irregular, discontinuous and non-periodic while the wave form of a musical note is regular, continuous and periodic. These wave forms are shown in figure given below:

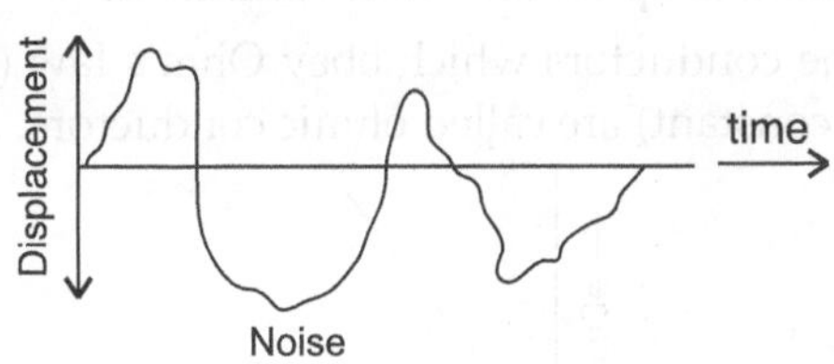

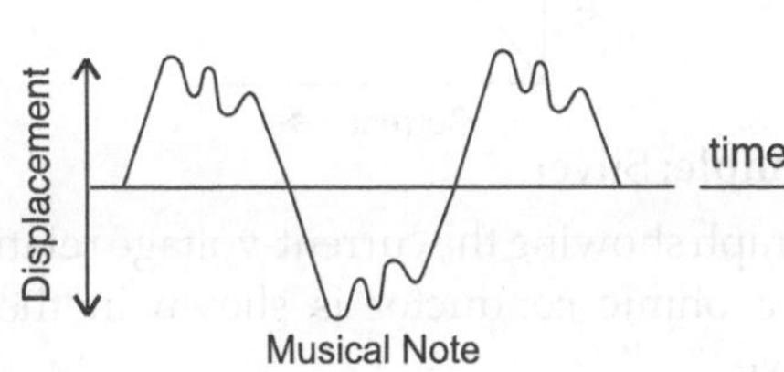

Chapter 7. Electricity

1. Draw a graph of potential difference (V) versus current (I) for an ohmic resistor. How can you find the resistance of the resistor from this graph ?

Ans.

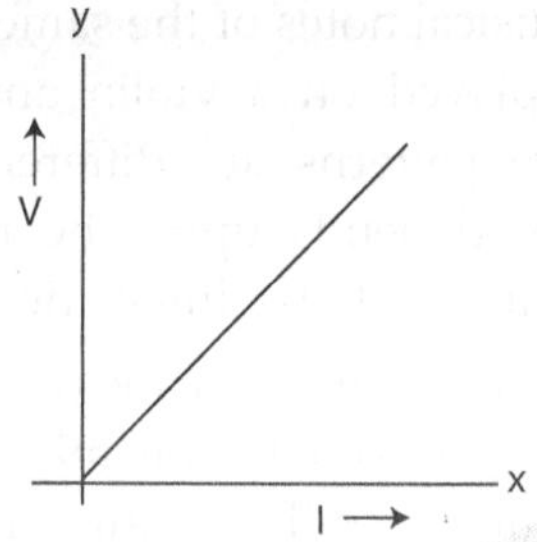

Graph of V versus I for an Ohmic conductor

We can find resistance by finding the slope of the graph.

2. Draw a neat diagram for the verification of Ohm's law by voltmeter-ammeter method. By another diagram show the relation between p.d. and current.

Ans.

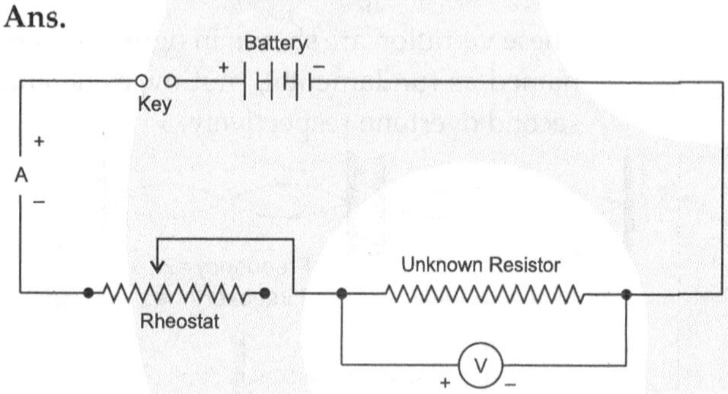

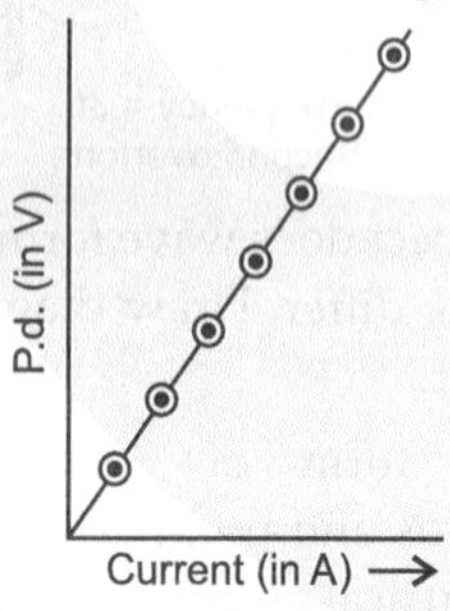

3. What are ohmic conductors? Give one example. Draw a graph showing the current-voltage relationship for an ohmic conductor.

Ans. The conductors which obey Ohm's law (*i.e.*, V/I is constant) are called ohmic conductors.

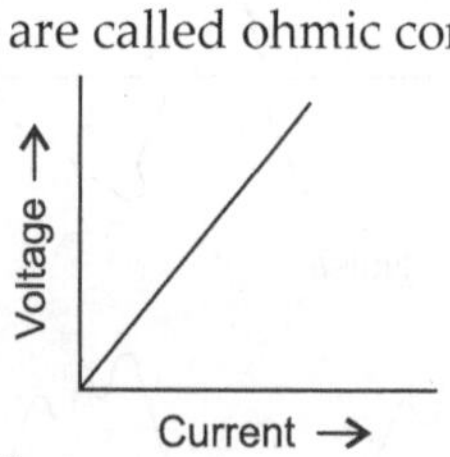

Example: Silver.

A graph showing the current-voltage relationship for a ohmic conductor is shown in the figure above.

4. What are non-ohmic conductors? Give one example. Draw a current-voltage graph for a non-ohmic conductor.

Ans. The conductors which do not obey Ohm's law (*i.e.*, V/I is not constant) are called non-ohmic conductors.

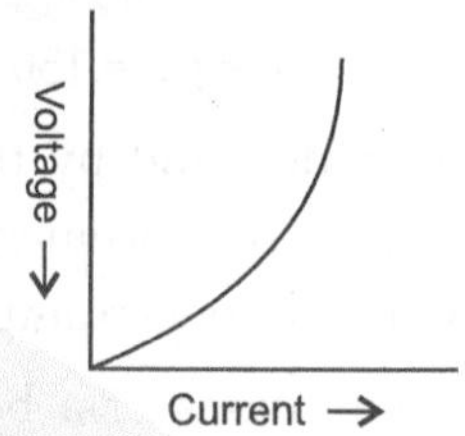

Example: Diode valve.

The current-voltage graph for a non-ohmic conductor is shown in figure above.

5. In the given figure, calculate the effective resistance between the points A and B.

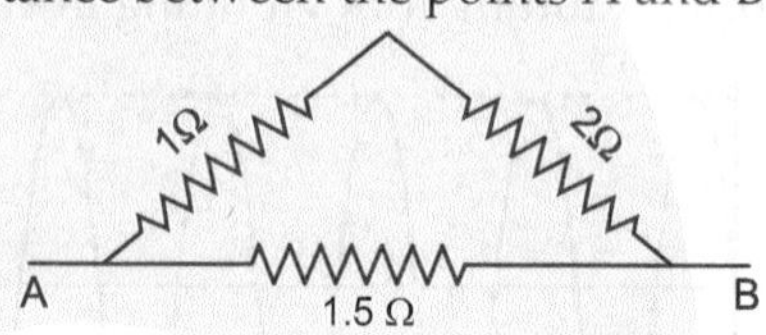

Sol. In the diagram, 2 Ω and 1 Ω resistances are in series. The equivalent resistance of the arm, R′ = 2 + 1 = 3 Ω.

Between the points A and B, the equivalent resistance R′ (= 3 Ω) is in parallel with 1.5 Ω

Hence, total effective resistance between A and B is

$$R = \frac{3 \times 1.5}{3 + 1.5} = \frac{4.5}{4.5}$$

$$= 1.0\ \Omega.$$ **Ans.**

6. A cell of e.m.f. 1.5 V and internal resistance 10 ohm is connected to a resistor of 5 ohm, with an ammeter in series (see fig.). What is the reading of the ammeter?

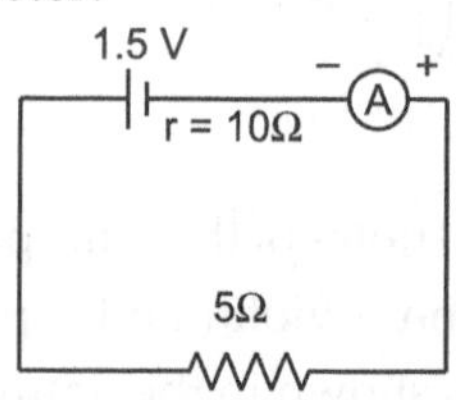

Sol.

$$E = I(R + r)$$

$$1.5 = I(5 + 10)$$

$$I = 0.1\ A.$$ **Ans.**

7. The V-I graph for a series combination and of a parallel combination of two resistors is as shown in the figure alongside:

Which of the two, A or B, represents the parallel combination? Give a reason for your answer.

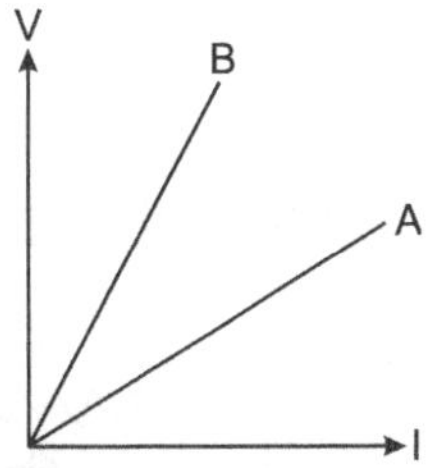

Ans. 'A' represents parallel combination because A is less steeper than B. So resistance of A is less than B and we know that resistance in parallel combination is less than the series combination.

8. Calculate the total resistance across AB :*

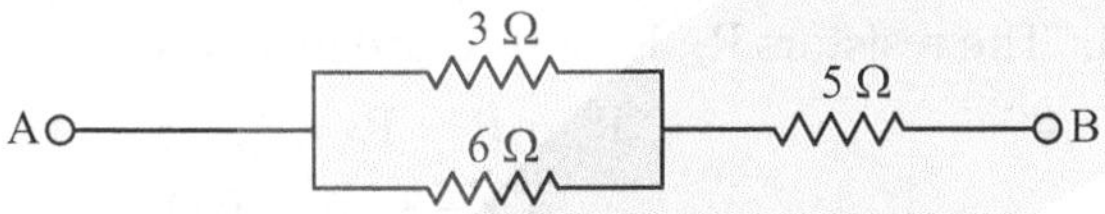

Ans. Equivalent resistance of 3 Ω and 6 Ω in parallel is,

$$R_1 = \frac{3 \times 6}{6+3} = \frac{18}{9} = 2\ \Omega$$

Total resistance across AB is $R_1 + R_2 = (5 + 2) = 7\ \Omega$.

9. Calculate the effective resistance across AB:*

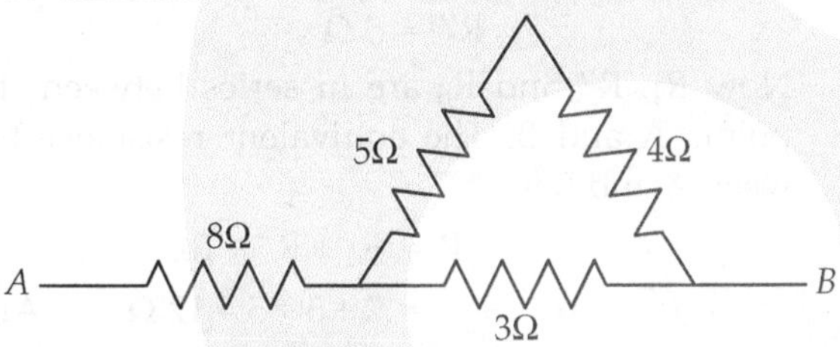

Ans. Given: 5 Ω and 4 Ω in series

$\therefore \quad R_1 = (5 + 4)\ \Omega = 9\ \Omega$

Diagram can be simplified as follows:

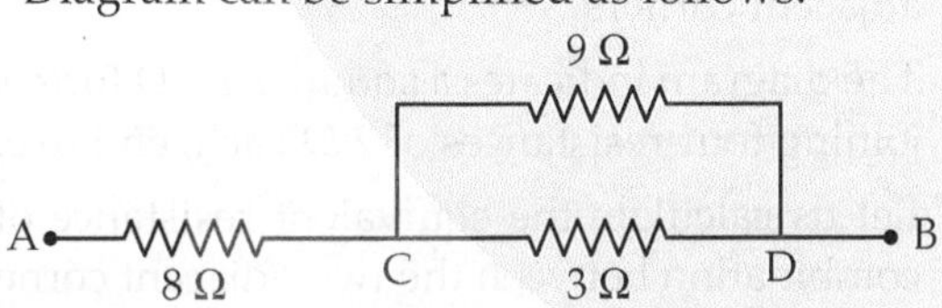

Now, 9 Ω and 3 Ω are in parallel.

$$\therefore \quad R_2 = \frac{9 \times 3}{9+3} = \frac{27}{12} = \frac{9}{4}\Omega$$

Now, 8 Ω and R_2 are in series

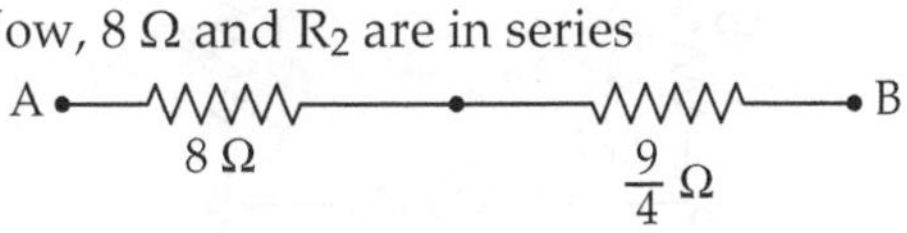

$$\therefore \quad R_3 = 8 + \frac{9}{4} = \frac{41}{4}\Omega = 10.25\ \Omega$$

10. You are provided with three resistors of resistance 1.0 Ω, 2.0 Ω and 3.0 Ω . How would you connect them to obtain the total effective resistance 1.5 Ω ? Draw diagram of the arrangement and check it by calculations.

Sol. Two resistors of resistance 1.0 Ω and 2.0 Ω are connected in series and then this combination is connected in parallel with the resistor of resistance 3.0 Ω. The arrangement is shown in fig. given below. Total resistance of series combination, R = 1 + 2 = 3 Ω.

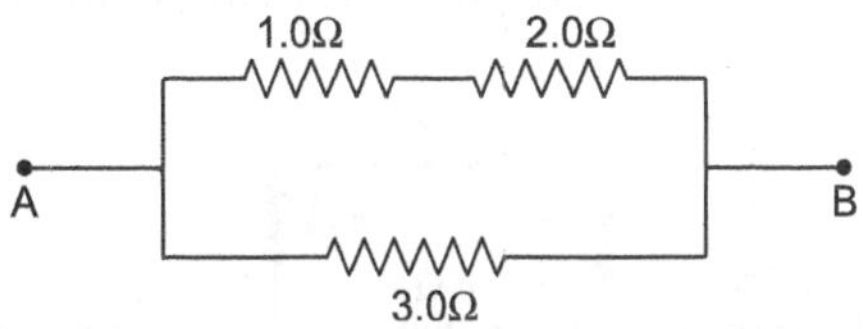

Total effective resistance of parallel combination,

$$R' = \frac{3 \times 3}{3+3} = 1.5\ \Omega. \qquad \textbf{Ans.}$$

11. Three resistors, each of 2 Ω are connected together so that their total resistance is 3 Ω. Draw a diagram to show this arrangement and check it by calculations.

Sol. The diagram of the arrangement is shown in figure below in which two resistors are in parallel and then this combination is in series with the third resistor.

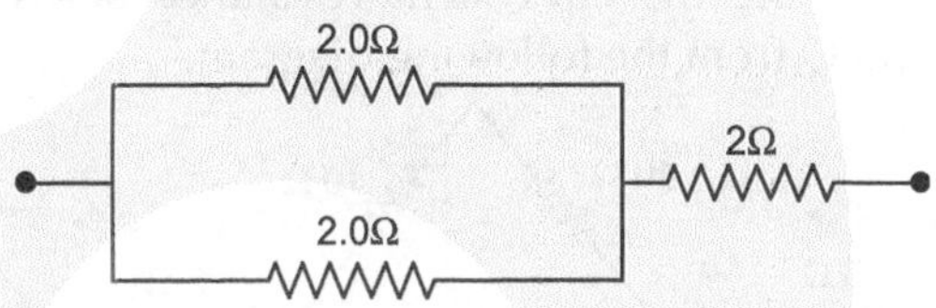

The equivalent resistance of parallel combination:

$$R = \frac{2 \times 2}{2+2} = \frac{4}{4} = 1\ \Omega.$$

Then total resistance of series combination (R') = 1 + 2 = 3 Ω. **Ans.**

12. Find the equivalent resistance between points A and B.

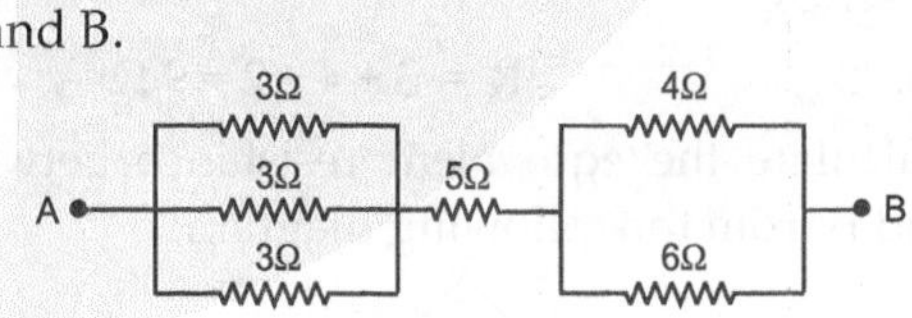

Sol.

$$\frac{1}{R_1} = \left(\frac{1}{3} + \frac{1}{3} + \frac{1}{3}\right)$$

$$\Rightarrow \quad \frac{1}{R_1} = 1$$

$$\Rightarrow \quad R_1 = 1\Omega$$

$$\Rightarrow \quad R_2 = 5\ \Omega$$

$$\Rightarrow \quad \frac{1}{R_3} = \frac{1}{4} + \frac{1}{6} = \frac{10}{24}$$

$$\Rightarrow \quad R_3 = \frac{24}{10}\Omega$$

* **Frequently asked previous years Board Exam Questions.**

∴ Equivalent resistance,

$$\text{Req.} = R_1 + R_2 + R_3 = 1 + 5 + \frac{24}{10}$$

$$R = 8{\cdot}4\ \Omega \qquad \textbf{Ans.}$$

13. Calculate the equivalent resistance between the points A and B for the following combination of resistors:

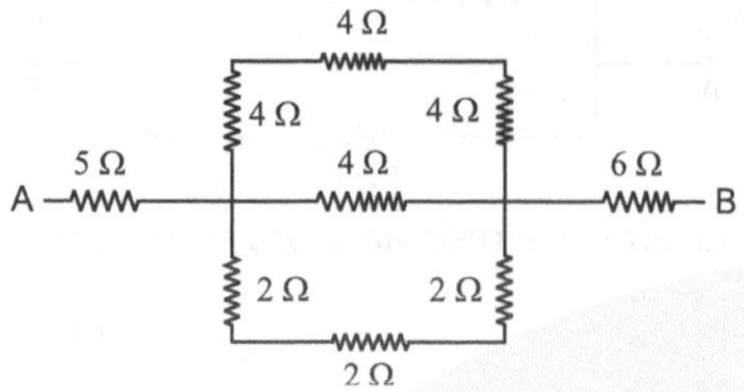

Sol.
$$\frac{1}{R_1'} = \frac{1}{12} + \frac{1}{4} + \frac{1}{6} = \frac{1+3+2}{12} = \frac{6}{12}$$

$$\frac{1}{R_1'} = \frac{1}{2}$$

$$R_1' = 2\Omega$$

Total resistance = 5 + 2 + 6 = 13 Ω. **Ans.**

14. Calculate the equivalent resistance between P and Q from the following diagram:

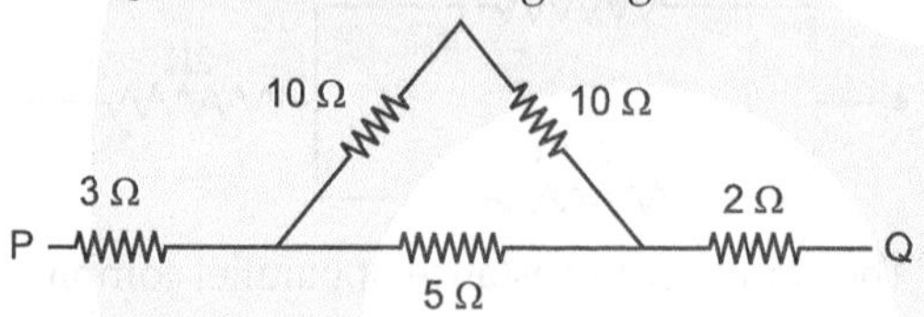

Sol.
$$\frac{1}{R_1} = \frac{1}{20} + \frac{1}{5}$$

$$\frac{1}{R_1} = \frac{1+4}{20} = \frac{5}{20}$$

$$R_1 = 4\ \Omega$$

Now, 3 Ω, 4 Ω and 2 Ω resistances are connected in series.

So, $R = 3 + 4 + 2 = 9\ \Omega$ **Ans.**

15. Calculate the equivalent resistance between A and B from the following diagram:

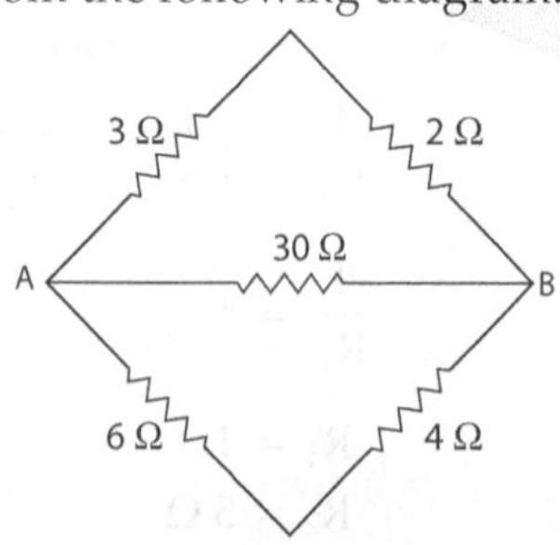

Sol.
$$\frac{1}{R} = \frac{1}{5} + \frac{1}{30} + \frac{1}{10}$$

$$\frac{1}{R} = \frac{6+1+3}{30}$$

$$\frac{1}{R} = \frac{10}{30} = \frac{1}{3}$$

$$R = 3\ \Omega. \qquad \textbf{Ans.}$$

16. Six resistances are connected together as shown in the figure. Calculate the equivalent resistance between the points A and B.

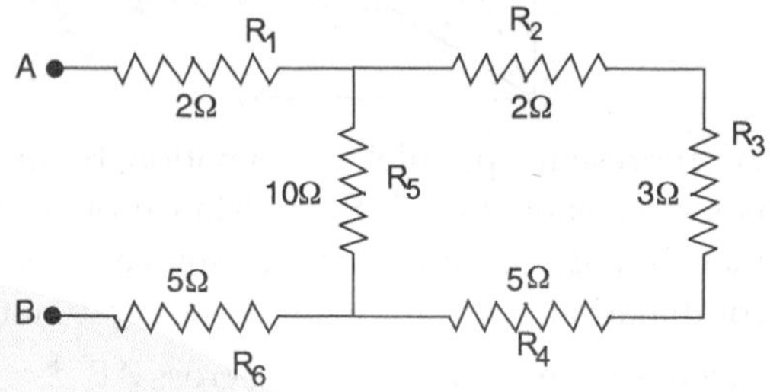

Sol. The resistors R_2, R_3 and R_4 are in series.

$$\therefore \quad R' = R_2 + R_3 + R_4$$
$$= 2 + 3 + 5 = 10\ \Omega$$

Now R′ and R_5 are in parallel.

$$\therefore \quad \frac{1}{R'} = \frac{1}{R'} + \frac{1}{R_5}$$

$$= \frac{1}{10} + \frac{1}{10} = \frac{2}{10} = \frac{1}{5}$$

$$\therefore \quad R'' = 5\ \Omega$$

Now R_1, R″ and R_6 are in series between the points A and B. The equivalent resistance between A and B is

$$R = R_1 + R'' + R_6$$
$$= 2 + 5 + 5 = 12\ \Omega \qquad \textbf{Ans.}$$

17. Four resistors of 2·0 Ω each are joined end to end, to form a square ABCD. Calculate the equivalent resistance of the combination between any two adjacent corners.

Sol. The diagram indicates a square ABCD formed by joining four resistances of 2 Ω each, end to end.

Let us calculate the equivalent resistance of the combination between the two adjacent corners C and D.

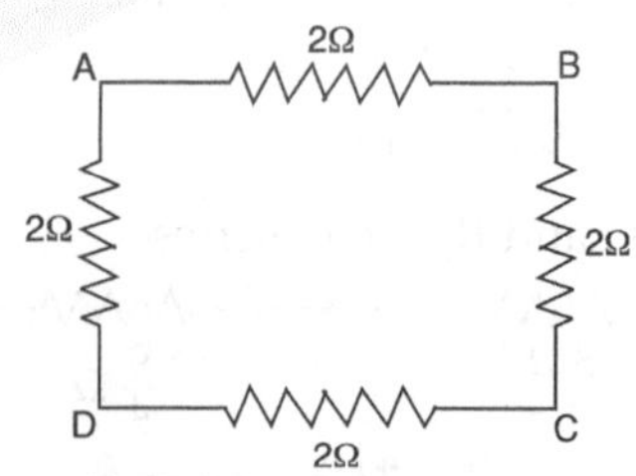

From figure, arms DA, AB and BC are joined in series.

So, their equivalent resistance = (2 + 2 + 2) Ω = 6 Ω

Now resistance of 6 Ω and 2 Ω are in parallel.

∴ The equivalent resistance of the combination is R.

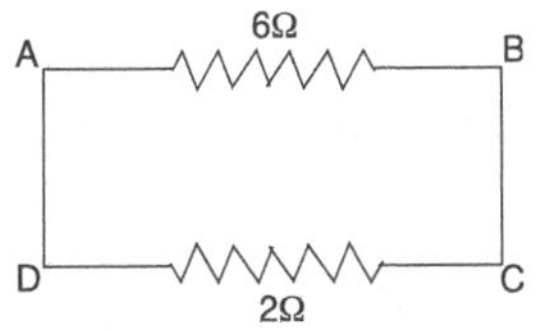

Then, $$\frac{1}{R} = \frac{1}{6}+\frac{1}{2}=\frac{4}{6}=\frac{2}{3}$$

∴ $$R = \frac{3}{2}\Omega$$

$$= 1{\cdot}5\ \Omega$$ **Ans.**

18. Three resistors of 6 Ω, 3 Ω and 2 Ω are connected together so that their total resistance is greater than 6 Ω but less than 8 Ω. Draw a diagram to show this arrangement and calculate its total resistance.

Sol. We connect the resistances of 3 Ω and 2 Ω in parallel to each other and then connect this parallel combination to the 6 Ω resistance in series. The required arrangement is as shown in figure.

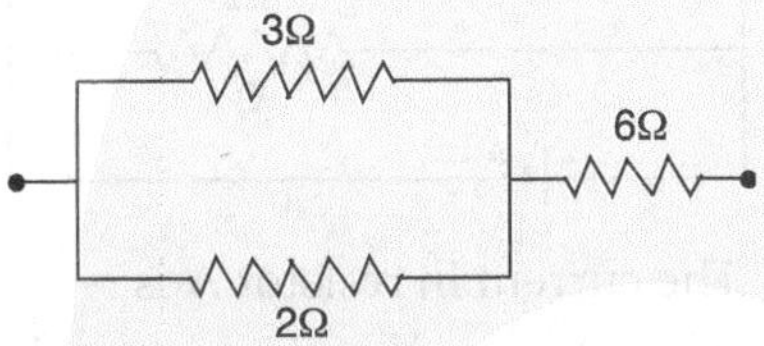

The equivalent resistance of 3 Ω and 2 Ω in parallel is given by

$$\frac{1}{R} = \frac{1}{3}+\frac{1}{2}=\frac{2+3}{6}=\frac{5}{6}$$

or $$R = \frac{6}{5}\Omega = 1.2\ \Omega.$$

Thus, the total resistance of the set up is (1.2 + 6) W = 7.2 Ω. **Ans.**

19. In the circuit shown below, calculate the equivalent resistance between the points (i) A and B, (ii) C and D.

Sol. (i) Between the points A and B: Three resistance 2 Ω, 2 Ω, 2 Ω are in series.

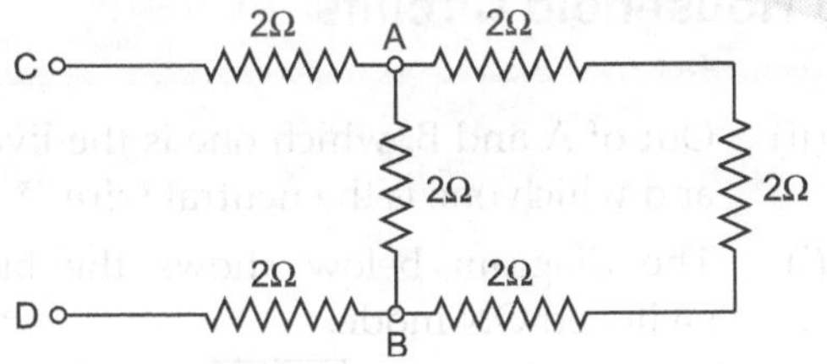

The equivalent resistance, R′= 2 + 2 + 2 = 6 Ω.

This is joined in parallel with a resistance 2 Ω.

The equivalent resistance,

$$R = \frac{2\times 6}{2+6}=\frac{12}{8}$$

$$= 1.5\ \Omega.$$ **Ans.**

(ii) Between the points C and D:

The above combination of equivalent resistance 1.5 Ω is in series with two resistances of 2 Ω each.

Total effective resistance between the points C and D, R = 1.5 + 2 + 2 = 5.5 Ω. **Ans.**

20. What is the equivalent resistance between the points X and Y for the given network?

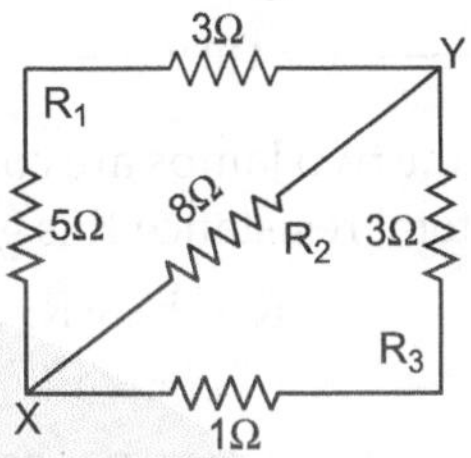

Sol. $$\frac{1}{R} = \frac{1}{R_1}+\frac{1}{R_2}+\frac{1}{R_3}$$

$R_1 = 5 + 3 = 8\ \Omega$

$R_2 = 8\ \Omega$

$R_3 = 1 + 3 = 4\ \Omega$

∴ $$\frac{1}{R} = \frac{1}{8}+\frac{1}{8}+\frac{1}{4}$$

$$\frac{1}{R} = \frac{1+1+2}{8}=\frac{4}{8}=\frac{1}{2}$$

$R = 2\ \Omega.$ **Ans.**

21. Six equal resistors of 1 ohm each are connected to form the sides of a hexagon ABCDEF. Calculate the resistance offered by the combination if the current enters at A and leaves it at D.

Sol. From figure it is clear that there are two sets of parallel resistances between A and D. Each group has three resistances of 1Ω each, in series. Equivalent resistance R_1 of one group (AB, BC and CD) = 1 Ω + 1 Ω + 1 Ω = 3 Ω.

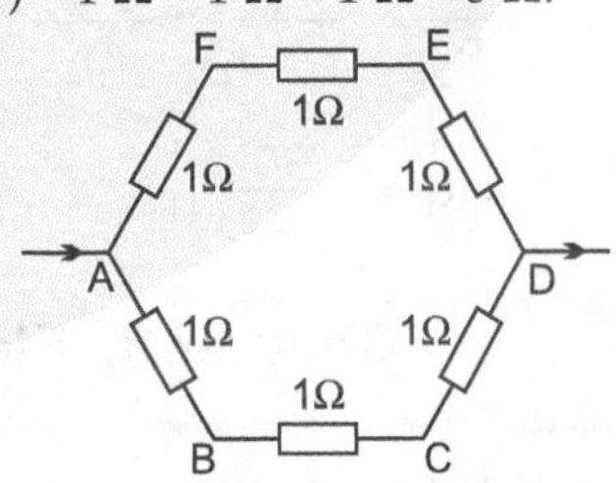

Similarly, equivalent resistance R_2 of the other group (AF, FE, ED)

$$= 1\ \Omega + 1\Omega + 1\Omega = 3\ \Omega$$

As the two groups are in parallel, we have

$$\frac{1}{R} = \frac{1}{R_1}+\frac{1}{R_2}=\frac{1}{3}+\frac{1}{3}=\frac{2}{3}$$

where R is the equivalent resistance of the whole combination.

∴ $$R = \frac{3}{2}\Omega \text{ or } 1.5\ \Omega.$$ **Ans.**

22. Two lamps of resistance 30 Ω and 20 Ω respectively are connected in series in a 110 V circuit. Calculate (i) the total resistance in the circuit (ii) the current in the circuit, and (iii) the voltage drop across each lamp.

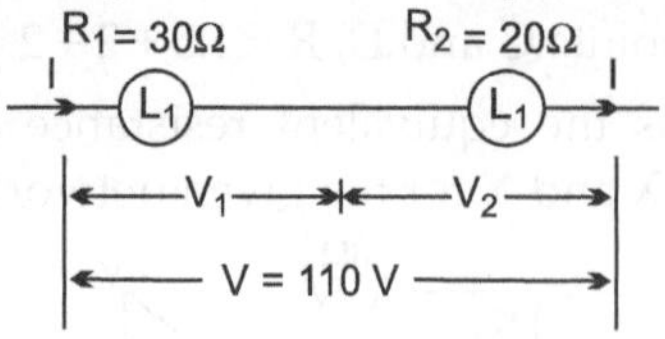

Sol. (i) Since the two lamps are connected in series, their total resistance R is given by:

$$R = R_1 + R_2$$
$$= 30\ \Omega + 20\ \Omega = 50\ \Omega.$$

(ii) We have, V = 110 V, R = 50 Ω

$\therefore$ By Ohm's law, $I = \frac{V}{R} = \frac{110}{50}$ A = 2.2 A.

(iii) We now have,

$$I = 2.2\ A$$

$\therefore$ $V1 = IR_1 = 2.2 \times 30\ V = 66\ V$

$\therefore$ Voltage drop across L_1 = 66 V

Similarly, $V_2 = IR_2 = 2.2 \times 20 = 44\ V$

$\therefore$ Voltage drop across L_2 = 44 V. **Ans.**

23. Calculate the equivalent resistance of the following combinations of resistors r_1, r_2, r_3, r_4.

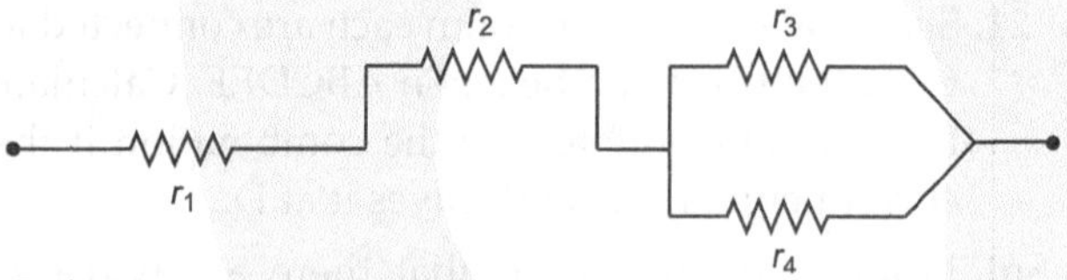

Sol. Since r_3 and r_4 are connected in parallel.

$\therefore$ Equivalent resistance R_1 of this combination is given by

$$\frac{1}{R_1} = \frac{1}{r_3} + \frac{1}{r_4} = \frac{r_4 + r_3}{r_3 r_4}$$

or $$R_1 = \frac{r_3 r_4}{r_3 + r_4}$$

Now r_1, r_2 and R_1 are in series

$\therefore$ Equivalent resistance R of the whole combination is

$$R = r_1 + r_2 + \frac{r_3 r_4}{r_3 + r_4}$$

$$= \frac{(r_1 + r_2)(r_3 + r_4) + r_3 r_4}{r_3 + r_4}\ \Omega.$$ **Ans.**

24. 2 Ω resistor A, 1 Ω resistor B and 4 Ω resistor C are connected in parallel. The combination is connected across a 2 V battery of negligible resistance. Draw a diagram of the arrangement and calculate: (i) The current in each resistor A, B and C, (ii) The current through battery.

Sol. The diagram of the arrangement is shown in figure given alongside:

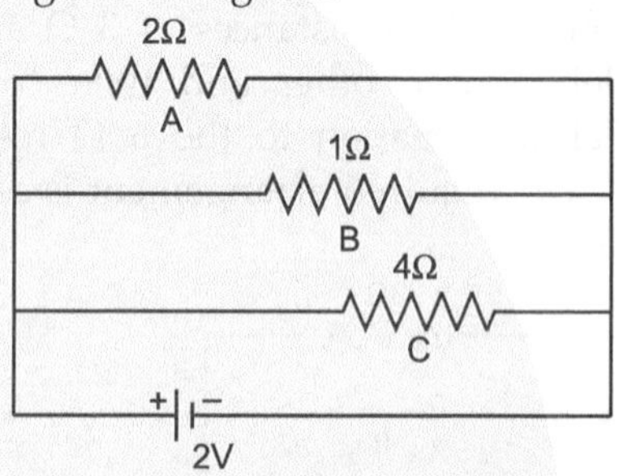

(i) The current in resistor A is

$$I_1 = \frac{V}{R_1} = \frac{2}{2} = 1.0\ A$$

The current in resistor B is

$$I_2 = \frac{V}{R_2} = \frac{2}{1} = 2.0\ A$$

The current in resistor C is

$$I_3 = \frac{V}{R_3} = \frac{2}{4} = 0.5\ A$$

(ii) The current through battery,

$$I = I_1 + I_2 + I_3$$
$$I = 1.0 + 2.0 + 0.5$$
$$= 3.5\ A.$$ **Ans.**

Chapter 8. Electric Power and Household Circuits

1. The diagram below shows a dual control switch circuit connected to a bulb.

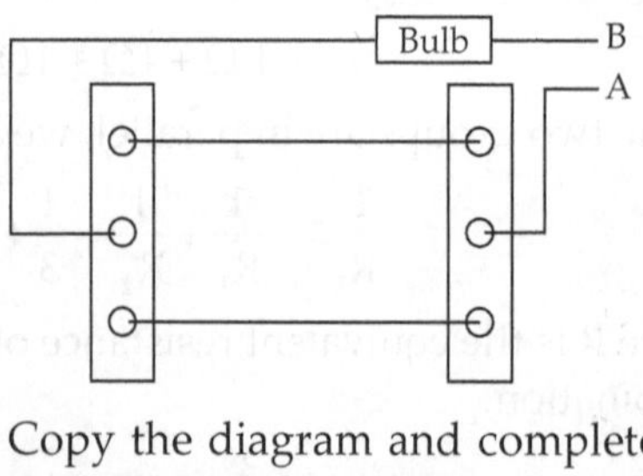

(i) Copy the diagram and complete it so that the bulb is switched ON.

(ii) Out of A and B, which one is the live wire and which one is the neutral wire ?*

Ans. (i) The diagram below shows the bulb in switched ON mode.

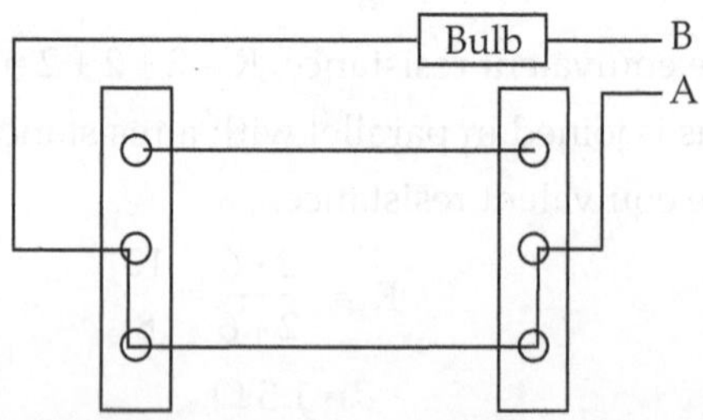

* **Frequently asked previous years Board Exam Questions.**

(ii) A is live wire, B is neutral wire.

2. Draw diagram of a three pin socket outlet and state the connections made to them.

Ans. A three pin socket outlet is shown in figure alongside.

E—Earth wire

N—Neutral wire

L—Live wire.

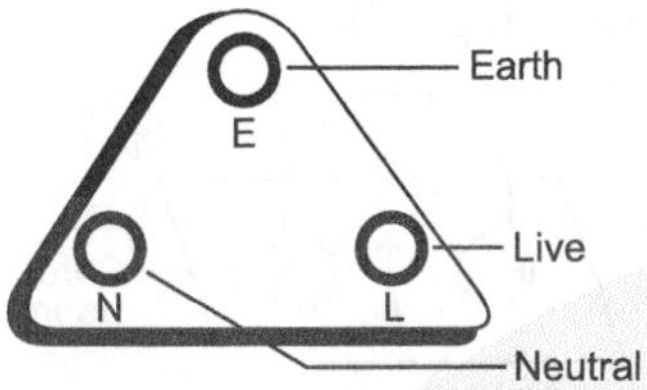

3. Draw a diagram of a three pin plug. Mark the pins used for live, neutral and earth connections as L, N and E respectively. To which part of the appliance is the pin marked as E connected. To which line L or N is the fuse connected?

Ans. The labelled diagram of a three pin plug is shown alongside. The pin marked as E is connected to the outer metallic case of the appliance. The fuse is connected to the live line L.

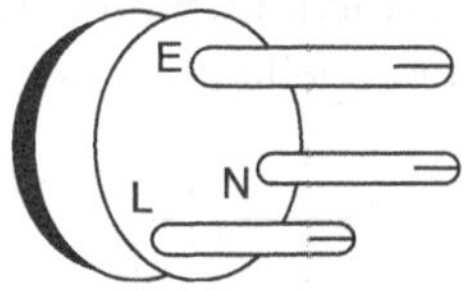

4. Draw a labelled diagram with necessary switch, regulator etc. to connect a bulb, a plug socket outlet and a fan with the mains. In what arrangement have you connected these to the mains ?

Ans. The labelled diagram of the circuit is given below :

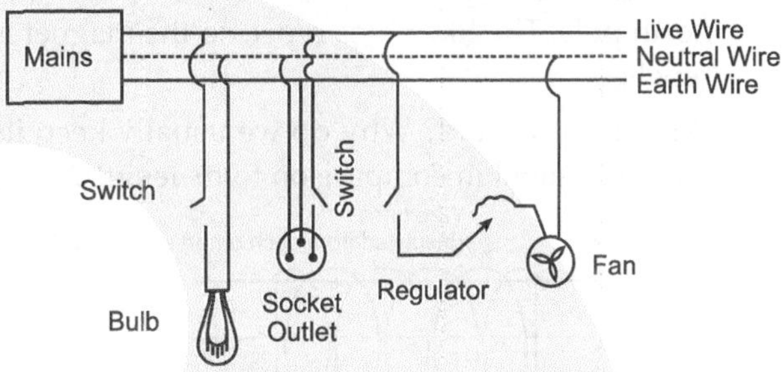

The different appliances have been connected to the mains in parallel.

Chapter 9. Magnetic Effect of Current

1. The diagram below shows a loop of wire carrying current I:

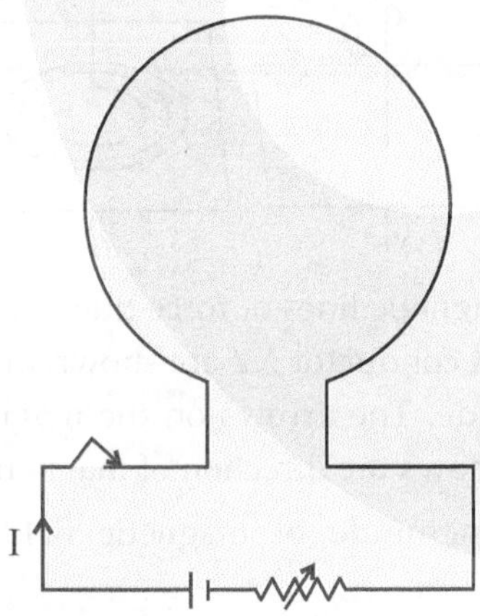

(i) What is the magnetic polarity of the loop that faces us ?

(ii) With respect to the diagram how can we increase the strength of the magnetic field produced by this loop?*

Ans. (i) South

(ii) By passing a stronger current.

2. A magnet kept at the centre of two coils A and B is moved to and fro as shown in the diagram. The two galvanometers show deflection.*

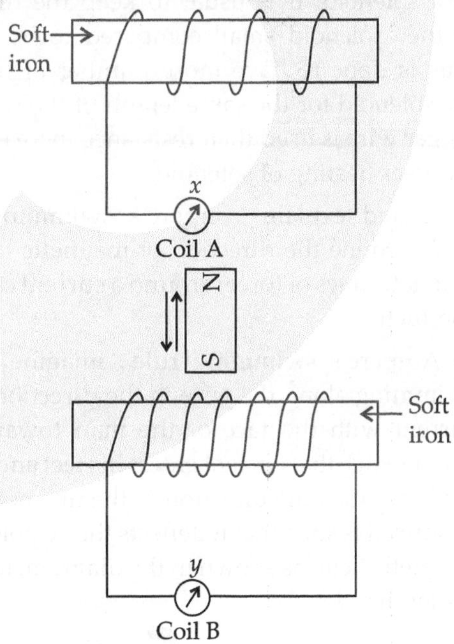

State with a reason whether:

$$x > y$$

or $$x < y.$$

[x and y are magnitudes of deflection]

Ans. $x < y$ because coil B has more number of turns hence there will be a greater change in magnetic flux linked with the coil B.

* **Frequently asked previous years Board Exam Questions.**

3. The adjacent diagram shows a small magnet placed near a solenoid. State whether the magnet is attracted or repelled, as the switch is pressed. Give a reason.

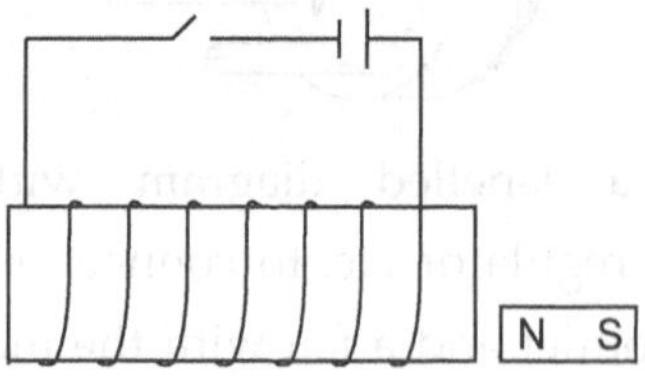

Ans. As the switch is pressed, the magnet will get repelled. The reason is that at the face of solenoid near the north pole of magnet, the current flows in anti-clockwise direction, so the face becomes north pole. The like poles repel, so the magnet is repelled.

4. What is a solenoid? Why do we usually keep its diameter small in comparison to its length?

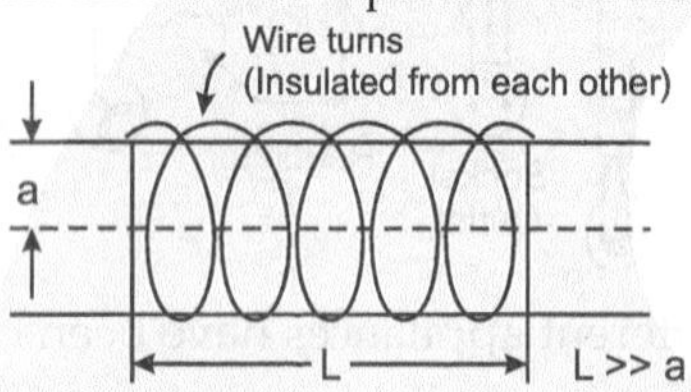

Ans. If a conducting wire (coated with an insulating paint or varnish) is closely wound on a (long) cylindrical core, the resulting coil is referred to as a solenoid. It is usual to keep the diameter of the solenoid small compared to its length. This is done to have more number of turns in the solenoid for the same length of the wire. If a longer wire is used then resistance increases and it causes heating of solenoid.

5. State and explain 'Ampere's swimming rule' to determine the direction of magnetic field (or magnetic lines of force) around a current carrying conductor.

Ans. In Ampere's swimming rule', imagine a man swimming along the wire in the direction of the current with the face of the man towards the needle, such that current enters his feet and leaves his head, then the direction of the magnetic field produced is such that it deflects the N-pole of the magnetic field, as shown in the diagram, towards his left hand.

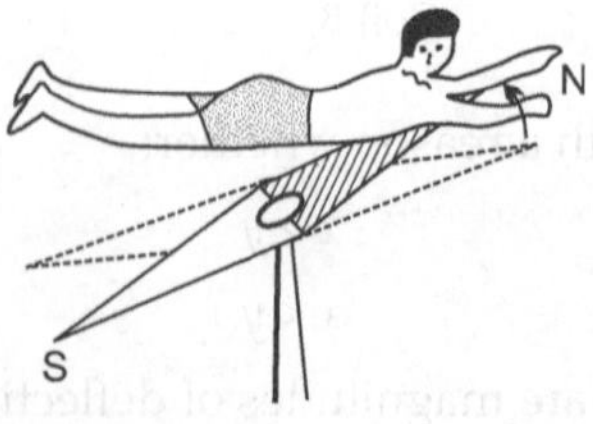

6. The diagram shows a current carrying coil passing through a cardboard sheet. Draw three magnetic lines of force on the board.

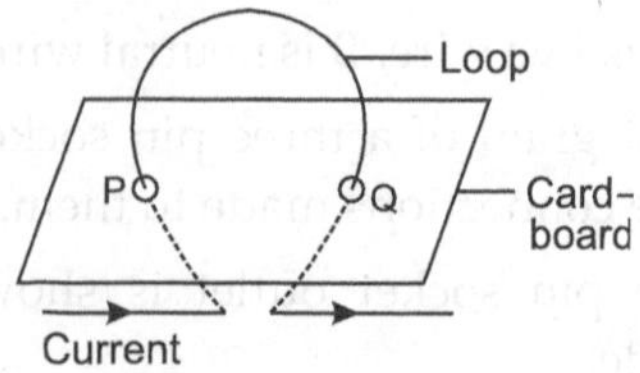

State two factors on which magnitude of magnetic field at the centre depends.

Ans. Figure shows the magnetic lines of force due to current carrying coil.

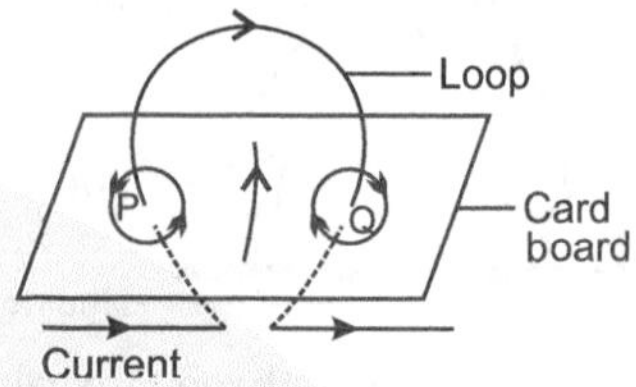

The magnitude of magnetic field at the centre of coil depend on: (i) the strength of current in the coil, and (ii) the number of turns in the coil.

7. In the diagram XY is a straight conductor carrying current in the direction marked by the arrow. The conductor is held vertical by passing it through a horizontal cardboard sheet. Draw three magnetic lines of force on the board and mark the direction of magnetic field in your diagram. State two factors on which magnitude of magnetic field at a point, depends.

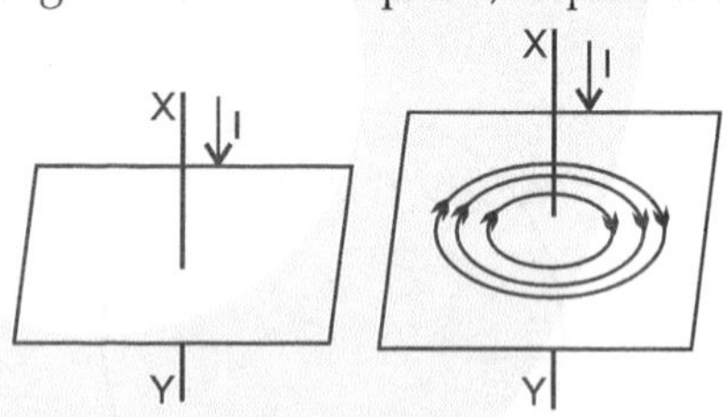

Ans. The magnetic lines of force due to current in the straight conductor XY are shown in figure given alongside. The arrows on the magnetic lines of force shows the direction of magnetic field.

The magnitude of magnetic field at a point depends on:

(i) The strength of current in the conductor, and

(ii) The distance of point from the conductor.

8. (i) A straight wire conductor passes vertically through a piece of cardboard sprinkled with iron filings. Copy the diagram and show the setting of iron filings when a current is passed through the wire in the upward direction and the cardboard is tapped gently. Draw arrows to represent the direction of the magnetic field lines.

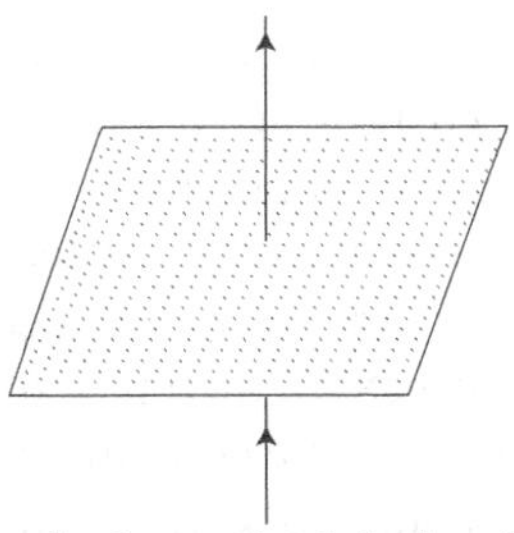

(ii) Name the law which helped you to find the direction of the magnetic field lines.

Ans. (i)

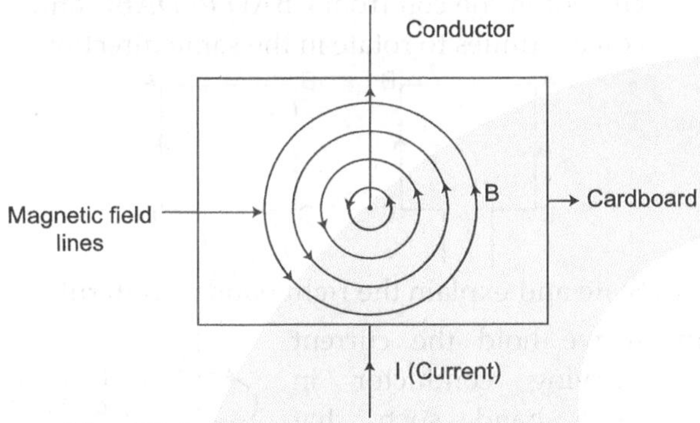

(ii) Right hand thumb rule.

9. Sketch the lines of force of the magnetic field of a solenoid. How does its field compare with that of a bar magnet?

Ans. The magnetic field of a solenoid is very similar to that of a bar magnet. This is shown in figure (i) and (ii) respectively, which shows the lines of force of the magnetic field of a current carrying solenoid and a bar magnet.

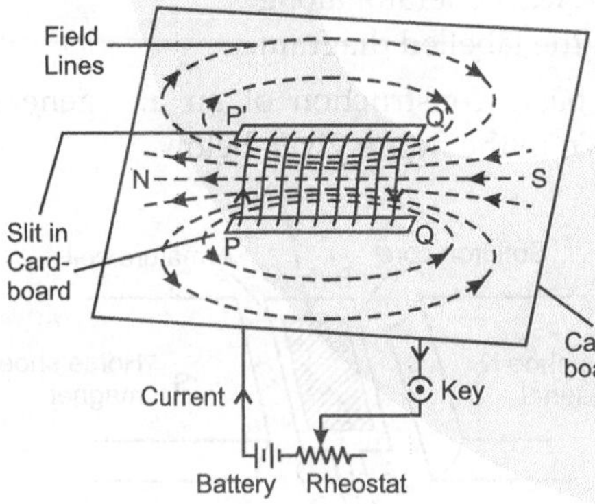

(i)

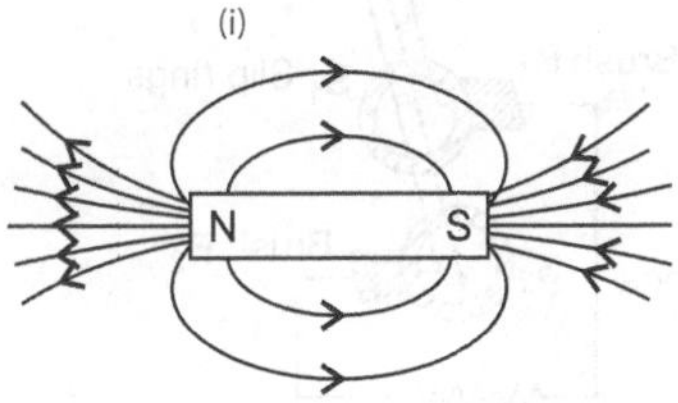

(ii)

10. What is an electromagnet ? Name and draw two common shapes of electromagnet.

Ans. An electromagnet is a temporary strong magnet made from a piece of soft iron, when current flows in the coil wound around it.

The two common shapes of electromagnet are:

(i) I-shape or bar magnet and (ii) U-shape or horse-shoe magnet.

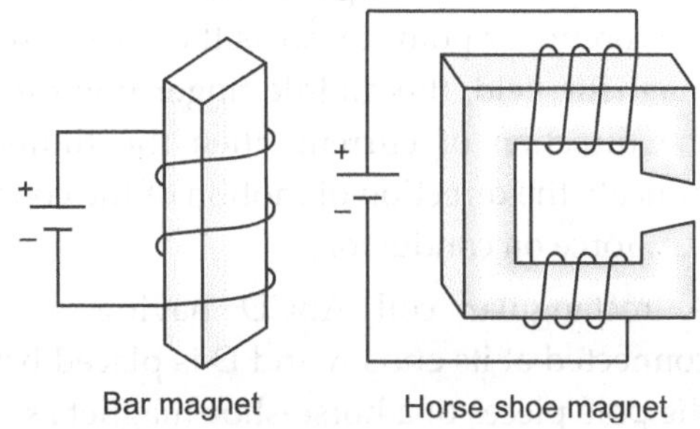

11. Draw a labelled diagram to make an electromagnet from a soft iron bar. Mark the polarity at its ends in your diagram. What precaution would you observe while making it?

Ans. The labelled diagram is shown in the figure given below:

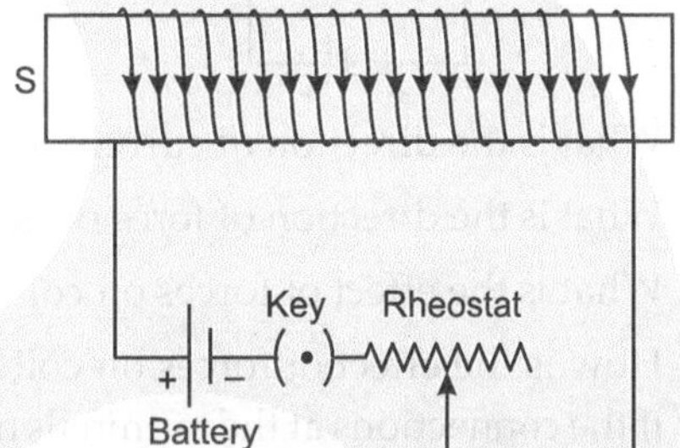

Precaution: The source of current must be the D.C. battery

12. You have been provided with a solenoid AB.

(i) What is the polarity at end A?

(ii) Give one advantage of an electromagnet over a permanent magnet.

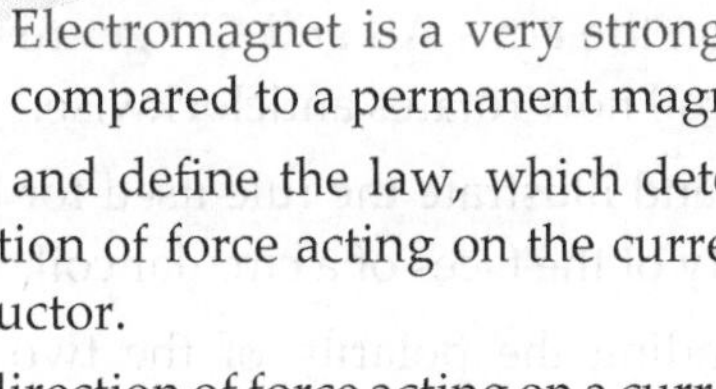

Ans. (i) North pole.

(ii) Electromagnet is a very strong magnet as compared to a permanent magnet.

13. State and define the law, which determines the direction of force acting on the current carrying conductor.

Ans. The direction of force acting on a current carrying conductor can be determined by the Fleming's left hand rule.

Fleming's left hand rule: Stretch the thumb, the forefinger and the middle finger of your left hand mutually at right angles to each other such that the forefinger point towards the direction of the magnetic field, the middle finger point towards the direction of current, then the thumb will indicate the direction of motion of the conductor (*i.e.*, force on conductor).

14. A rectangular coil ABCD having a battery connected at its ends A and D is placed between the pole pieces of a horse-shoe magnet as shown in the figure:

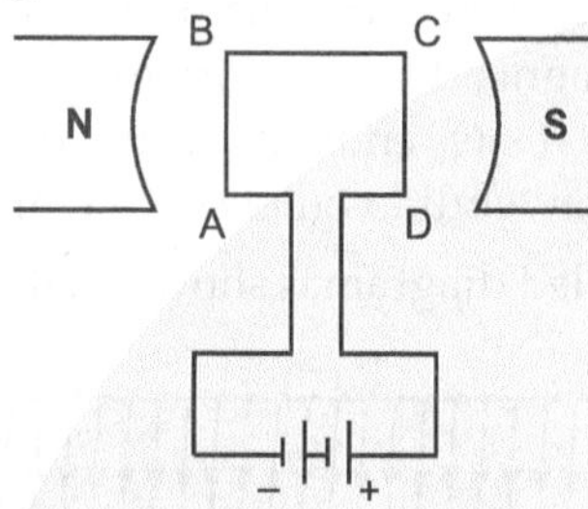

(i) What is the direction of current in the coil ?

(ii) What is the direction of force on each arm ?

(iii) What is the effect of forces on coil ?

(iv) How is the effect of forces on coil changed if the connections at the terminals of battery are interchanged ?

Ans. (i) The current in coil is in the direction DCBA.

(ii) Using Fleming's left hand rule, force on arm AB is outward at right angles to the plane of coil. No force acts on arm BC. On the arm CD the force is inwards perpendicular to the plane of coil. No force acts on the arm DA.

(iii) The forces on the arm AB and CD form a couple and the coil rotates clockwise.

(iv) On interchanging the connections at the terminals of battery, the direction of current in the coil is reversed, and also the forces on the arms AB and CD get reversed. The coil now rotates anticlockwise.

15. State and illustrate the rule used for finding the polarity of the faces of a circular coil.

Ans. For finding the polarity of the two faces of a current carrying circular coil, the following rule applies:

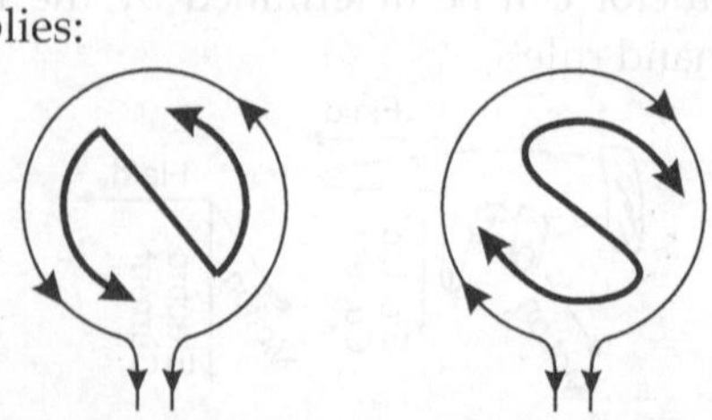

When an observer, looking at the circular coil, finds the current to be flowing in the anti-clockwise direction, then face of the coil behaves like the N-pole of the equivalent magnet. On the other hand, if the current is seen to flow in the clockwise direction, then face of the coil behaves like the S-pole of the equivalent magnet.

16. State the function of a split ring in D.C. motor.

Ans. When the coil of a D.C. motor turns through 180∞, the split ring reverses the direction of the current in the coil from CBAD to DABC and the coil continues to rotate in the same direction.

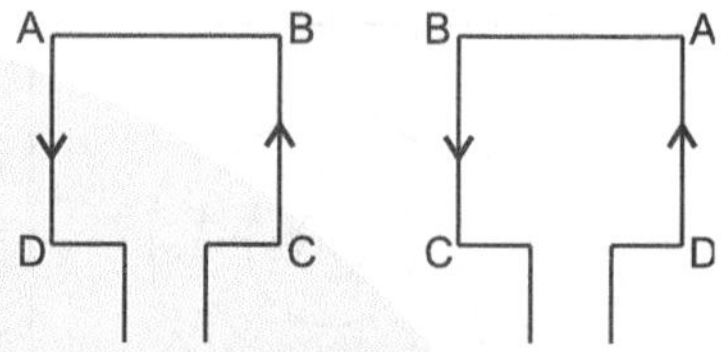

17. State and explain the right hand thumb rule.

Ans. If we hold the current carrying conductor in right hand such that the thumb points in the direction of flow of current, then the fingers encircle the wire in the direction of the magnetic field lines.

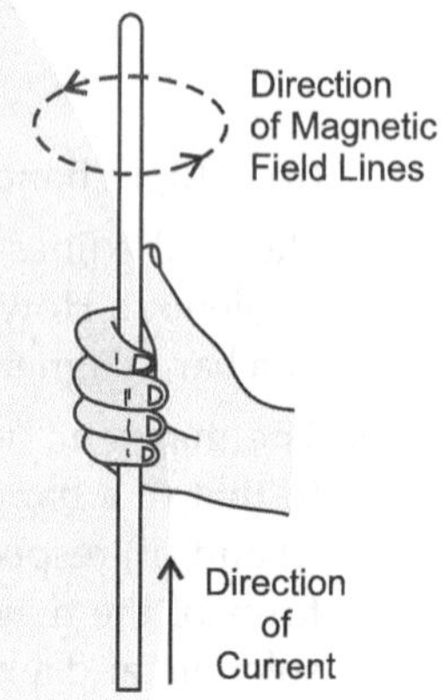

18. Explain the construction of an A.C. generator along with the labelled diagram.

Ans. The basic construction of an a.c. generator is shown in the figure given below:

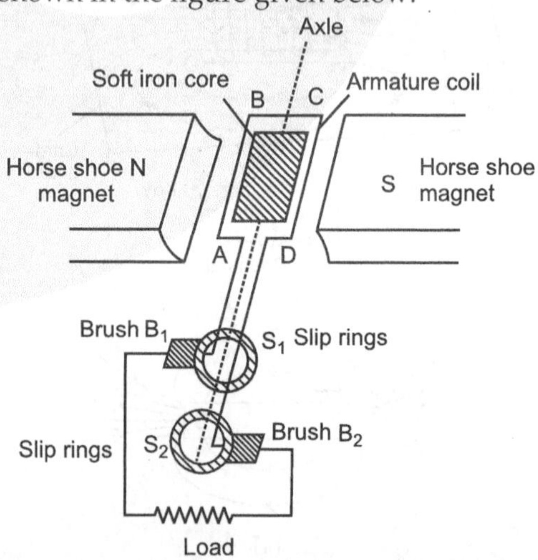

The main parts of an a.c. generator are:

(i) **A horse shoe electromagnet NS:** It is a strong magnet with poles N and S.

(ii) **Armature coil ABCD:** There is a soft iron core on which a coil ABCD having a large number of turns of insulated copper wire is wound. The coil is rotated rapidly in the magnetic field between the poles of the field magnet, by an axle fixed to it.

(iii) **Slip rings S_1 and S_2:** There are two coaxial metallic rings S_1 and S_2 which rotate along with the coil. One end of the coil is connected to the slip ring S_1 and the other end of the coil is connected to the slip ring S_2. Each ring slips against a brush which is stationary.

(iv) **Brushes B_1 and B_2:** There are two brushes B_1 and B_2 made of carbon, which press gently against the slip rings S_1 and S_2 respectively. The other ends of brushes act as terminals of the generator across which the external circuit containing load is connected. The brushes B_1 and B_2 do not rotate along with the coil and the slip rings.

19. State change that you would make to convert an A.C. generator into a D.C. generator. Illustrate the change by a diagram.

Ans. As shown in the diagram, the only change is that both the slip-rings of an A.C. generator are replaced by a split-ring (X–Y) (a ring split in two halves) which acts as a commutator or current-reverser.

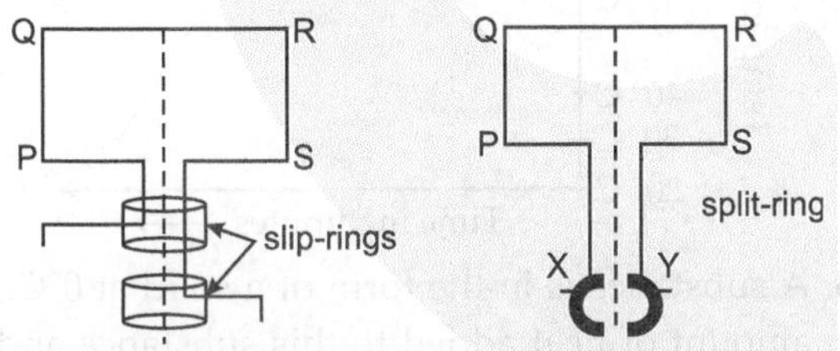

A.C. Generator D.C. Generator

The other parts in both are the same. By this device the current in the external circuit becomes unidirectional.

20. Draw a labelled diagram to show the various parts of a step-up transformer and step down transformer.

Ans. Step-up transformer is used to change low voltage alternating e.m.f. to high voltage alternating e.m.f. and step-down transformer is used to change high voltage alternating e.m.f. to low voltage alternating e.m.f. The essential parts are shown in the given diagram:

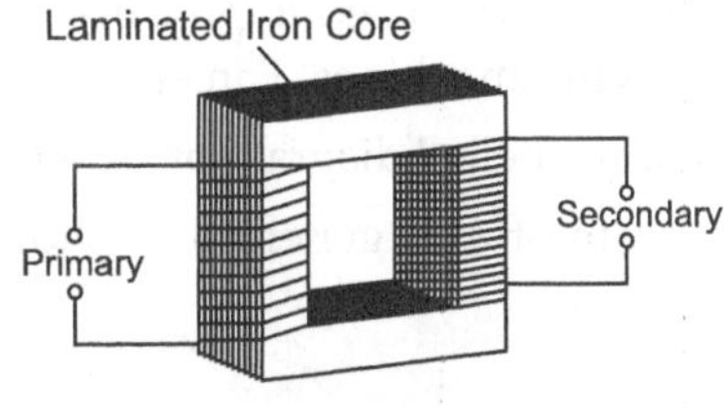

Step Up

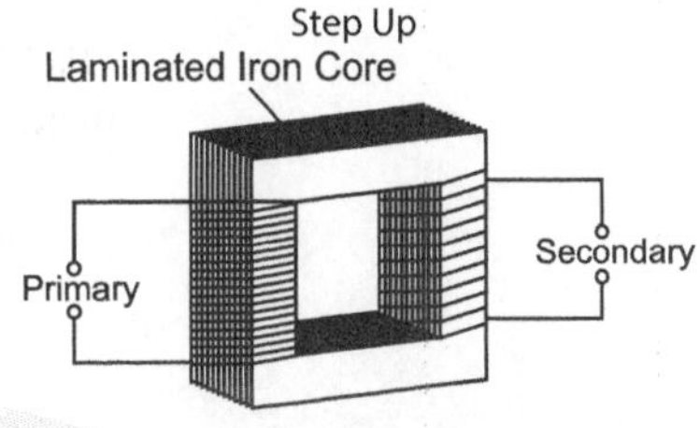

Step Down

21. The diagram below shows the core of a transformer and its input and output connections.*

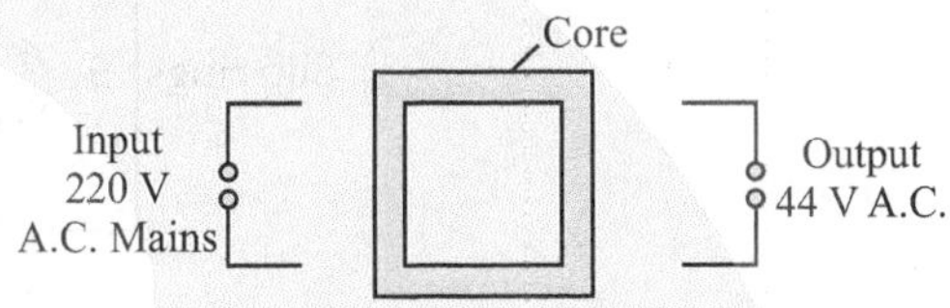

(i) State the material used for the core.

(ii) Copy and complete the diagram of the transformer by drawing input and output coils.

Ans. (i) The material used for core is laminated soft iron.

(ii)

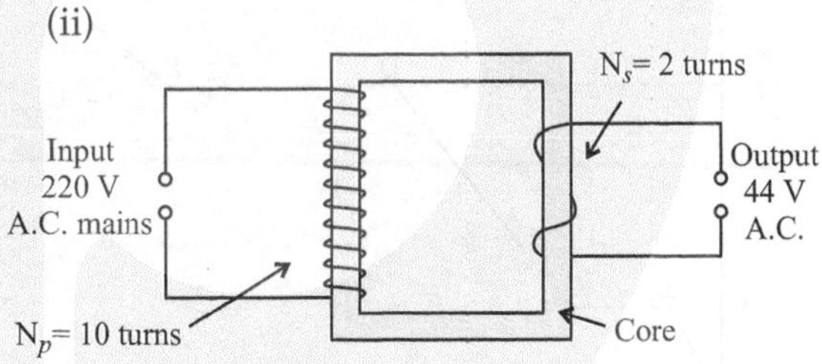

$$\text{Turns ratio} = \frac{N_s}{N_p} = \frac{E_s}{E_p} = \frac{44}{220} = \frac{1}{5}$$

Hence, number of turns in secondary coil : primary coil is equal to 1 : 5

22. (i) Draw a neat labeled diagram of a d.c. motor.

(ii) Write any one use of a d.c motor.*

Ans. (i) Diagram of a d.c. motor.

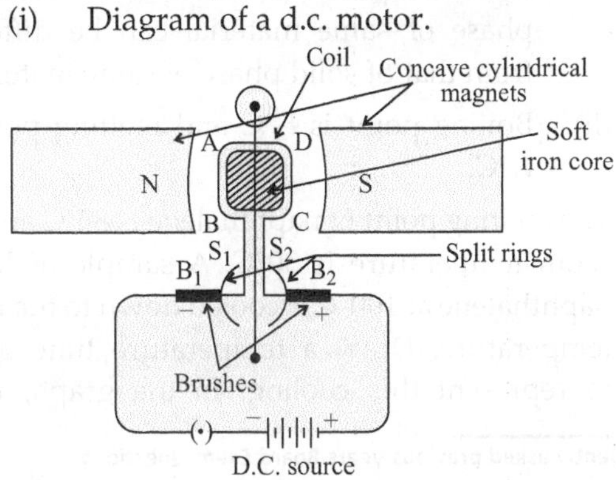

* Frequently asked previous years Board Exam Questions.

Figure Based Short Answers

(ii) D.C. motor is used to produce rotational motion by using electricity in devices such as spinning and weaving machines, Fan etc.

23. Draw a neat labelled diagram of an A.C. generator.*

Ans. The diagram of A.C. generator is shown below.

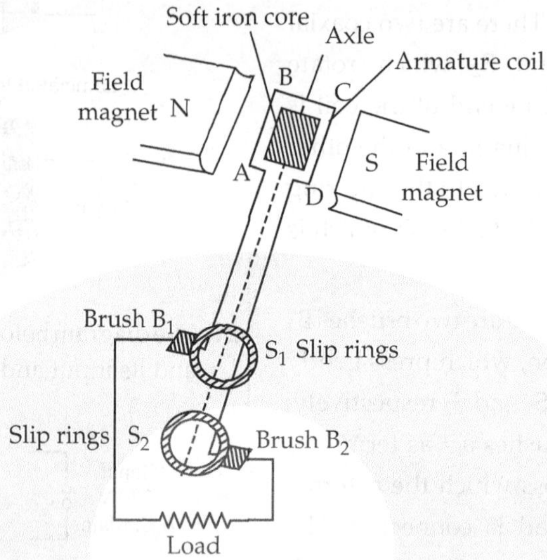

Chapter 10. Specific Heat Capacity and Latent Heat

1. The diagram below shows the change of phases of a substance on a temperature vs time graph on heating the substance at a constant rate.

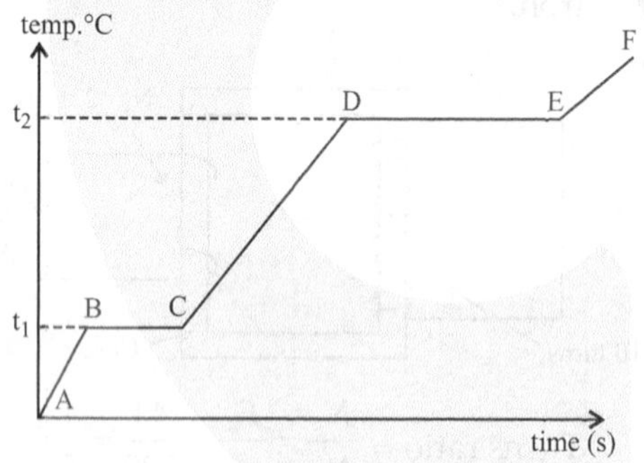

(i) Why is the slope of CD less than slope of AB?

(ii) What is the boiling and melting point of the substance?*

Ans. (i) The slope of CD is less than slope of AB because specific heat capacity of liquid phase of same material can be different from that of solid phase of same material.

(ii) Boiling point is t_2°C and melting point is t_1°C.

2. The melting point of naphthalene is 80°C and the room temperature is 30°C. A sample of liquid naphthalene at 100°C is cooled down to the room temperature. Draw a temperature time graph to represent this cooling. In the graph, mark the region which corresponds to the freezing process.*

Ans. BC represents the freezing process in the graph.

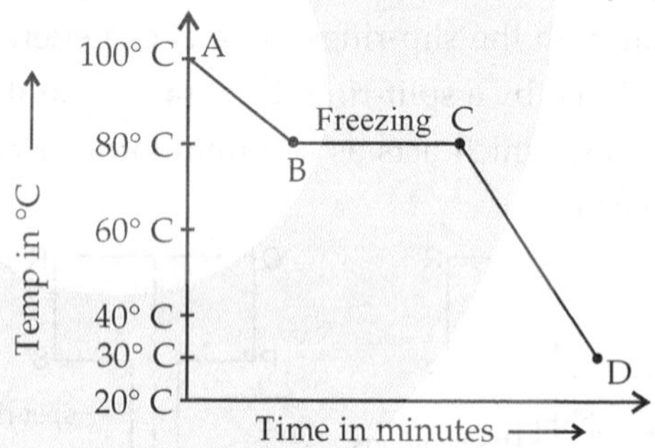

3. A substance is in the form of a solid at 0°C. The amount of heat added to this substance and the temperature of the substance are plotted on the following graph:

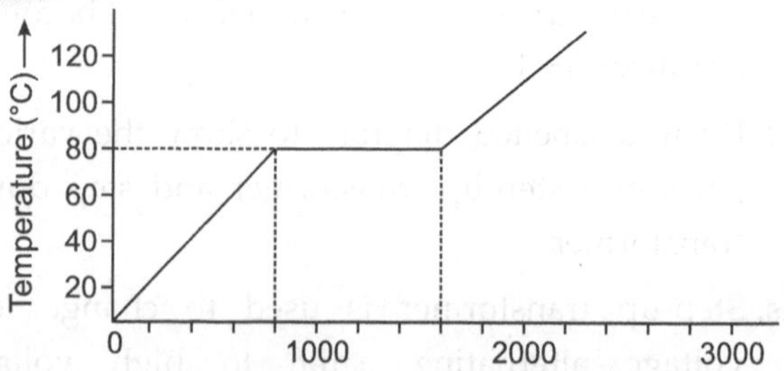

If the specific heat capacity of the solid substance is 500 J/kg °C, find from the graph, the mass of the substance.

* Frequently asked previous years Board Exam Questions.

Sol. From graph heat supplied is 800 joule during change in temperature from 0∞ to 80 °C.

Heat = Mass × Specific heat capacity × Rise in temperature

$800 = m \times 500 \times 80$

$m = \frac{1}{50}$ kg

$= 20$ g. **Ans.**

Chapter 11. Radioactivity and Nuclear Energy

1. The diagram given below shows a radioactive source S place in a thick lead walled container. The radiations given out are allowed to pass through a magnetic field. The magnetic field (shown as ×) acts perpendicular to the plane of paper inwards. Arrows show the paths of the radiations A, B and C.

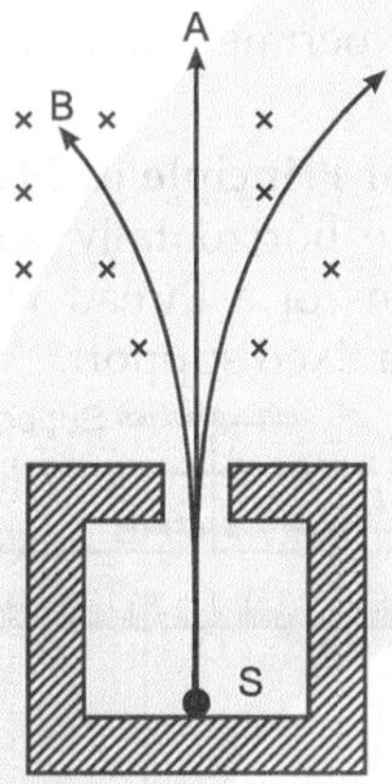

Name the radiations labelled A, B and C.

Ans. The radiations labelled A, B and C are gamma radiations, alpha particles and beta particles respectively.

2. The figure given below shows a radioactive source S placed in a thick walled lead container. The radiations given off pass through a magnetic field acting in direction perpendicular to the plane of paper inwards as shown by X. Copy the diagram and show the path of radiations. Explain why the source is kept in a thick-walled lead container.

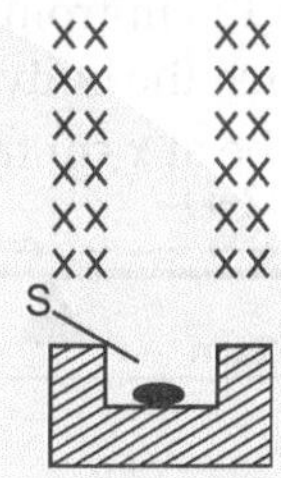

Ans. The given figure shows the path of radiation in the magnetic field. The source is kept in a thick-walled lead container so that its radiations may not come out in the undesired direction otherwise they will harm the person working on it.

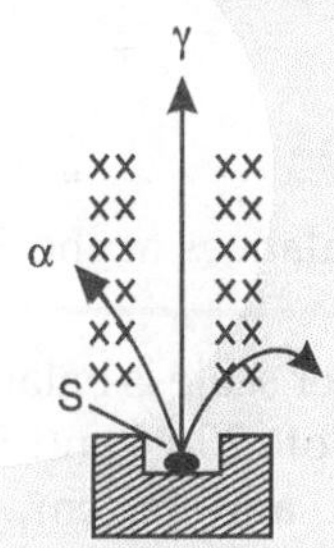

Figure Based Long Answers | Set 4 |

Chapter 1. Force, Work, Power and Energy

1. Two unlike forces (parallel) of 10 N and 25 N act at a distance of 12 cm from each other. Find the point about which the body will balance.

Ans. Let the balance be at x cm from the 25 N force.

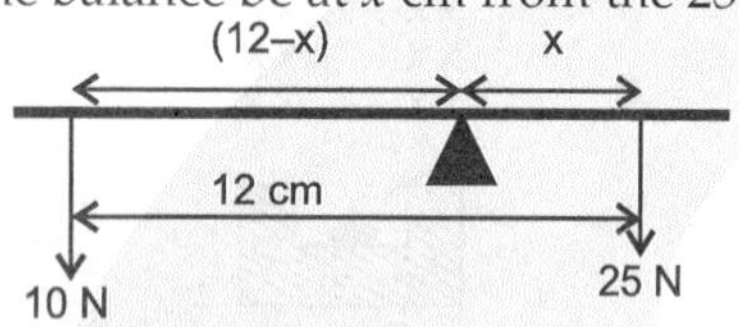

$$\text{Anti-clockwise moment} = 10\,(12 - x)$$
$$\text{Clockwise moment} = 25\,x$$

Applying principle of moments,

$$\Rightarrow \quad 10\,(12 - x) = 25\,x$$
$$\Rightarrow \quad 120 - 10\,x = 25\,x$$
$$\Rightarrow \quad 35\,x = 120$$
$$\Rightarrow \quad x = \frac{120}{35}$$
$$= 3{\cdot}43 \text{ cm.}$$

∴ The body balances at the 3·43 cm mark from the 25 N force. **Ans.**

2. A uniform meter scale is balanced at 40 cm mark, when weights of 20 gf and 5 gf are suspended at 5 cm mark and 75 cm mark respectively. Calculate weight of meter scale.

Ans.

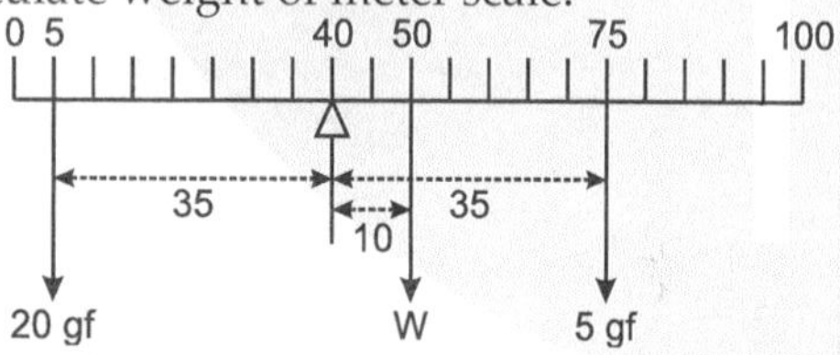

A meter scale is uniform, therefore its weight acts at 50 cm mark.

Taking moments about 40 cm mark.

Moment in clockwise direction,

$$= W \times 10 \text{ cm} + 5 \text{ gf} \times 35 \text{ cm}$$
$$= 10 \text{ W cm} + 175 \text{ gf cm.}$$

Moment in anticlockwise direction

$$= 20 \text{ gf} \times 35 \text{ cm}$$
$$= 700 \text{ gf cm.}$$

applying principle of moments,

$$10 \text{ W cm} + 175 \text{ gf cm} = 700 \text{ gf cm}$$
$$\therefore \quad 10 \text{ W cm} = (700 - 175) \text{ gf cm}$$
$$\therefore \quad W = \frac{525}{10} \text{ gf} = 52{\cdot}5 \text{ gf.}$$ **Ans.**

3. Describe an experiment to verify the principle of moments.

Ans. Verification of Principle of Moments: Suspend a meter scale horizontally with its mid-point at O, by means of a thread with its other end connected to a fixed support.

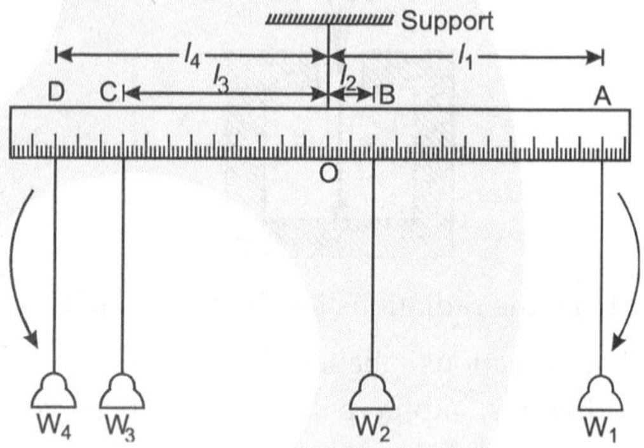

Now suspend some weights on both sides of the mid-point and adjust their distances in such a way that scale again becomes horizontal.

Let W_1, W_2 and W_3, W_4 be the weights suspended and l_1, l_2, l_3 and l_4 are respective lengths.

The weights W_1 and W_2 tend to turn the scale clockwise while the weights W_3 and W_4 tend to turn the scale anticlockwise.

Total clockwise moments

$$= W_1 \times l_1 + W_2 \times l_2$$

Total anticlockwise moments

$$= W_3 \times l_3 + W_4 \times l_4$$

In equilibrium, when the scale is horizontal, it is found that :

Total clockwise moments = Total anticlockwise moments

$$W_1 l_1 + W_2 l_2 = W_3 l_3 + W_4 l_4$$

This verifies the principle of moments.

4. A uniform meter rule rests horizontally on a knife-edge at the 60 cm mark when a mass of

10 gram is suspended from one end. At which end must this mass be suspended ? What is the mass of the rule?

Ans. Let M gram be the mass of the rule. Since density of the rule is uniform, its weight M g will act at its mid-point *i.e.*, at the 50 cm mark (Fig.).

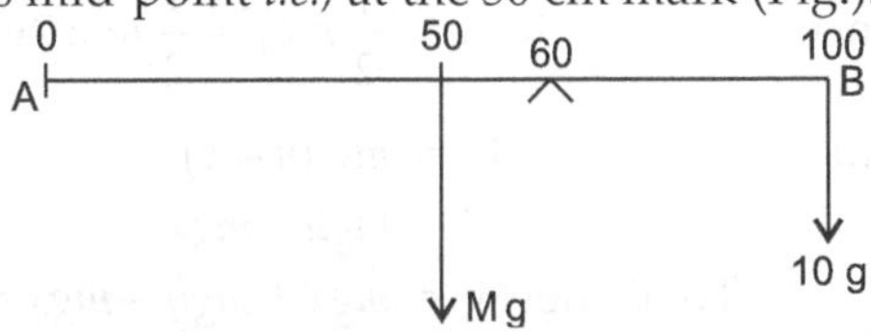

The weight of the rule produces an anticlockwise moment about the knife-edge O. In order to balance it, 10 gram mass is to be suspended at the end B, to produce a clockwise moment.

From the principle of moments,

Anticlockwise moment = Clockwise moment

$$M \times (60 - 50) = 10 \times (100 - 60)$$

or $$M \times 10 = 10 \times 40$$

$\therefore$ $$M = 40 \text{ gram}$$ **Ans.**

5. A uniform metre scale can be balanced at the 70·0 cm mark when a mass of 0·05 kg is hung from the 94·0 cm mark.

(i) Draw a diagram of the arrangement.

(ii) Find the mass of the metre scale.

Ans. (i) Diagram of the given arrangement is shown below:

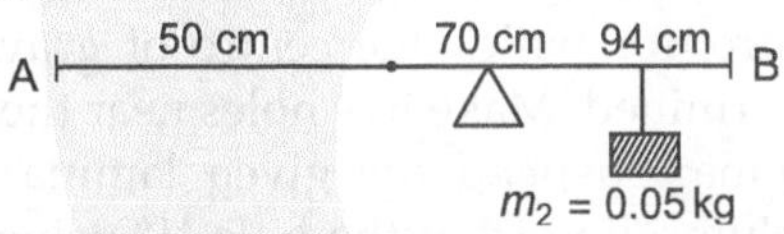

(ii) As the given meter scale is a uniform scale, so its centre of gravity lies at 50 cm. Let mass of metre scale be m_1 kg and its weight (m_1 kgf) will act at 50 cm mark.

By principle of moments,

$$m_1x_1 = m_2x_2$$

$$m_1 \times (70 - 50) = 0{\cdot}05 \times (94 - 70)$$

$$m_1 = \frac{0.05 \times 24}{20}$$

$$= 0{\cdot}06 \text{ kg.}$$

$$= 60 \text{ g.}$$ **Ans.**

6. A block of mass 30 kg is pulled up a slope, as shown in the diagram with a constant speed, by applying a force of 200 N parallel to the slope. A and B are initial and final positions of block.

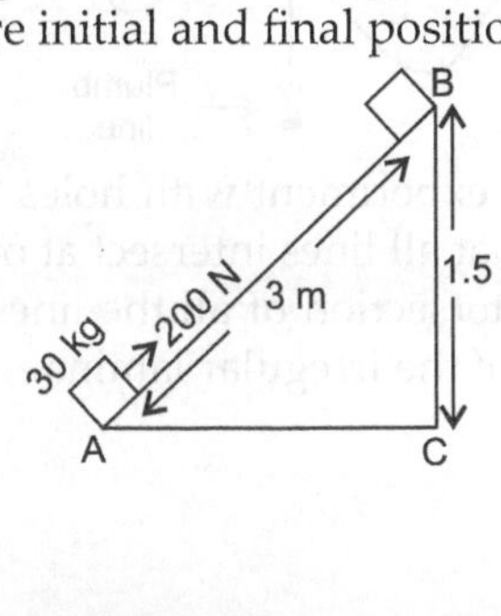

(i) Calculate the work done by force in moving the block from A to B.

(ii) Calculate P.E. gained by block.

[Take $g = 10 \text{ ms}^{-2}$]

Ans. (i) Work done in moving the block from A to B,

$$= F \times s$$

$$= 200 \text{ N} \times 3 \text{ m} = 600 \text{ J}$$ **Ans.**

(ii) Gain in potential energy

$$= mgh = 30 \times 10 \times 1{\cdot}5$$

$$= 450 \text{ J.}$$ **Ans.**

7. A wheel of diameter 1·6 m is shown in the figure with axle at O. A force F = 4 N is applied at B in the direction shown in figure. Calculate the moment of force about (i) Centre O, and (ii) Point A.

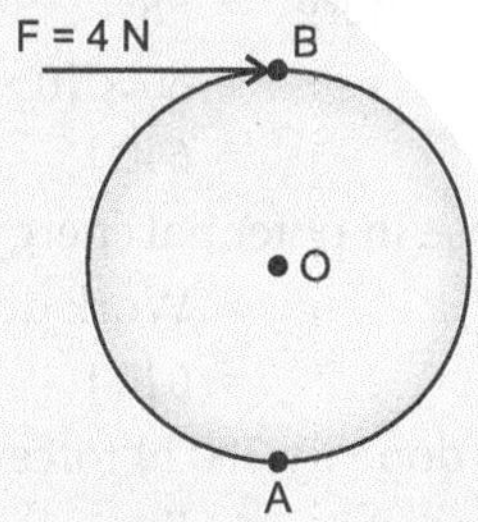

Ans. (i) Given: F = 4 N, $BO = \frac{1.6}{2}$ m = 0·8 m

Moment of F about O = F × BO

= 4 × 0·8 Nm (Clockwise)

= 3·2 Nm (Clockwise)

(ii) Given : F = 4 N, BA = 1·6 m

Moment of F about A = F × BA

= 4 × 1·6 Nm (Clockwise)

= 6·4 Nm (Clockwise)

8. A truck driver can load oil drums into the back of the truck by pushing them up a sloping plank, or by lifting them directly. Each drum has a mass of 80 kg, the plank is 3 m long, and the back of the truck is 0·8 m above the ground.

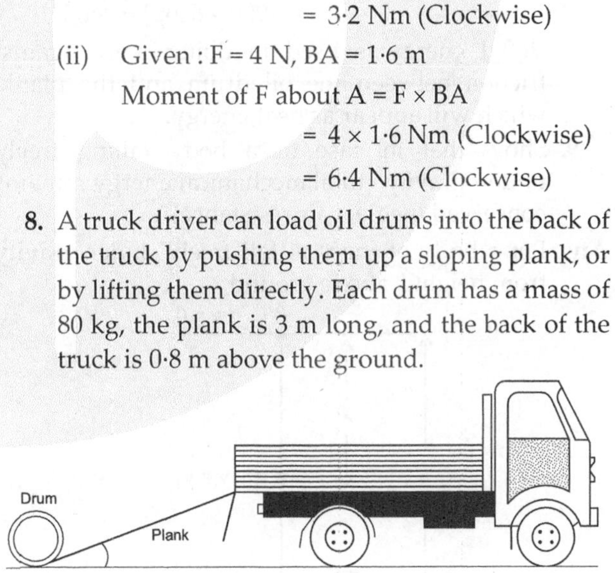

(i) Calculate the weight of one oil drum (Take $g = 10 \text{ ms}^{-2}$).

(ii) How much energy would be required in lifting the drum into the truck without the plank ?

(iii) How does the potential energy increase, when the oil drum is lifted directly into the truck without the plank ?

(iv) If a force of 300 N is needed to push the oil drum along the plank 3 m long to load it on the truck, calculate the work done.

(v) Account for the difference in work done by the force and the increase in potential energy of the oil drum.

Ans.

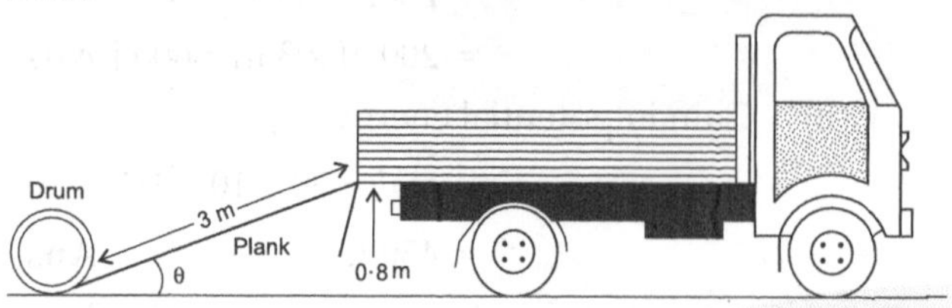

(i) Weight of one oil drum

$= mg$

$= 80 \times 10\ N = 800\ N$

(ii) Energy $= mgh$

$= 80 \times 10 \times 0{\cdot}8\ J$

$= 640\ J$

(iii) Increase in potential energy

= Work done in (ii)

$= 640\ J$

(iv) Work done by 300 N force

$= F \times s$

$= 300 \times 3\ J$

$= 900\ J$

(v) Difference between work done and gain in potential energy

$= (900 - 640)\ J = 260\ J$

260 J energy is used in doing work against friction between the oil drum and the plank, which will appear as heat energy.

9. Show that in case of a body falling freely under gravity, total mechanical energy remains conserved (neglect air resistance).

Ans. Let a body of mass m fall freely under gravity from height h above ground.

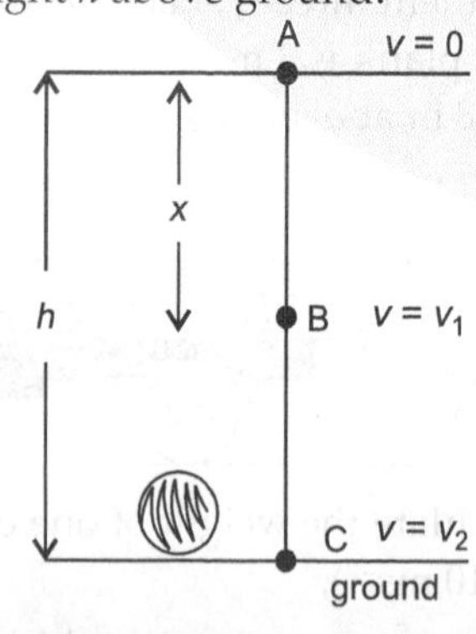

Let A, B and C be the positions of body.

Let x be the distance fallen from A to B.

At position A:

K.E. $= 0$ (body is at rest)

and P.E. $= mgh$

$\therefore$ Total energy $= 0 + mgh = mgh$ ...(i)

At position B:

Let v_1 be the velocity of body, then $u = 0$, $S = x$.

From equation $v^2 = u^2 + 2aS$

$v_1^2 = 0 + 2gx = 2gx$

Since K.E. $= \frac{1}{2} mv_1^2 = \frac{1}{2} m \times 2gx = mgx$

and P.E. $= mg(h - x)$

$= mgh - mgx$

$\therefore$ Total energy $= mgx + mgh - mgx = mgh$ ...(ii)

At position C :

Let velocity of body be v_2, then $u = 0$, $S = h$.

From equation $v^2 = u^2 + 2gS$

$v_2^2 = 0 + 2gh = 2gh$

Since K.E. $= \frac{1}{2} mv_2^2 = \frac{1}{2} m \times 2gh = mgh$

and P.E. $= 0$ (body at ground)

$\therefore$ Total energy $= mgh + 0 = mgh$ ...(iii)

$\therefore$ From (i), (ii) and (iii) it is clear that sum of mechanical energy remains same at any point in the path of free fall of a body.

10. Describe an experiment to determine the centre of gravity of an irregular lamina by the method of balance using a plumb line.

Ans. Let A be an irregular lamina as shown in figure, for which the centre of gravity is to be determined. Make five holes near the edge of the lamina. Suspend the given lamina along with a plumb line from the hole '1' using a nail and clamp it horizontally on a retort stand. Lamina must be free to oscillate on the nail about the point of suspension. When the plumb line and nail are stationary, draw a line on the card board, along the thread of the plumb line.

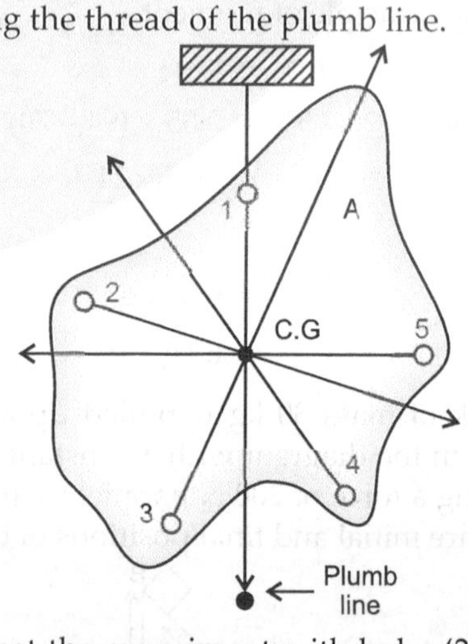

Repeat the experiment with holes '2', '3', '4' etc. It is seen that all lines intersect at one point. This point of intersection of all the lines is the centre of gravity of the irregular lamina.

Chapter 2. Simple Machines

1. (i) How do you define mechanical advantage of a machine?

   (ii) Calculate mechanical advantage of the lever shown in figure.

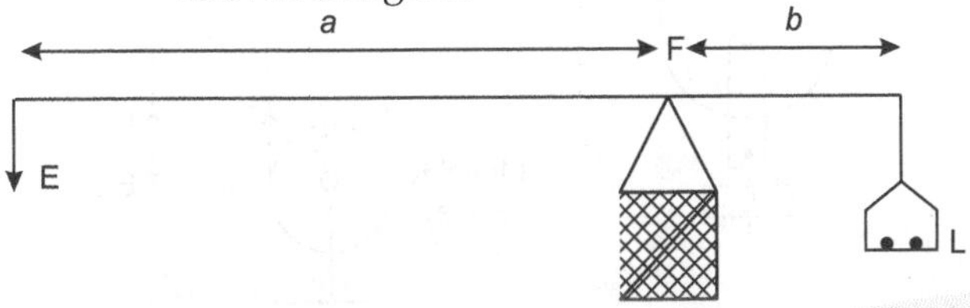

Ans. (i) Mechanical advantage is the ratio of the load lifted to the effort applied.

$$M.A. = \frac{L}{E}$$

(ii) Clockwise moment $= b \times L$

Anticlockwise moment $= a \times E$

In equilibrium by principle of moments,

$$b \times L = a \times E$$

or $$\frac{L}{E} = \frac{a}{b}$$

But $$\frac{L}{E} = M.A.$$

$$\therefore \quad M.A. = \frac{L}{E} = \frac{a}{b} =$$

2. A uniform lever of length 1 m is kept in equilibrium when supported at the length of 60 cm and a mass M is suspended at a length of 90 cm as shown in the figure. State with reasons, whether the weight of the lever is greater than, less than or equal to the weight of mass M. Also, calculate the mechanical advantage.

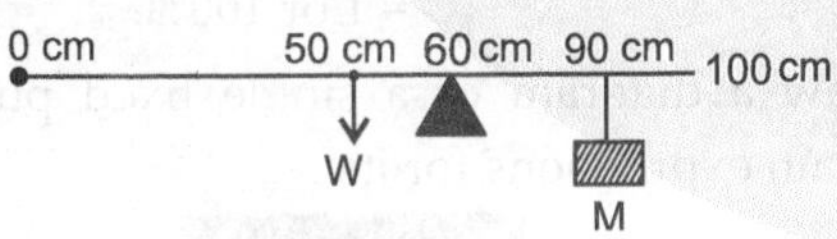

Ans. Let M be the load.

$\therefore$ Load arm $= 90 - 60 = 30$ cm

Since weight of the lever will act at centre of gravity which is the midpoint of the lever.

$\therefore$ Effort arm $= 60 - 50 = 10$ cm

Let weight of the lever be W.

By the principle of moments,

$$L \times d_L = E \times d_E$$

$$M \times 30 = W \times 10$$

$$W = 3M$$

Since weight of the lever is three times that of M.

$\therefore$ Weight of the lever is greater than the weight of M.

$$M.A. = \frac{\text{Load}}{\text{Effort}}$$

$$= \frac{M}{W} = \frac{M}{3M}$$

$$= \frac{1}{3}$$

3. What is a lever?

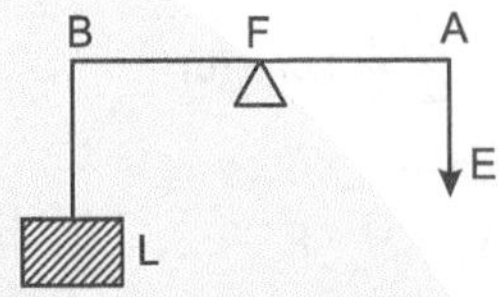

Ans. A lever is a special case of the application of the law of moments. It is a rigid bar, straight or bent, which is capable of rotation about a fixed point or a fixed line, called axis of rotation. This fixed point or the axis of rotation is called the fulcrum. An effort E is applied at the point A of the lever to overcome a resistance (or load) at another point B, as shown in the diagram.

4. The diagram shows the use of a lever.

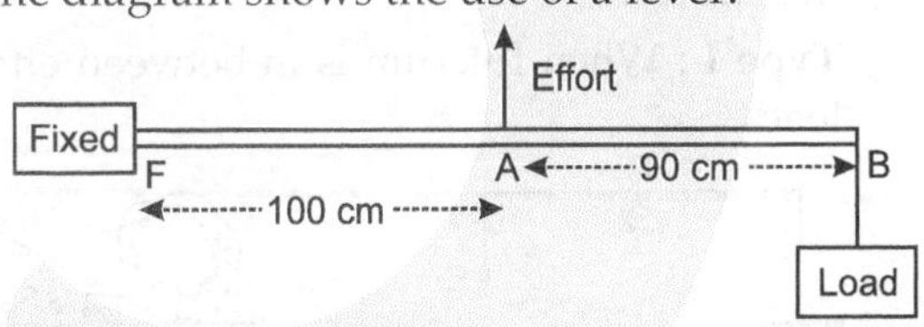

   (i) State the principle of moments as applied to the above lever.

   (ii) Which class of lever is this ? Give an example of this class of lever.

   (iii) If FA = 100 cm, AB = 90 cm, calculate the effort required to lift the load of 20 N.

Ans. (i) In equilibrium position of lever, sum of the clockwise moments about the fulcrum is equal to the sum of the anticlockwise moments about the fulcrum.

(ii) Lever of class III, example: foreceps or a pair of sugar tongs.

(iii) By the principle of moments,

$$E \times 100 = 190 \times 20$$

$$\therefore \quad E = 38 \text{ N.}$$ **Ans.**

5. Draw a system of pulleys using one fixed pulley and three movable pulleys. State the value of (i) M.A. and (ii) V.R. of this pulley system in the ideal case.

Ans.

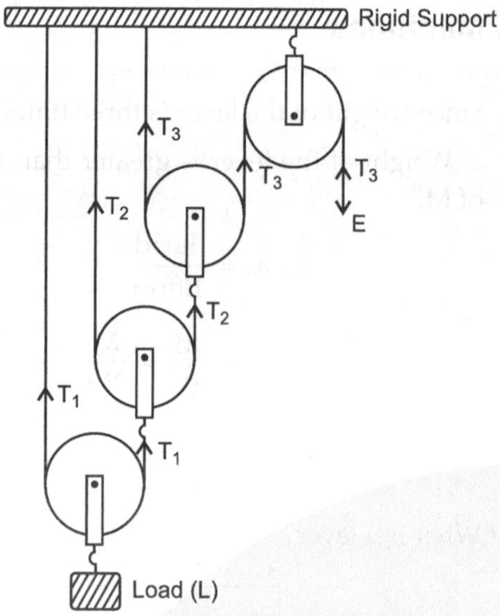

In ideal case:

(i) M.A. = $2^3 = 8$

(ii) V.R. = $2^3 = 8$

(In general, if there are *'n'* movable pulley connected to a fixed pulley, then in the ideal case: M.A. = 2^n and V.R. = 2^n)

6. State the types (or kinds) of lever, and give one example of each kind.

Ans. There are three kinds of lever, depending upon the different positions of the load, fulcrum and the effort.

Type I : When fulcrum is in between effort and load.

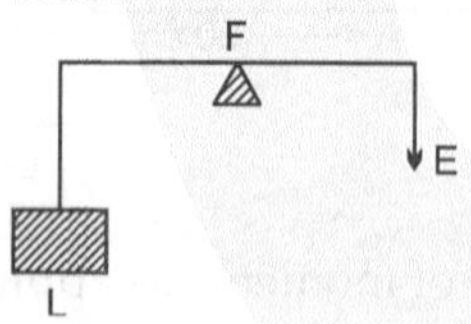

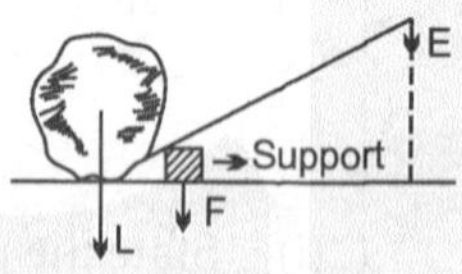

Crowbar

Type II : When the load is in between fulcrum and the effort.

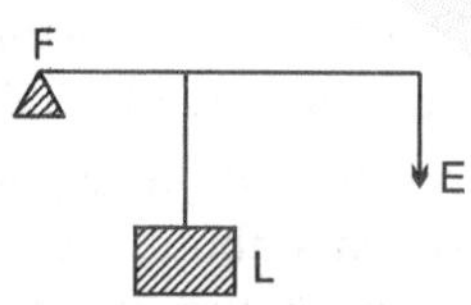

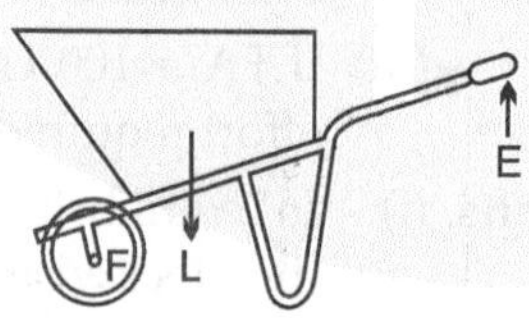

Wheel Barrow

Type III : When the effort lies in between the fulcrum and load.

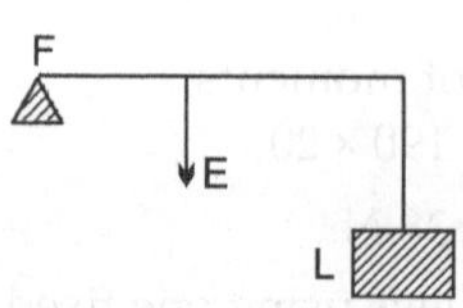

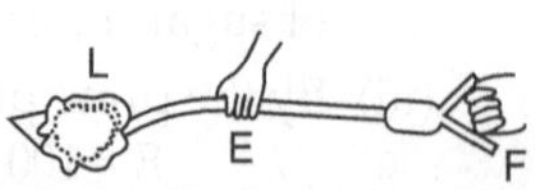

Spade

7. Draw the diagram of a single movable pulley and obtain its mechanical advantage, velocity ratio and efficiency.

Ans. Since it is inconvenient to apply the effort in an upward direction, a single fixed pulley is used to change the direction of effort.

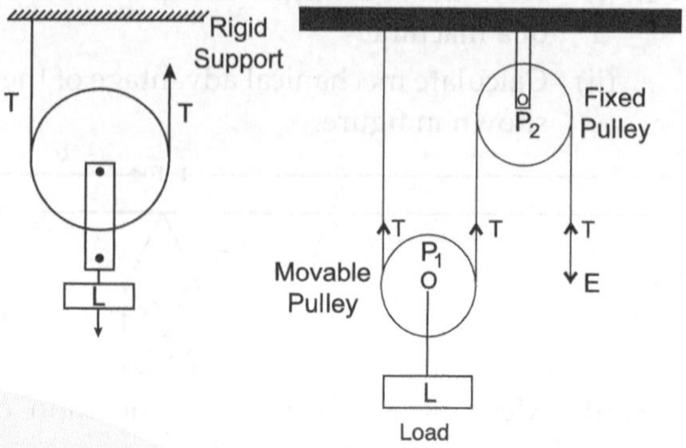

In equilibrium, L = T + T = 2T

E = T

Mechanical advantage

$$= \frac{\text{Load (L)}}{\text{Effort (E)}} = \frac{2T}{T} = 2$$

As the effort is pulled down through a distance d_E, the two segments of the thread carrying the movable pulley and the load goes up by $d_E/2$.

Hence, $d_L = d_E/2$

$$\text{Velocity ratio (V.R.)} = \frac{d_E}{d_L}$$

$$= \frac{d_E}{\frac{d_E}{2}} = 2.$$ **Ans.**

$$\text{Efficiency} = \frac{\text{M.A.}}{\text{V.R.}}$$

$$= \frac{2}{2}$$

= 1 or 100%. **Ans.**

8. Draw a diagram of a single fixed pulley and obtain expressions for its :

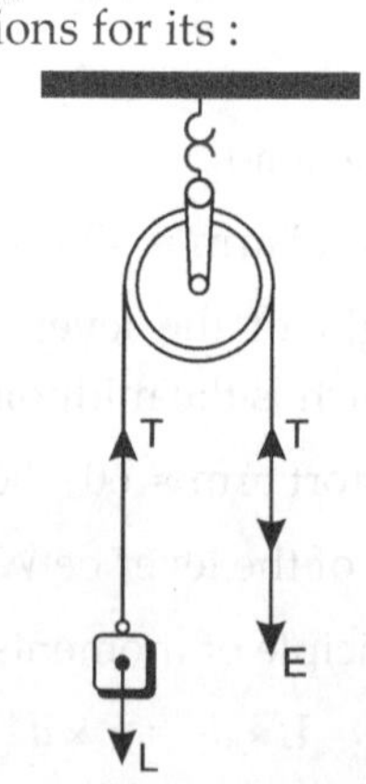

(i) Mechanical advantage, (ii) Velocity ratio, and (iii) Efficiency, in the ideal case.

Ans. The diagram of a single fixed pulley is shown alongside. If T is the tension in each strand of the string and in the ideal case string is massless and there is no friction in the pulley bearings, then in equilibrium, E = T and L = T

(i) Mechanical advantage $= \frac{L}{E} = \frac{T}{T} = 1$

(ii) If the effort E moves a distance d downwards, the load L moves the same distance d upwards.

So, velocity ratio $= \frac{d_E}{d_L} = \frac{d}{d} = 1$

(iii) Efficiency $= \frac{\text{M.A.}}{\text{V.R.}} = \frac{1}{1} = 1$ (or 100%).

9. The pulley system shown in the figure is to be used to lift a load W. If the man applying the effort cannot apply a force exceeding 1000 N, what is the maximum load that can be lifted? The actual load that the man is finally able to lift turns out to be 2700 N. What are the values of the actual M.A. obtained and the efficiency of the actual setup ?

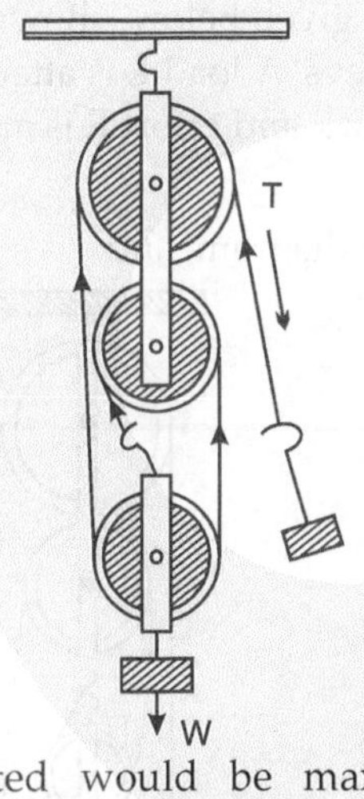

Ans. The load lifted would be maximum when conditions are ideal. Since this setup uses three strands of string, the load gets raised only through 1/3rd of the distance through which the effort moves. Hence,

$$\text{Velocity ratio} = \frac{\text{Distance moved by the effort}}{\text{Distance moved by the load}}$$

$$= \frac{1}{1/3} = 3.$$

Under ideal conditions,

V.R. = M.A.

Hence, Ideal M.A. of the setup

= 3.

∴ Maximum load that can be lifted

= Effort × 3 = 1000 N × 3

= 3000 N. Ans.

Under actual conditions, M.A.

$$= \frac{\text{Load}}{\text{Effort}}$$

$$= \frac{2700 \text{ N}}{1000 \text{ N}}$$

$= 2.7$

∴ Efficiency $= \frac{\text{M.A.}}{\text{V.R.}} = \frac{2.7}{3}$

$= 0{\cdot}9 = 90\%$. Ans.

10. The alongside figure shows the combination of a movable pulley P_1 with a fixed pulley P_2 used for lifting up a load W.

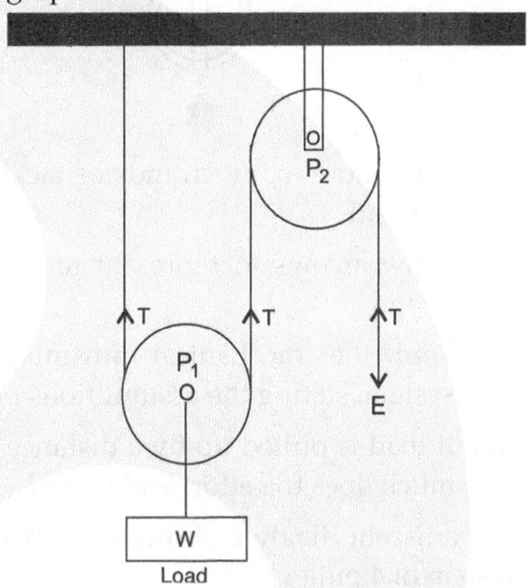

(i) State the function of the fixed pulley P_2.

(ii) If the free end of the string moves through a distance x, what is the distance by which the load W is raised.

(iii) Calculate the force to be applied at C to just raise the load W = 20 kgf, neglecting the weight of the pulley P_1 and friction.

Ans. (i) It is quite difficult to apply effort in the upward direction, if fixed pulley P_2 is not used. The fixed pulley changes the direction of effort from upward to downward direction, making the application of the effort more convenient and easier.

(ii) The distance travelled by load is half the distance moved by effort $= x/2$. Ans.

(iii) Here, W = 20 kgf

In equilibrium, W = 2T and E = T

∴ Effort applied, $E = \frac{W}{2} = \frac{20}{2} = 10$ kgf.

Ans.

11. Draw a diagram showing a block and tackle system of 4 pulleys.

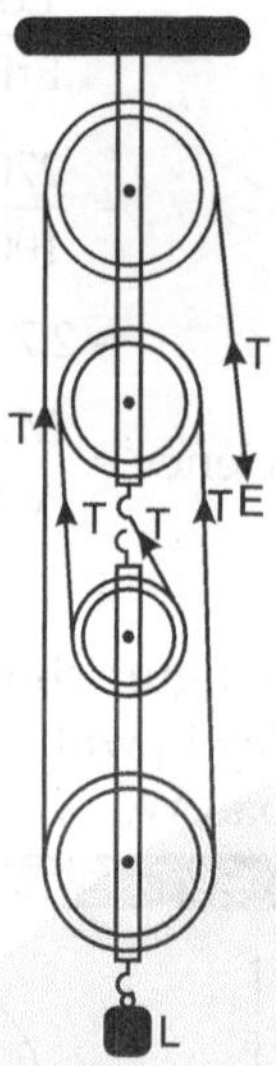

(i) State, how many strands of tackle support the load?

(ii) Draw arrows to represent tension in each strand.

(iii) Find the mechanical advantage of the system, stating the assumptions made.

(iv) If load is pulled up by a distance 1 m, how much does the effort end move?

Ans. The adjacent diagram shows a block and tackle system of 4 pulleys.

(i) Four strands of tackle support the load.

(ii) The tension in each strand is shown by the arrow marked as T.

(iii) If we neglect the friction of the pulleys and weight of the pulleys in the lower block, then :

$$L = 4T$$

and $$E = T$$

$$\therefore \quad M.A. = \frac{L}{E} = \frac{4T}{T} = 4. \qquad \text{Ans.}$$

(iv) If load is pulled up by a distance 1 m, the effort end moves by 4 m.

12. A block and tackle system of pulleys has a velocity ratio 5.

(i) Draw a labelled diagram of the system indicating clearly the points of application and directions of load and effort.

(ii) What is the value of the mechanical advantage of the given pulley system if it is an ideal pulley system ?

Ans. (i) The required diagram is given below:

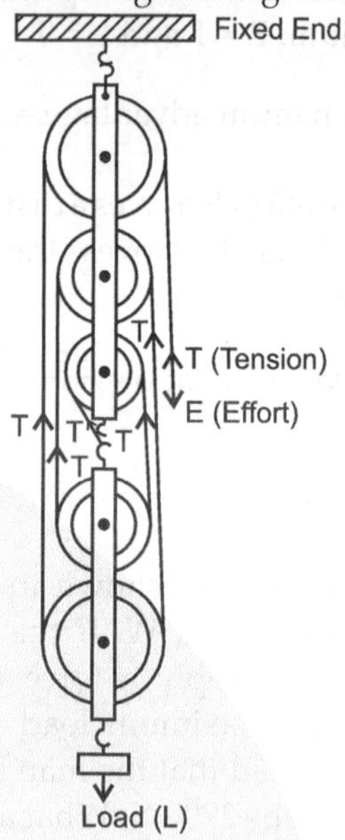

(ii) M.A. = 5 for an ideal pulley system.

13. Diagram given below, shows an arrangement of four pulleys. A load L is attached to the movable lower block and effort E is applied at free end of the string.

Copy the diagram; and

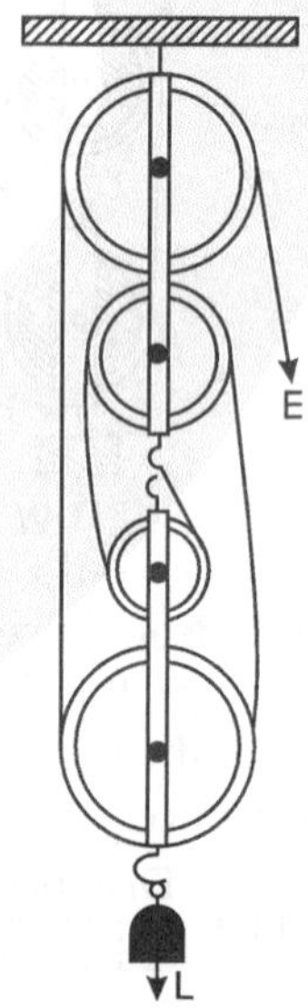

(i) Draw arrows to indicate tension in each part of the string.

(ii) Calculate the mechanical advantage of the system.

Ans. The following diagram shows a block and tackle system of 4 pulleys.

(i) The tension in each string is shown by the arrow marked as T.

(ii) If we neglect the friction of the pulleys and weight of the pulleys in the lower block, then

$$L = 4\,T$$

and $$E = T$$

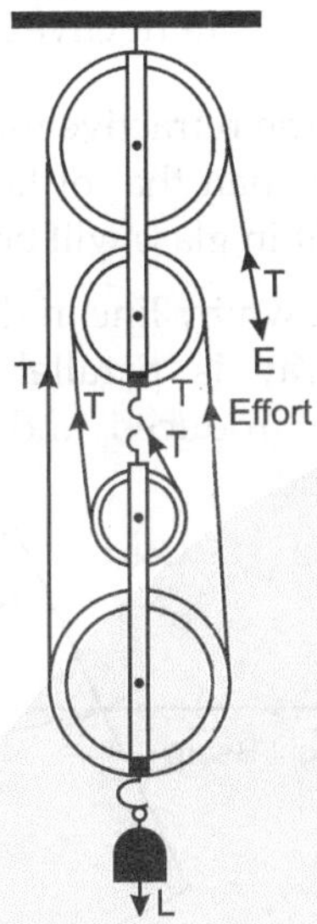

$\therefore$ Mechanical advantage

$$= \frac{L}{E} = \frac{4T}{T} = 4$$ **Ans.**

Chapter 3. Refraction of Light

1. How will you verify the laws of refraction or how the refractive index of glass is determined in the laboratory?

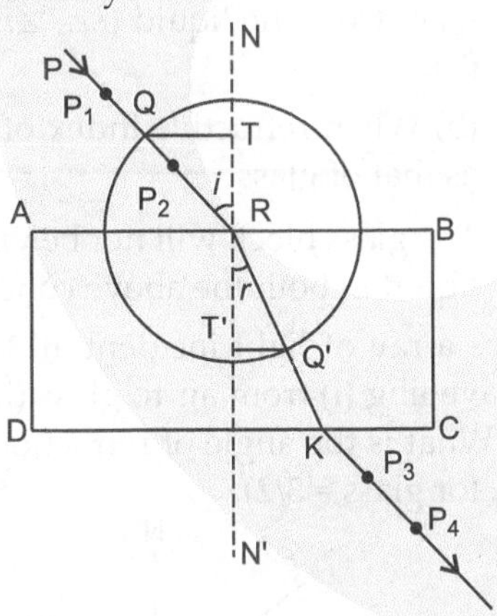

Ans. Place a white sheet of paper on a drawing board. Put a rectangular slab of glass on this white sheet and trace out its boundary ABCD. At any point R on the edge AB, draw a normal NN′. Draw an incident ray PR at any angle say 30°. Fix two pins P_1 and P_2 vertically on the line PR and at a distance not less than 3 cm from each other. Looking through the other side of the block, fix two pins, P_3 and P_4 vertically on the white sheet, such that all the four pins P_1, P_2, P_3 and P_4 appear to be in a straight line as seen through the glass block. Mark the pin points with a fine pencil, remove the block and join the points P_3 and P_4 by the straight line RK.

PR represents the incident ray and RK represents the refracted ray. ∠PRN and ∠KRN′ are respectively angles of incidence and refraction.

With R as centre and convenient radius, draw a circle cutting the incident ray at Q and the refracted ray at Q′.

From Q and Q′, draw perpendiculars QT and Q′T′ on the normal NN′

Now $\sin i = \dfrac{QT}{QR}$

and $\sin r = \dfrac{Q'T'}{Q'R}$

$$\mu = \frac{\sin i}{\sin r} = \frac{QT}{QR} \cdot \frac{Q'R}{Q'T'} = \frac{QT}{Q'T'}$$

since QR = Q′R being radii of the same circle.

Repeat this experiment for other angles of incidence, say 45° and 60° and find μ.

It is found that for different values of i the ratio.

$$\frac{\sin i}{\sin r} = \text{constant.}$$

Thus, it is verified that,

(i) $\dfrac{\sin i}{\sin r}$ is a constant and is equal to ${}_a\mu_g$.

(ii) The incident ray, the refracted ray and normal lie in the same plane.

2. The diagram shows a point source P inside a water container. Three rays A, B and C starting from P are shown up to the water surface. Show in the diagram the path of these rays after striking the water surface. The critical angle for water-air pair is 48°.

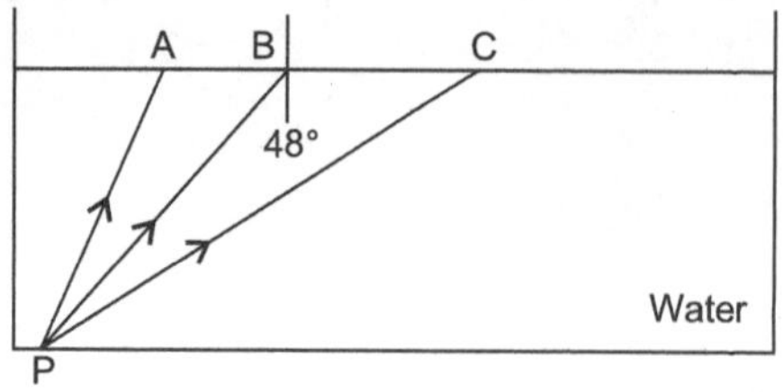

Ans.

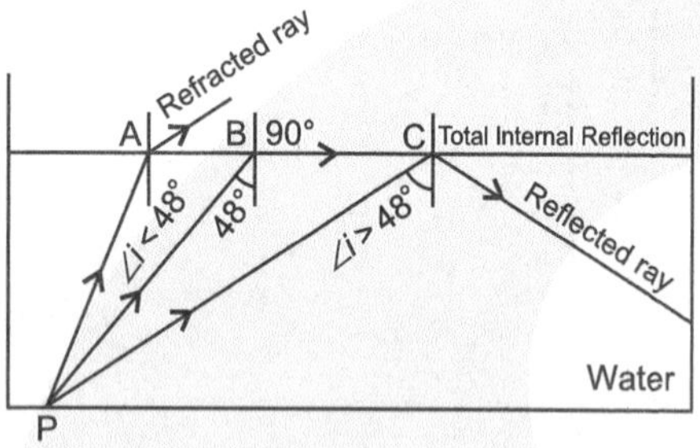

3. The diagram given alongside shows the path of a ray of light through a rectangular glass block placed in a liquid of uniform density.

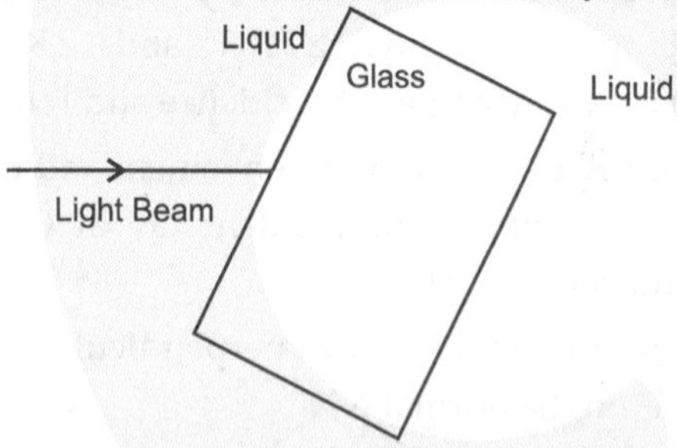

(i) (a) Does the light speed up or slow down in glass. (b) Give reason for your answer.

(ii) What is the angular deviation of the emergent ray from the glass block with respect to the incident ray?

(iii) State two conditions under which the light ray moving from liquid to glass passes straight without bending.

(iv) Will the glass be visible when the ray passes undeviated?

Ans. (i) (a) The speed of light, travelling from liquid to glass, slows down.

(b) Speed of light in liquid

$$= \frac{\text{Speed of light in air}}{\text{Refractive index of liquid}}$$

$$= \frac{3 \times 10^8}{\text{Refractive index of liquid}}$$

Speed of light in glass

$$= \frac{\text{Speed of light in air}}{\text{Refractive index of glass}}$$

$$= \frac{3 \times 10^8}{\text{Refractive index of glass}}$$

Since the refractive index of glass is usually higher than that of liquid, hence the speed of light in glass will be less.

(ii) As shown by line in the diagram, the emergent ray is parallel to the incident ray (when produced) and so there is no angular deviation.

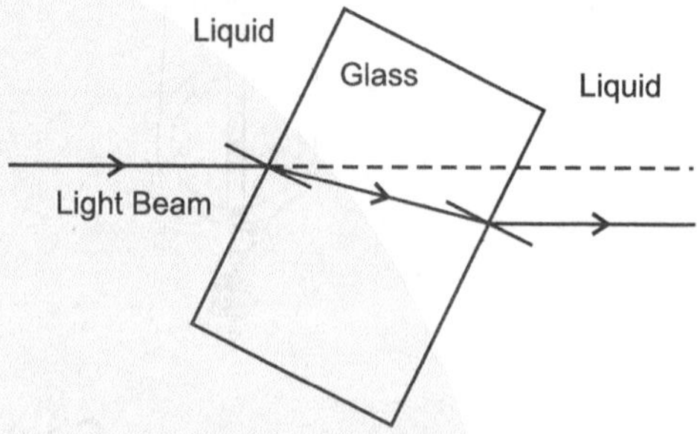

(iii) The light beam moving from liquid to glass passes undeviated under the following two conditions :

(a) When the light ray falls normally on glass from the liquid (*i.e.*, $\angle i = 0°$, then $\angle r = 0°$).

(b) When refractive index of liquid is same as that of glass.

(iv) The glass block will not be visible inside the liquid in both the above conditions.

4. Trace a ray of light incident at 30° on a surface if travelling (i) from air to glass (ii) from glass to air. What is the angle of refraction in each case ? (R.I. for glass = 3/2).

Ans. (i)

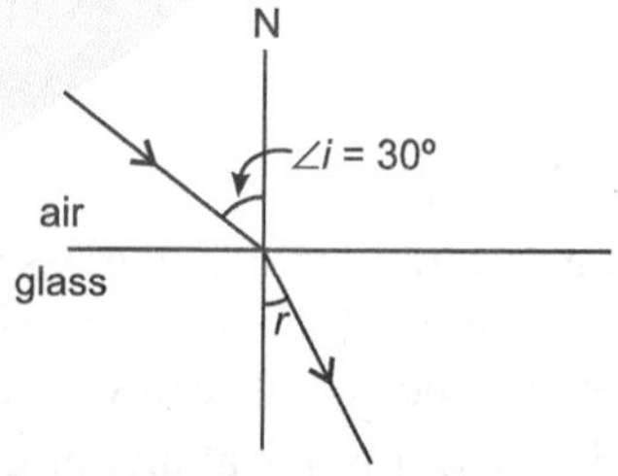

$${}_a\mu_g = \frac{\sin i}{\sin r}$$

$$\therefore \quad \frac{3}{2} = \frac{\sin 30°}{\sin r}$$

$$\therefore \quad \sin r = \frac{2}{3} \times \sin 30°$$

$$\therefore \quad \sin r = \frac{2}{3} \times \frac{1}{2}$$

$$= \frac{1}{3}$$

$\therefore \quad \sin r = 0.333$

$\therefore \quad r = 19° \; 30'.$ **Ans.**

(ii) For a ray of light travelling from glass to air, R.I. is denoted as ${}_g\mu_a$.

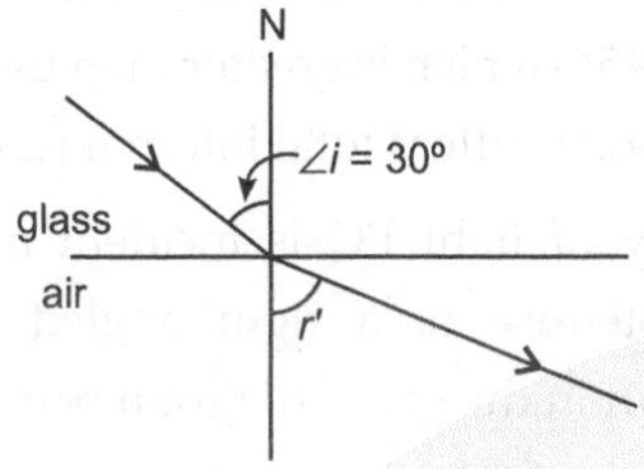

$${}_g\mu_a = \frac{1}{{}_a\mu_g}$$

$$\therefore \quad \frac{\sin i}{\sin r'} = \frac{1}{{}_a\mu_g}$$

$$\sin r' = {}_a\mu_g \times \sin i$$

and $\quad {}_a\mu_g = \frac{3}{2}$

$$\therefore \quad \sin r' = \frac{3}{2} \times \frac{1}{2}$$

$$= \frac{3}{4}$$

$\therefore \quad \sin r' = 0.7500$

$\therefore \quad r' = 48° \; 36'.$ **Ans.**

5. A monochromatic point source of light X is viewed by an observer through a prism P as shown in the figure. Complete the diagram to show the image formed by the prism and as seen by the observer. Label the image by the letter I.

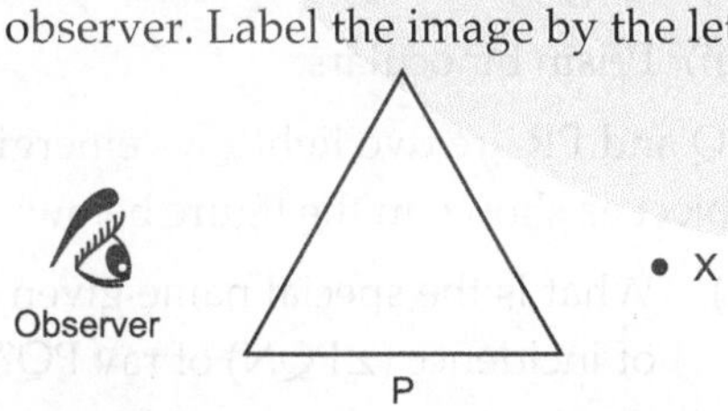

Ans. The completed ray diagram is shown in the figure.

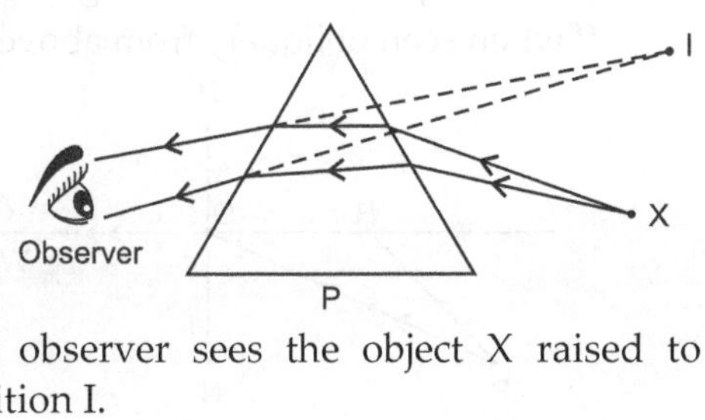

The observer sees the object X raised to the position I.

6. Draw a diagram of a prism and label:

(i) the base,

(ii) the refracting surfaces,

(iii) the refracting edge,

(iv) the refracting angle in it.

Ans. Figure below shows a prism, in which BEFC is the base, BADE and CADF are the refracting surfaces, AD is the refracting edge and ∠BAC or ∠EDF is the refracting angle of prism which has been marked A.

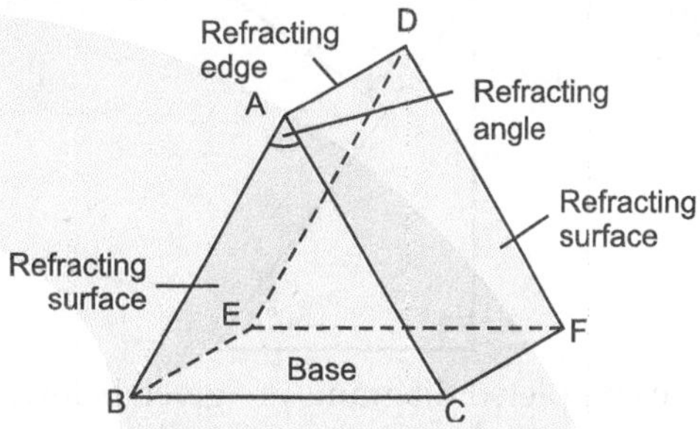

7. (i) A ray of light passes through a right angled prism as shown in the figure. State the angles of incidence at the faces AC and BC.

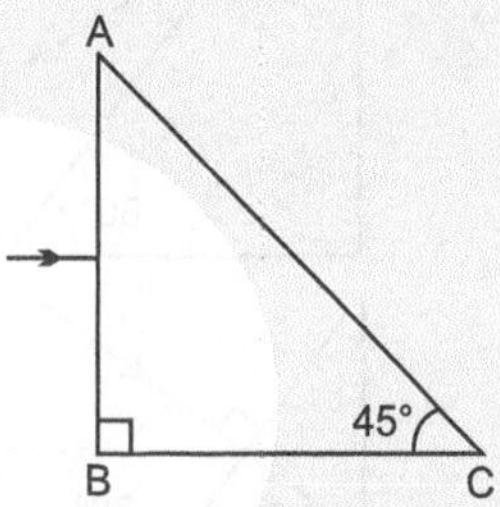

(ii) Calculate the velocity of light in a glass block of refractive index 1·5. (Velocity of light in air = 3×10^8 m/s)

Ans. (i) At AC, angle of incidence is 45°

At BC, angle of incidence is 0°.

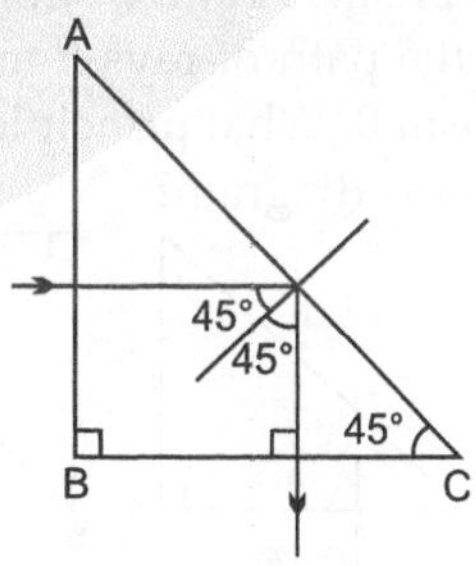

(ii) $${}_a\mu_g = \frac{\text{Speed of light in air } (v_{\text{air}})}{\text{Speed of light in glass } (v_{\text{glass}})}$$

$$1{\cdot}5 = \frac{3 \times 10^8}{v_{\text{glass}}}$$

$$\therefore \quad v_{\text{glass}} = \frac{3 \times 10^8}{1{\cdot}5} \text{ ms}^{-1} = 2 \times 10^8 \text{ ms}^{-1}.$$ **Ans.**

8. Complete the following diagrams to show the path of the ray of single colour in each diagram as it enters in and emerges out of the prism. Mark the angles wherever necessary. (Critical angle for glass = 42°).

(i)

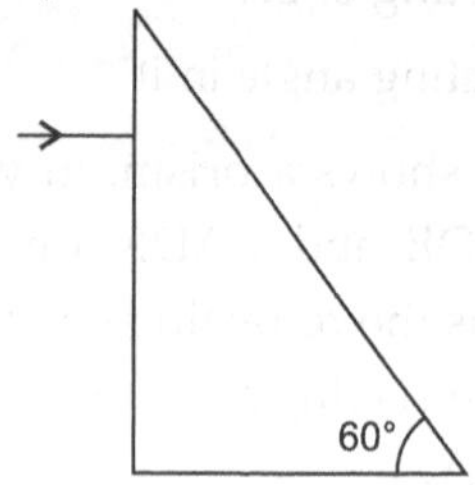

(ii)

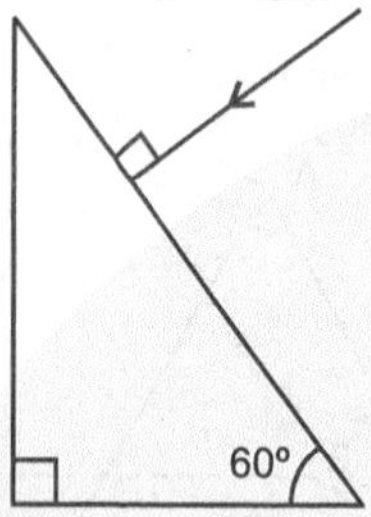

Ans. The completed diagrams are given below.

(i)

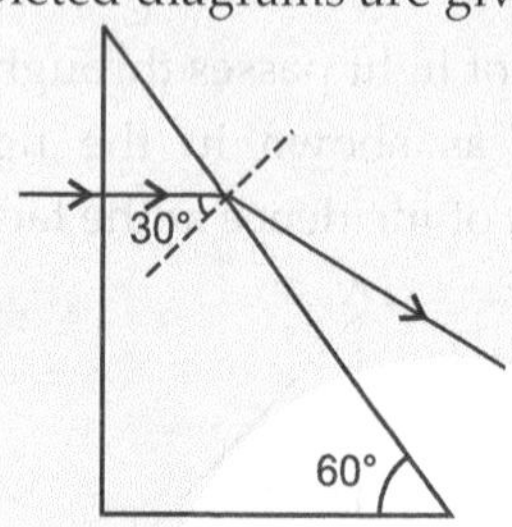

(ii)

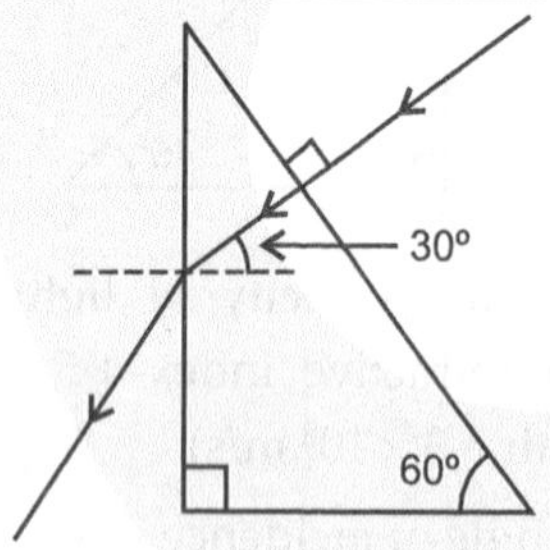

9. The adjacent diagram shows two right-angled isosceles prisms A and B. Complete the diagram to show the path of rays P and Q emerging out of the prism B. What principles have you used to complete the diagram?

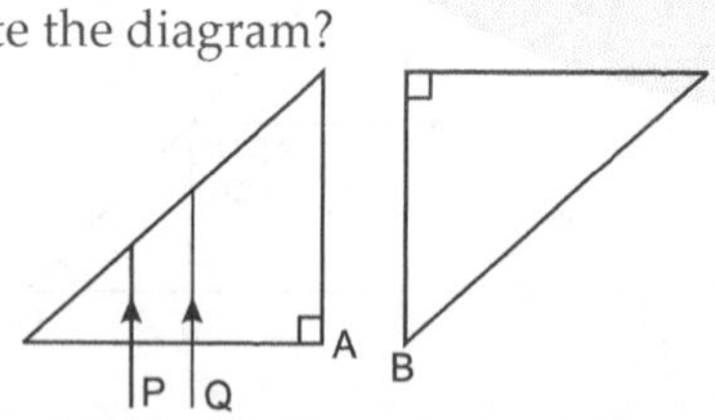

Ans. The completed ray diagram is shown in figure given below.

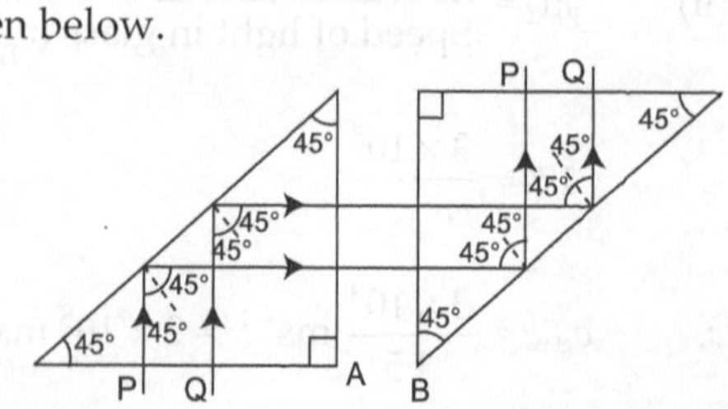

The following two principles are used to complete the diagram:

(i) A ray of light incident normally on the interface of two media passes undeviated.

(ii) A ray of light travelling from glass, when strikes the interface of two media (glass and air) at an angle of incidence, equal to 45° (which is greater than the critical angle = 42°) suffers total internal reflection.

10. A ray of light PQ is incident normally on the hypotenuse of a right angled prism ABC as shown in the given diagram below:

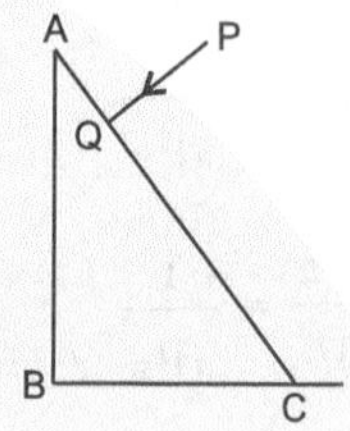

(i) Copy the diagram and complete the path of the ray PQ till it emerges from the prism.

(ii) What is the value of the angle of deviation of the ray?

(iii) Name an instrument where this action of the prism is used.

Ans. (i)

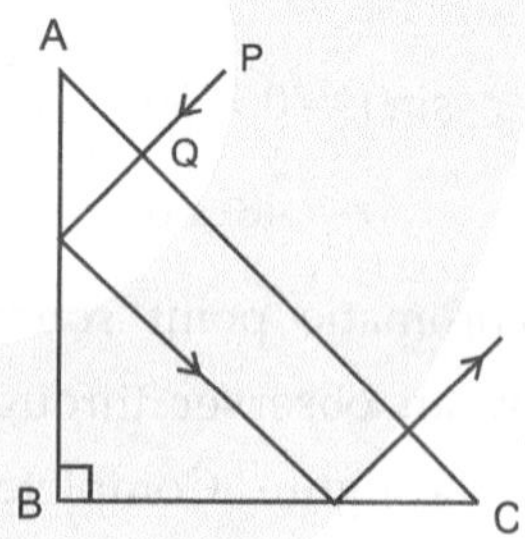

(ii) Angle of deviation of the ray = 180°.

(iii) Prism binoculars.

11. PQ and PR are two light rays emerging from the object as shown in the figure below:

(i) What is the special name given to the angle of incidence (∠PQN) of ray PQ?

(ii) Copy the ray diagram and complete it to show the position of the image of the object P when seen obliquely from above.

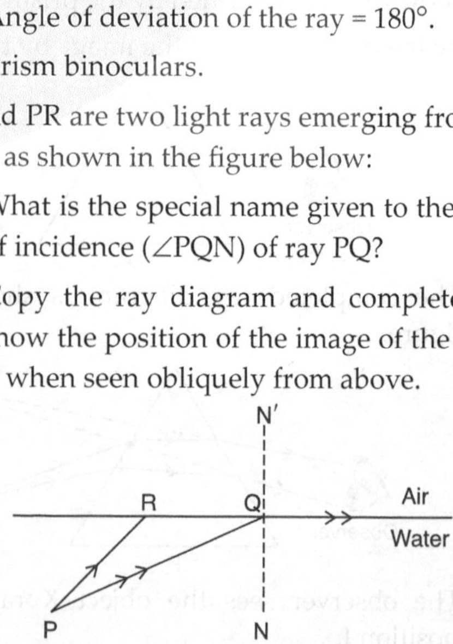

(iii) Name the phenomenon that occurs if the angle of incidence ∠PQN is increased still further.

Ans. (i) Critical angle

(ii) The completed diagram is given below.

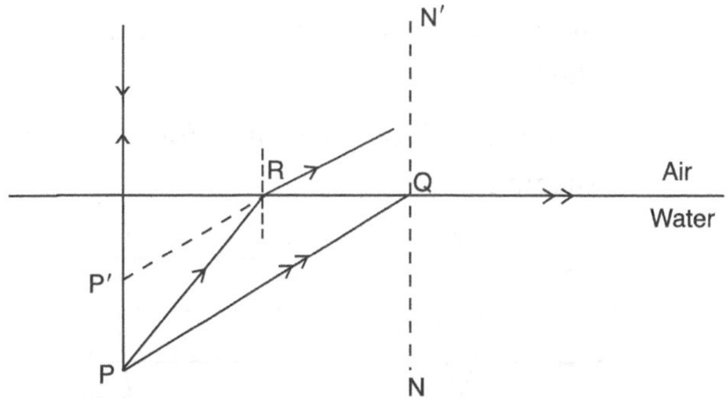

(iii) Total internal reflection of light.

Chapter 4. Refraction through Lenses and Optical Instruments

1. Name the types of lens.

Ans. There are two types of lenses:

(i) Concave or diverging lens which may be either double concave, plano-concave or convexo-concave.

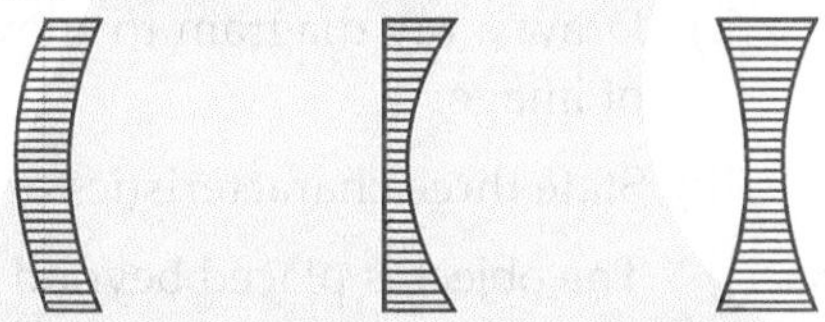

Convexo-Concave Plano-Concave Bi-Concave

(ii) Convex or converging lens which may be either double convex, plano-convex, or concavo-convex.

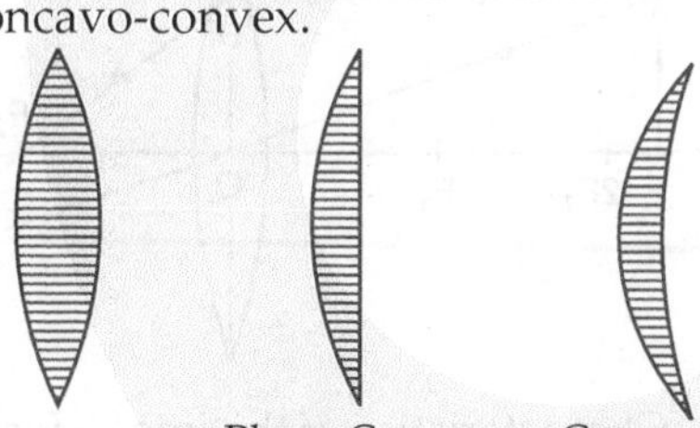

Bi-Convex Plano-Convex Concavo-Convex

2. Define (i) Principal axis, and (ii) Optical centre of a lens.

Ans. (i) **Principal axis:** It is the straight line passing through both the centres of curvatures, C_1 and C_2 and also through its optical centre O.

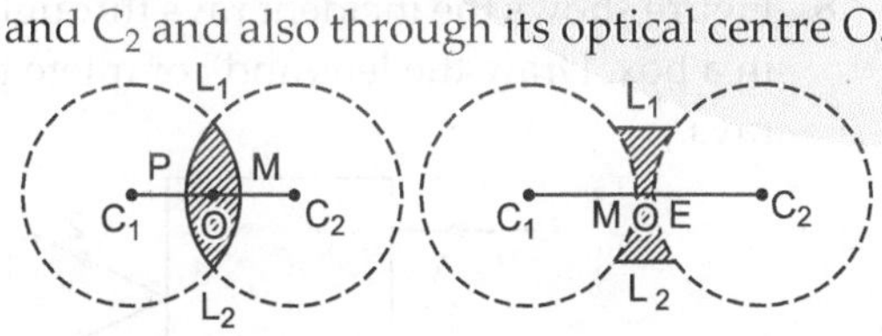

(ii) **Optical centre:** It is the centre of the lens through which the principal axis passes. Any ray of light passes through it undeviated, provided the lens is thin.

3. The given diagram shows a lens as a combinations of one glass block and two prisms. Complete the ray diagram and show the part of the incident ray AB after passing through the lens.

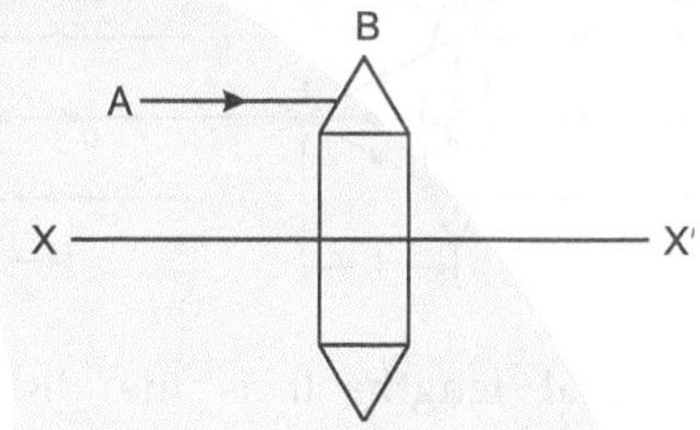

(i) Name the lens formed by this combination.

(ii) What is the line XX′ called?

(iii) Mark the focus F.

Ans. The completed ray diagram is shown.

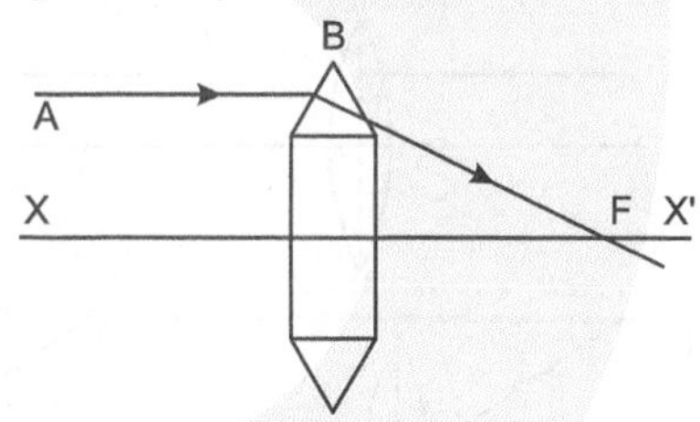

(i) The lens formed by the combination is the convex lens.

(ii) The line XX′ is called the principal axis.

(iii) The focus has been marked by the letter F in the diagram.

4. Define the principal focus of a (i) Convex lens and (ii) Concave lens.

Ans. (i) For a convex lens, principal focus is a point on the principal axis of a convex lens to which light rays parallel to the principal axis converge after passing through the lens.

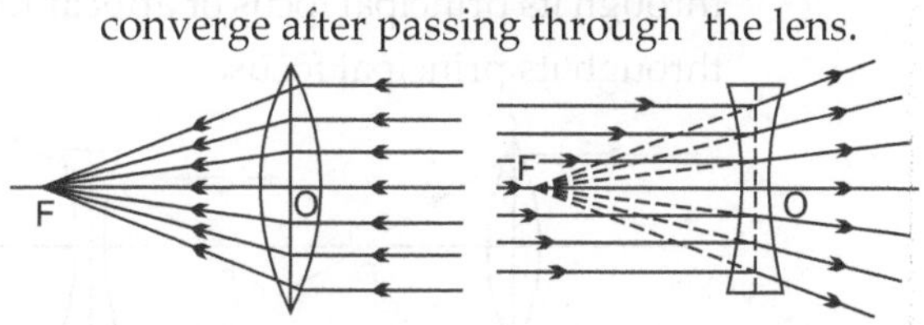

(ii) For a concave lens, principal focus is a point on the principal axis at which the rays of light originally parallel to the principal axis, appear to come from it after refraction by the lens.

5. (i) Copy and complete the following ray diagrams.

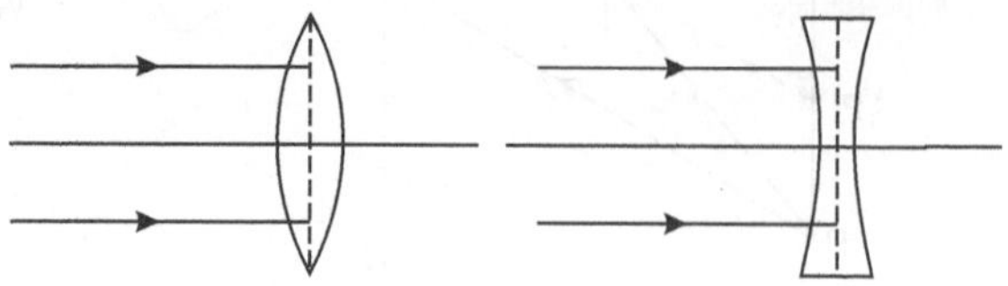

(i) (ii)

(ii) Define focal length.

Ans. (i)

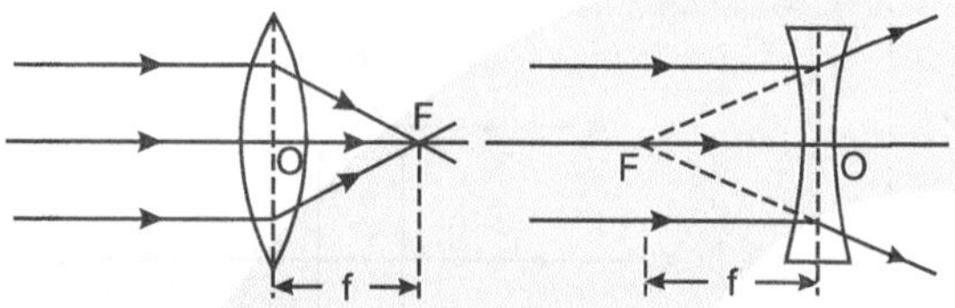

(ii) **Focal length:** It is the distance of the principal focus of the lens from its centre.

6. Complete the following diagram and state what happens to the ray of light after refraction through the lens.

(i)

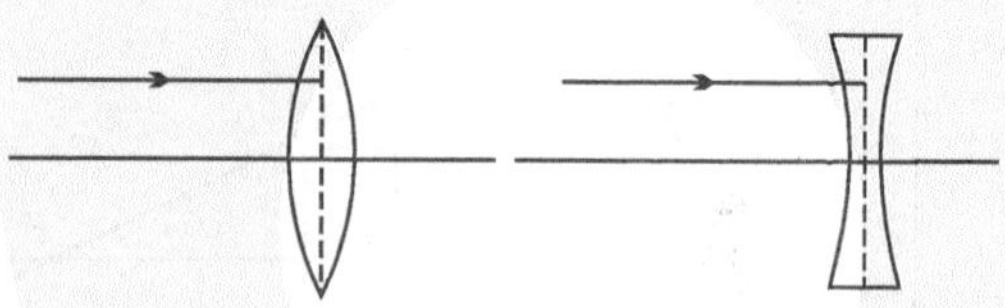

(ii)

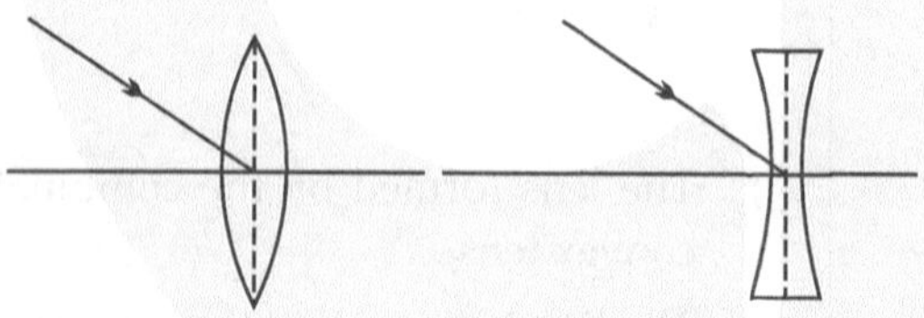

(iii)

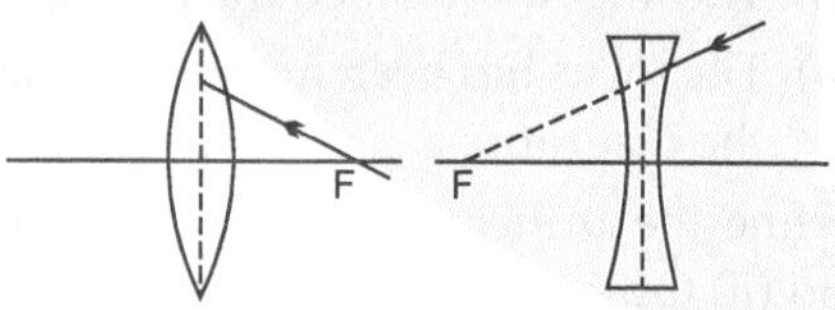

Ans. (i) A ray of light parallel to the principal axis, after refraction by the lens either passes through its principal focus or appear to pass through its principal focus.

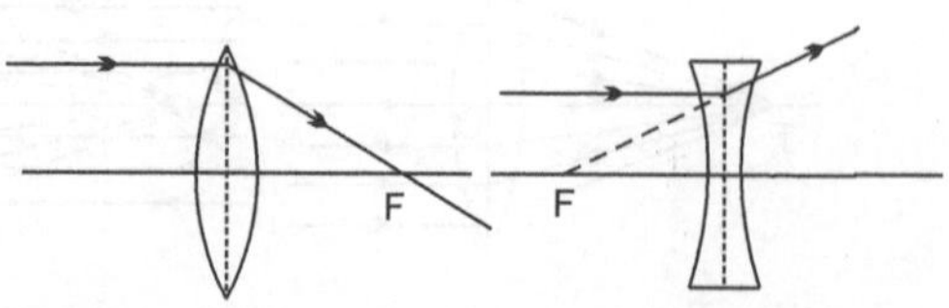

(ii) A ray of light goes undeviated through the optical centre of a lens, whether convex or concave.

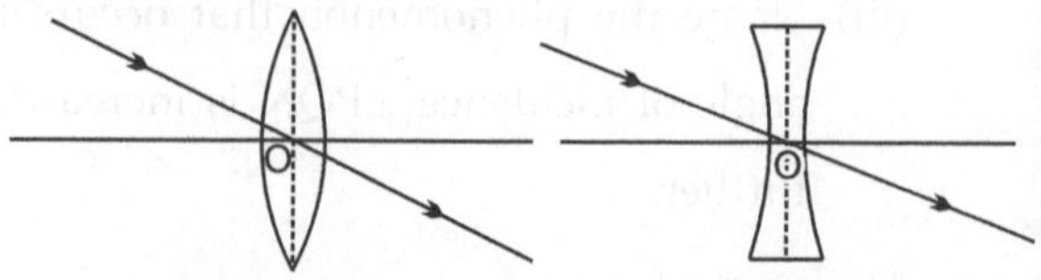

(iii) A ray of light that passes through the principal focus or appears to pass through the principal focus, goes parallel to its principal axis.

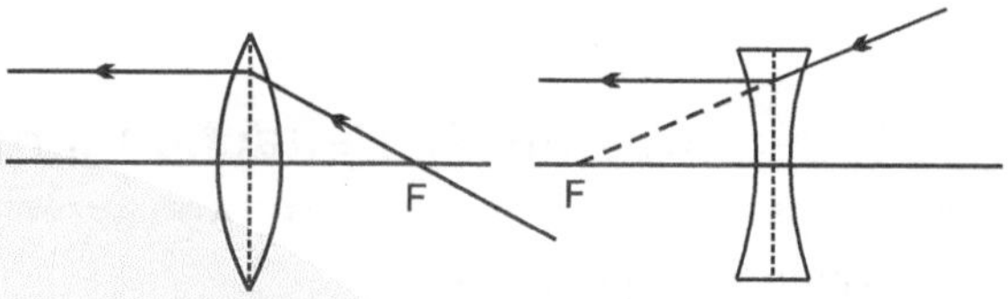

7. A convex lens forms an image of an object placed in front of it, between F_2 and $2F_2$ of the lens.

(i) Where is the object placed?

(ii) Draw a ray diagram to show the formation of image.

(iii) State three characteristics of the image.

Ans. (i) The object is placed beyond $2F_1$.

(ii) The required ray diagram is given below:

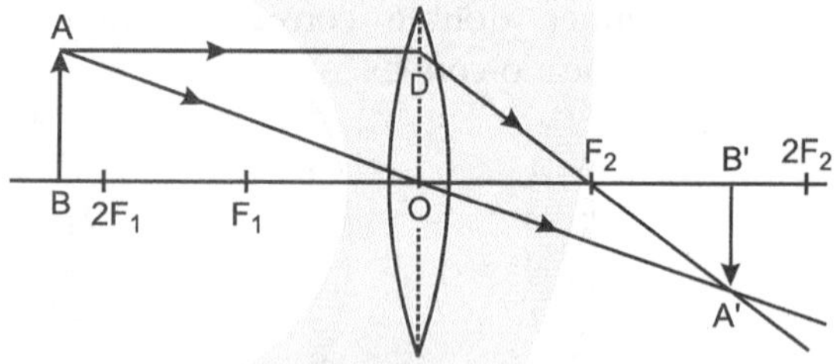

(iii) Characteristics of the image formed:

(a) Real.

(b) Inverted.

(c) Diminished.

8. Figure shows the incident rays through lens kept in a box. Draw the lens and complete the path of rays.

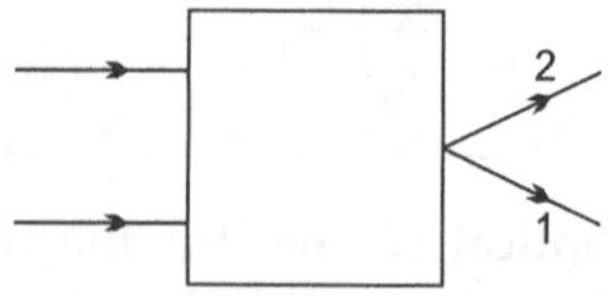

Ans. The completed diagram is given below:

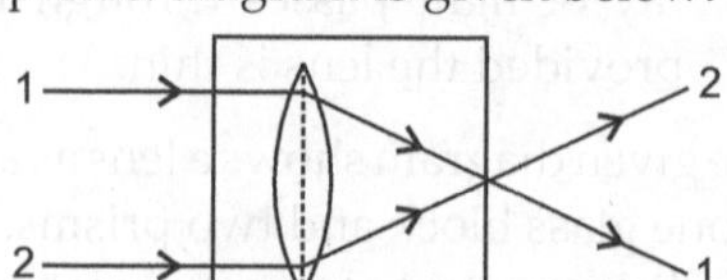

9. Complete the diagram to show the formation of image of the object AB.

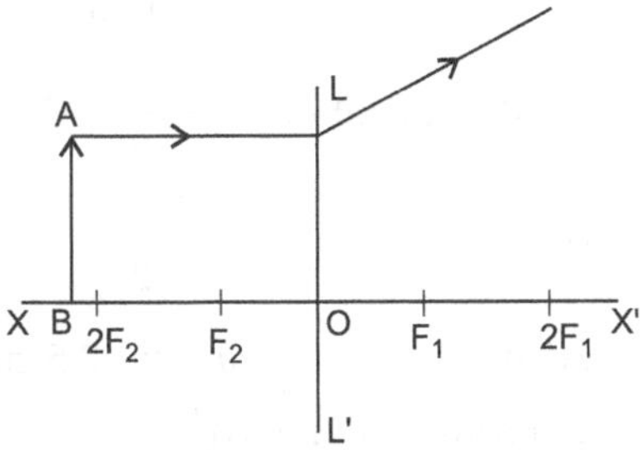

(i) Name the lens LL′.

(ii) Where is the image of the object AB formed?

(iii) State three characteristics of the image.

Ans.

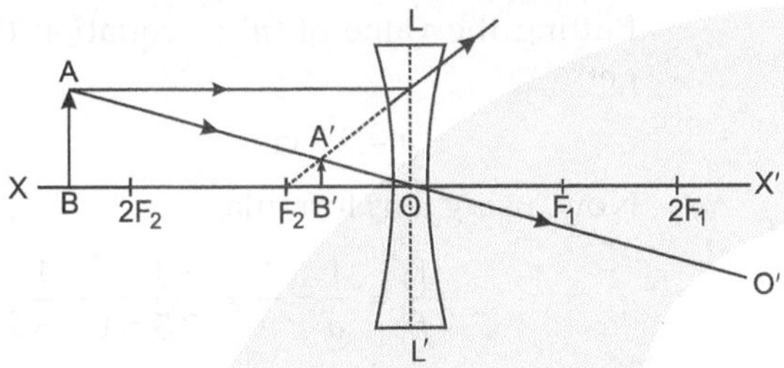

(i) The lens is diverging (or concave) lens.

(ii) The image of the object is A′B′ which is formed between the optical centre and focus of the lens, in front of it.

(iii) The image is (a) virtual, (b) upright, and (c) diminished.

10. Draw a ray diagram to illustrate the formation of an enlarged and upright image of an object by a lens. Name the lens used. Where should the object be placed relative to the lens ? Will the image be real or virtual ? Name a device where this action of lens is used.

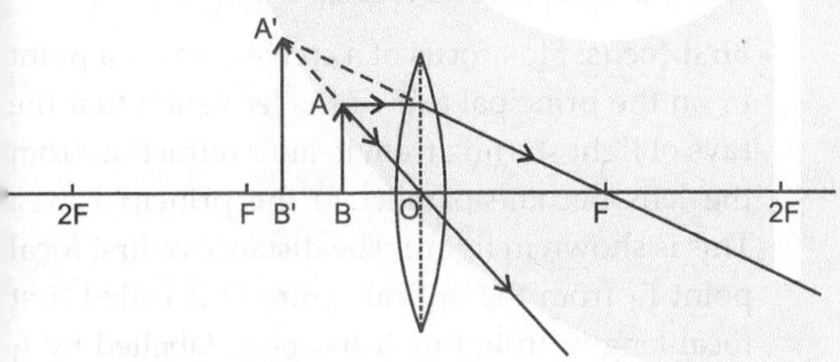

Ans. The ray diagram showing the formation of an enlarged and upright image of an object by a lens is shown in the figure alongside. For the object AB, the image is A′B′. The lens used is convex (converging).

The object should be placed between optical centre O and focus F of the lens, infront of it.

The image is virtual.

This action of lens is used in a simple magnifier (or a reading glass).

11. In the given figure AB is the object and CD is the image formed.

(Given : AB = 2 cm, CD = 1·5 cm and BC = 8 cm).

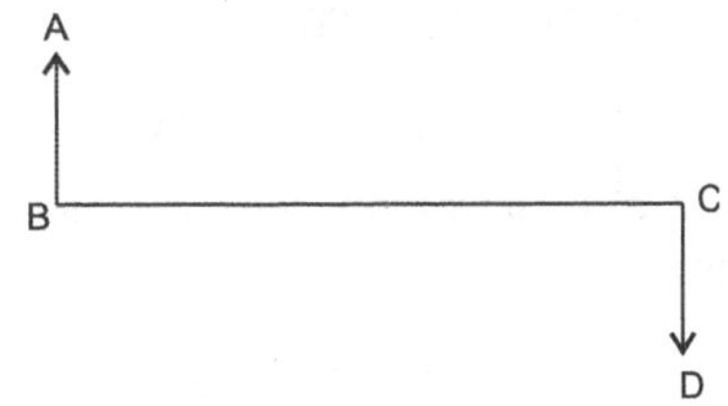

Answer the following:

(i) Nature of lens used.

(ii) Position of lens.

(iii) Position of focus.

(iv) Focal length of lens.

Ans. (i) Convex or converging lens is used, since the image formed is inverted (*i.e.*, real).

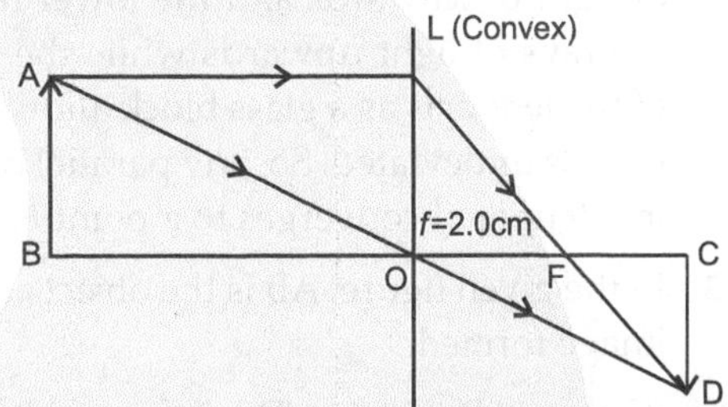

(ii) Position of lens is at point O.

(iii) Position of focus is point F.

(iv) Object size, $u = 2$ cm

Image size, $v = 1{\cdot}5$ cm

$$\text{Magnification, } m = \frac{-v}{u} = \frac{-1{\cdot}5}{2} = \frac{-3}{4}$$

$$\Rightarrow \quad v = \frac{3}{4}u \qquad \text{...(i)}$$

Now, $u + v = 8$ cm (given)

From equation (i),

$$u + \frac{3}{4}u = 8$$

$$\frac{7u}{4} = 8 \Rightarrow u = \frac{32}{7} \qquad \text{...(ii)}$$

Putting the value of 'u' in equation (i),

$$v = \frac{3}{4} \times \frac{32}{7} = \frac{24}{7} \qquad \text{...(iii)}$$

Using lens formula,

$$\frac{1}{f} = \frac{1}{v} - \frac{1}{u}$$

$$= \frac{7}{24} - \left(\frac{-7}{32}\right)$$

$$\Rightarrow \quad \frac{7}{24} + \frac{7}{32} = \frac{49}{96}$$

$$\Rightarrow \quad f = 1{\cdot}959 \approx 2 \text{ cm.}$$ **Ans.**

12. How does a convex lens act as a 'converging lens'?

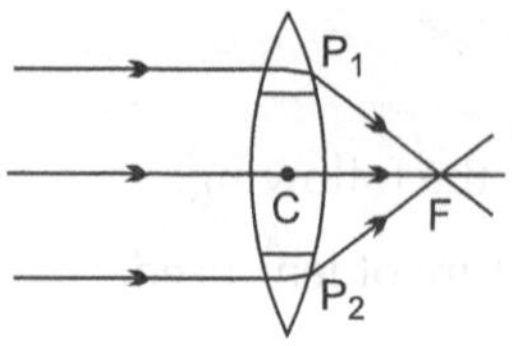

Ans. A convex lens may be considered as being made of a large number of prisms, such that in its upper half, prisms have their bases downwards and in its lower half, the prisms have their bases upwards. Due to this, the upper half part deviates the rays of light downwards and the lower half deviates the rays of light upwards while the central part of the lens acts as a glass block allowing the light to pass undeviated. So, any parallel beam of light incident on it converges to a point F.

13. In the given figure, AB is the object and CD is the image formed.

(Given : AB = 2 cm, CD = 1 cm and BC = 2·5 cm).

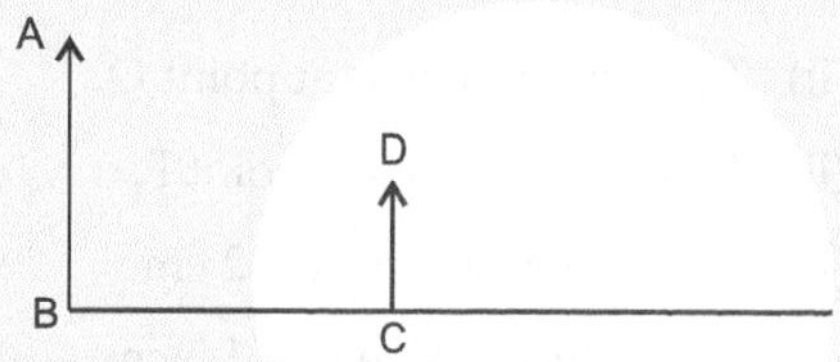

Answer the following:

(i) Nature of lens used.

(ii) Position of lens.

(iii) Position of focus.

(iv) Focal length of lens.

Ans. (i) A concave lens is used, since the image formed is erect and diminished (*i.e.*, it is virtual image).

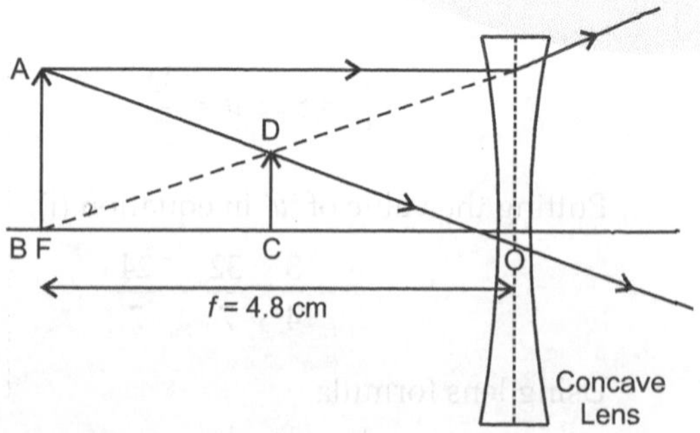

(ii) Position of lens is at point O.

(iii) Position of focus is at point F.

(iv) Object size, u = 2 cm

Image size, v = 1 cm

$$\therefore \text{ Magnification, } m = \frac{v}{u} = \frac{1}{2}$$

$$\text{or} \quad v = \frac{u}{2} \quad \ldots\text{(i)}$$

Also, $u = v + 2{\cdot}5$ cm. (Given)

$\therefore$ From equation (i),

$$u = \frac{u}{2} + 2{\cdot}5$$

$$\Rightarrow \quad u = 5 \text{ cm}$$

Putting the value of 'u' in equation (i), we get

$$v = 2{\cdot}5 \text{ cm}$$

Now, using lens formula,

$$\frac{1}{f} = \frac{1}{v} - \frac{1}{u} = \frac{-1}{2{\cdot}5} - \left(-\frac{1}{5}\right)$$

On solving, $f = -5$ cm **Ans.**

14. Define the terms principal foci and focal lengths as applied to a convex lens, and show them with the help of proper diagrams.

Ans.

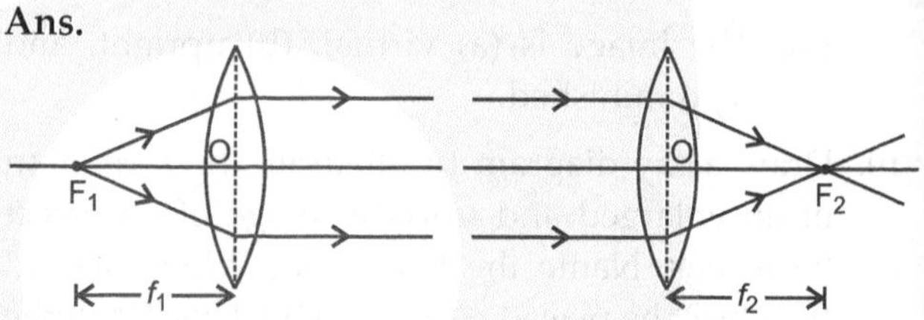

A convex lens has two principal foci: (i) first focus, and (ii) second focus.

First focus: First focus of a convex lens is a point F_1 on the principal axis of the lens such that the rays of light starting from it, after refraction from the lens become parallel to the principal axis. This is shown in figure. The distance of first focal point F_1 from the optical centre O is called first focal length. In figure, it has been labelled by f_1 (*i.e.*, $OF_1 = f_1$)

Second focus: Second focus of a convex lens is a point F_2 on the principal axis of the lens such that the rays of light incident parallel to the principal axis of the lens, after refraction from the lens pass through it. This is shown in figure. The distance of second focal point F_2 from the optical centre O is called second focal length. In figure it has been labelled by f_2 (*i.e.*, $OF_2 = f_2$).

15. Diagram shows an object AB placed on the principal axis of a convex lens placed in air. F_1 and F_2 are the two foci of the lens.

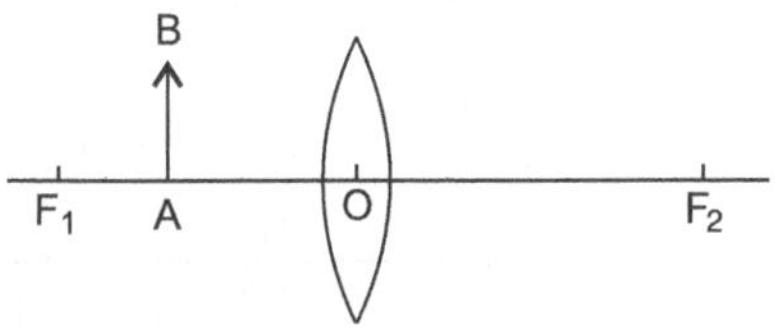

Copy the diagram:

(i) Draw a ray of light starting from B and passing through O. Show the same ray after refraction by lens.

(ii) Draw another ray from B which passes through F_2 after refraction by the lens.

(iii) Locate the final image.

(iv) Is the image real or virtual?

Ans. (i), (ii), (iii) see the diagram given below.

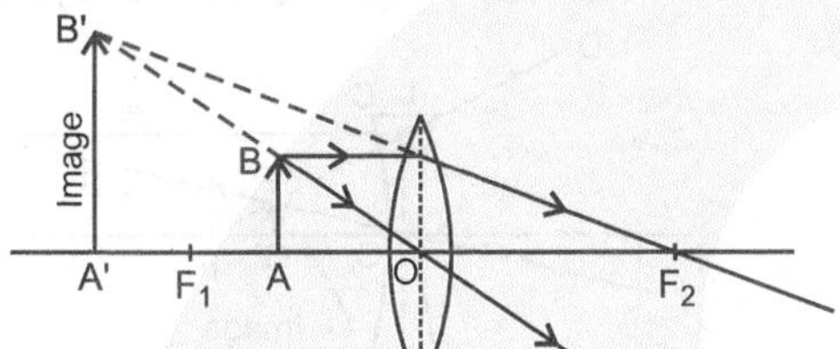

(iv) Image is virtual.

16. In the following diagram, L_1 and L_2 are the two convex lens placed at separation equal to the sum of focal lengths of the two lenses. A and B are the two rays of light incident on the lens L_1. Complete the path of rays till they emerge out of the lens L_2.

What principles have you used in completing the diagram?

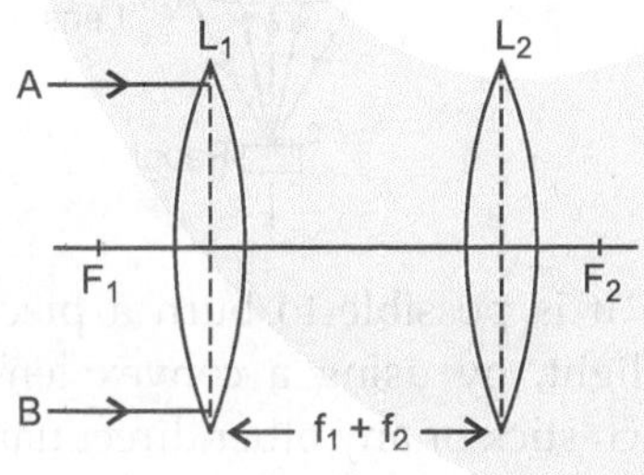

Ans. The complete ray diagram is shown alongside. The following two principles have been used in completing the diagram:

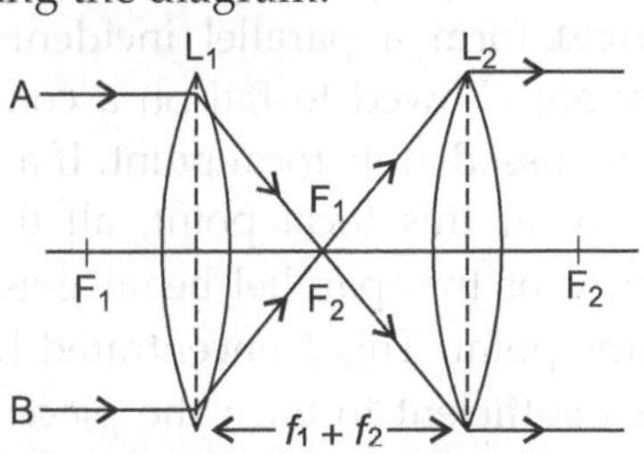

(i) A ray of light incident parallel to the principal axis, after refraction from the convex lens passes through the focus.

(ii) A ray of light incident on the convex lens through its first focus, after refraction becomes parallel to the principal axis of the lens.

17. An erect, magnified and virtual image is formed, when an object is placed between the optical centre and principal focus of a lens.

(i) Name the lens.

(ii) Draw a ray diagram to show the formation of the image with the above stated characteristics.

Ans. (i) The lens is convex.

(ii)

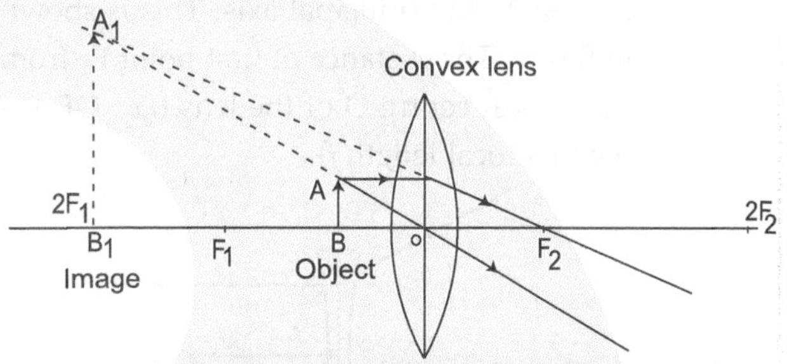

Ray Diagram of image formation

18. An object AB is placed between $2F_1$ and F_1 on the principal axis of a convex lens as shown in the diagram. Copy the diagram and using three rays starting from point A, obtain the image of the object formed by the lens.

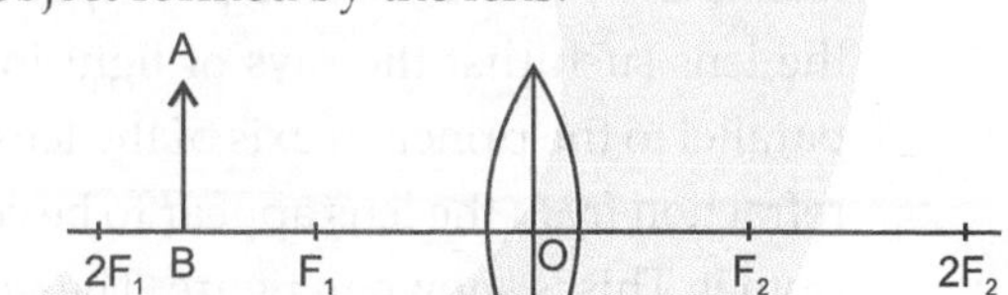

Ans.

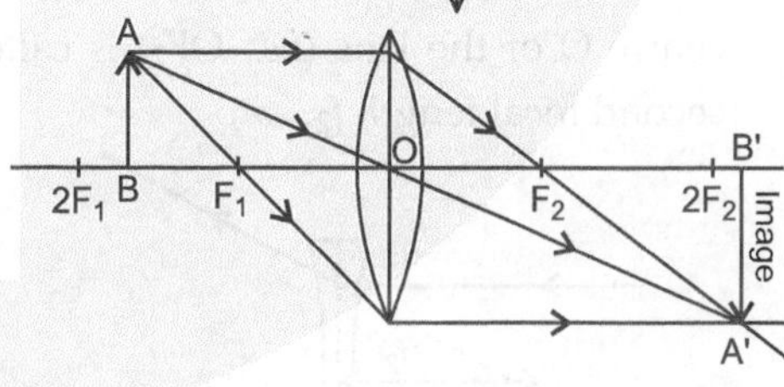

19. How does a concave lens act as a 'diverging lens'?

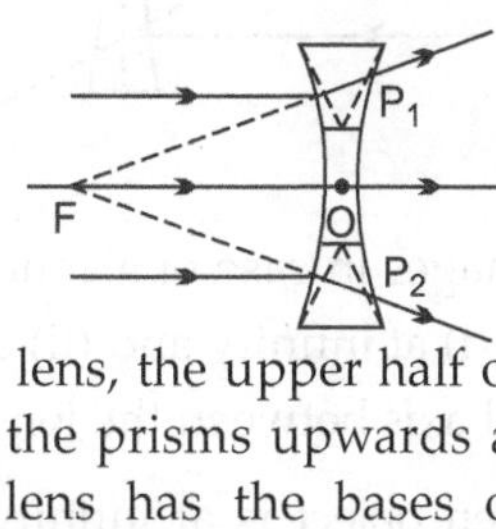

Ans. In a concave lens, the upper half of the lens has the bases of the prisms upwards and the lower half of the lens has the bases of the prisms downwards. Therefore, in this lens in the upper half, the rays of light suffer divergence in the upwards direction and in the lower half, the rays diverge in the downward direction. On being produced backwards by dotted lines, these rays appear to diverge from a point F.

20. Define the term principal foci as applied to a concave lens. Illustrate your answer with the aid of proper diagrams. Show the focal length of the lens in diagram.

Ans. A concave lens has two principal foci: (i) first focus and (ii) second focus.

(i) **First focus:** First focus of a concave lens is the point F_1 on the principal axis of the lens such that the rays of light appearing to meet at it, after refraction from the lens become parallel to the principal axis. This is shown in figure. The distance of first point F_1 from the optical centre O of the lens (*i.e.*, OF_1) is the first focal length f_1.

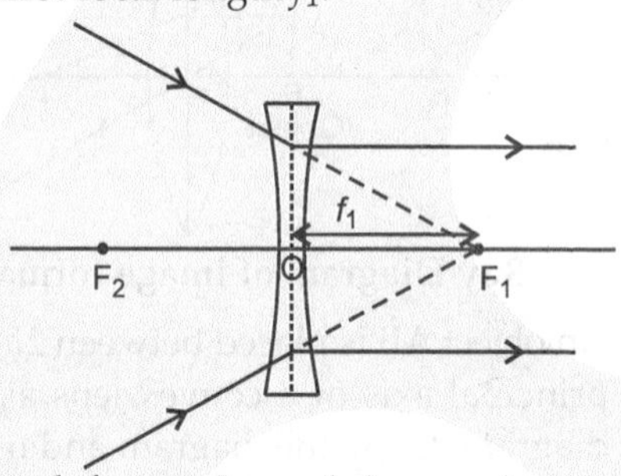

(ii) **Second focus:** Second focus of a concave lens is the point F_2 on the principal axis of the lens such that the rays of light incident parallel to the principal axis of the lens, after refraction from the lens appear to be coming from it. This is shown in figure. The distance of second focal point F_2 from the optical centre O of the lens (*i.e.*, OF_2) is called the second focal length f_2.

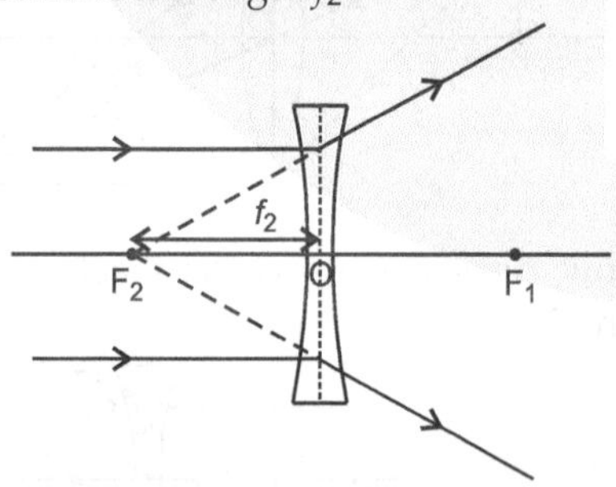

21. Draw images in case of a concave lens when the object is (i) at infinity and (ii) at any point on the principal axis between the lens and infinity.

Ans. (i) When object is at infinity, the incident ray is taken parallel to the principal axis. As shown in the figure, the image is virtual and is formed at the focus F_1. It is diminished (small) in size.

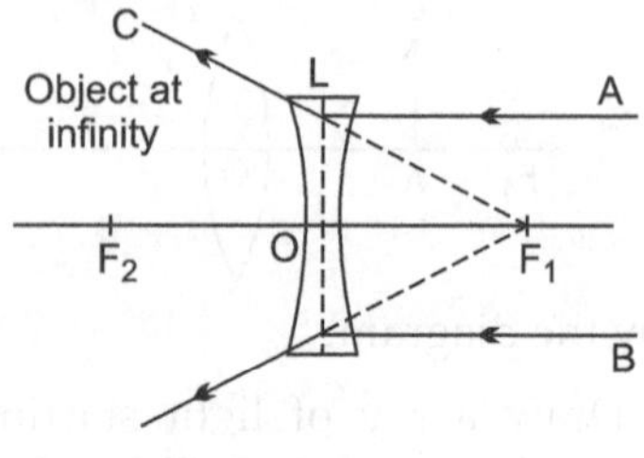

(ii) Here the object PQ is between the concave lens L and infinity and is at the principal axis. Two incident rays A and B are taken, A after refraction goes in the direction CD while the other ray B goes straight through the optical centre O. DC when produced backwards meets the ray B at P′, which is the virtual image P′Q′ of the object PQ.

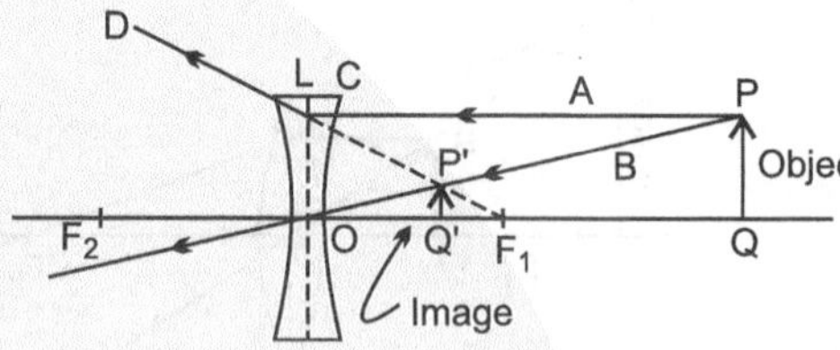

The image is virtual, upright and diminished and is on the side of the object PQ.

22. Can one burn a piece of paper in daylight by just using a convex lens instead of a match stick or any direct flame? Support your answer with the help of an appropriate ray diagram.

Ans.

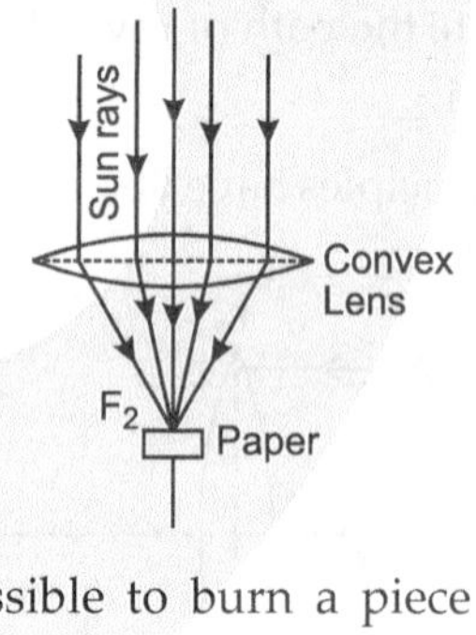

Yes, it is possible to burn a piece of paper, in daylight, by using a convex lens instead of a match stick or any other direct flame.

We know that a convex lens converges a beam of parallel incident rays at its focal point. The rays of the Sun (practically from an infinite distance source) form a parallel incident beam. When they are allowed to fall on a convex lens, they get focussed at its focal point. If a piece of paper is kept at this focal point, all the (large) heat energy of this parallel beam gets concentrated at this point. This (concentrated heat energy) is often sufficient to burn the piece of paper. The required ray diagram is as shown in figure.

23. What is a magnifying glass (or a simple microscope)? Draw a ray diagram to show how a convex lens may be used as a magnifying glass.

Where should the object be placed? State three characteristics of the image formed.

Ans.

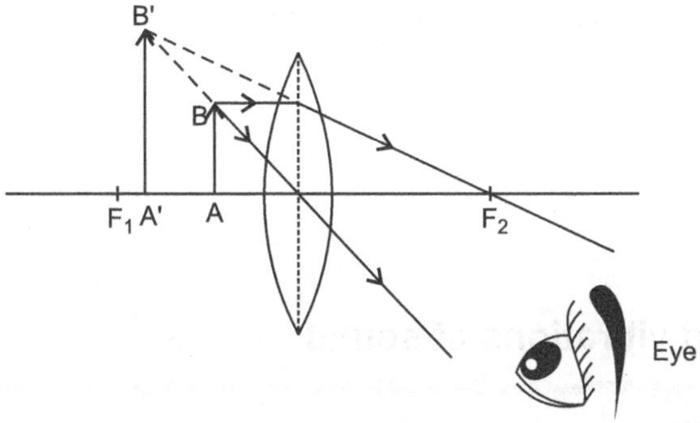

A magnifying glass is a convex (or converging) lens of short focal length mounted in a plastic or wooden frame provided with a handle. It is used to see small objects or to read small letters.

The ray diagram showing the use of a convex lens as a magnifying glass is given in the above figure.

The object should be placed infront of the lens between its optical centre and focus.

24. Study the given diagram and answer the following:

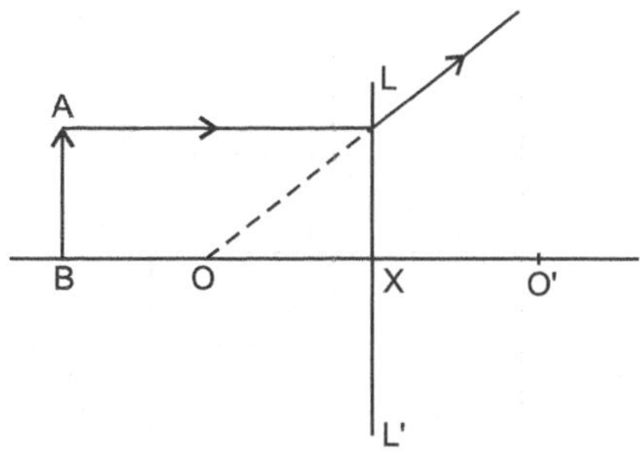

(i) Name the lens LL'.

(ii) What are the point O, O' called.

(iii) Complete the diagram to form the image of the object AB.

Ans. (i) The lens is a concave lens.

(ii) Point O is the 2nd focal point. Point O' is the 1st focal point.

(iii) The completed ray diagram is shown in the figure below :

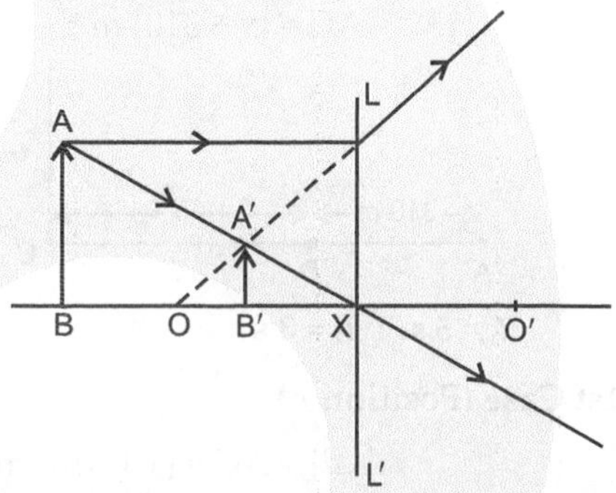

Chapter 5. Spectrum

1. With the help of two prisms show the dispersion and recombination of white light.

Ans.

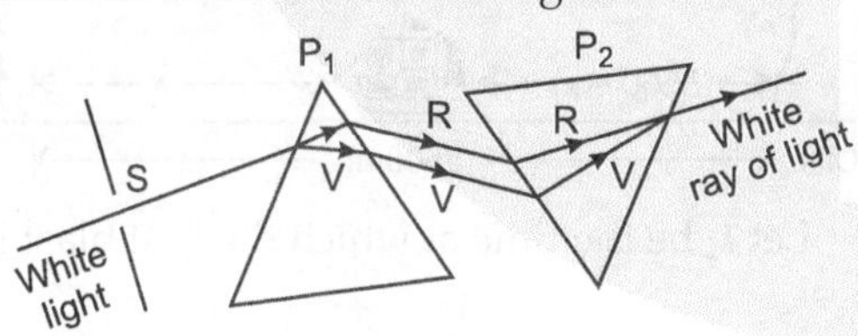

Before making any comment, the following simple experiment should be performed. The required arrangement is shown in the diagram. P_1 and P_2 are two glass prisms exactly similar and placed at some distance from each other such that their refracting surfaces are parallel to each other and the base of the prism P_2 is upwards and vertex downwards. S is a narrow slit through which a ray of white light is incident on the refracting face of the prism P_1. It is dispersed by it into its seven constituent colours.

When this dispersed light falls on the refracting surface of the prism P_2, having its base upwards so that the rays are deviated in a direction different from the first, we see on the screen only a white spot of light. It is evident that the second prism P_2 recombines these colours to form white light acting as a 'recombination prism' while the first prism acts as a 'dispersing prism'.

2. An imaginary material reflects red light, scatters green light, transmits blue light and absorbs yellow, orange and violet light. A cube of this material is shown in the given diagram. The surface is illuminated with a beam of white light. What colour will this cube appear (i) from above, (ii) from below and (iii) from the sides?

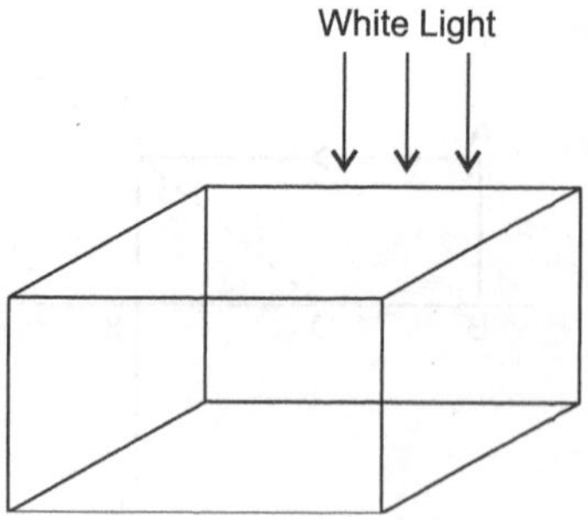

Ans. When viewed from above – Red.
When viewed from below – Blue.
When viewed from sides – Green.

Chapter 6. Echoes and vibrations of sound

1. A man fires a gun and hears its echo after 5 s. The man then moves 310 m towards the hill and fires his gun again. This time he hears the echo after 3 s. Calculate the speed of sound.

Ans. Let the speed of sound = v ms^{-1}

Let $BC = x$ m

$\therefore \quad AC = (x + 310)$ m

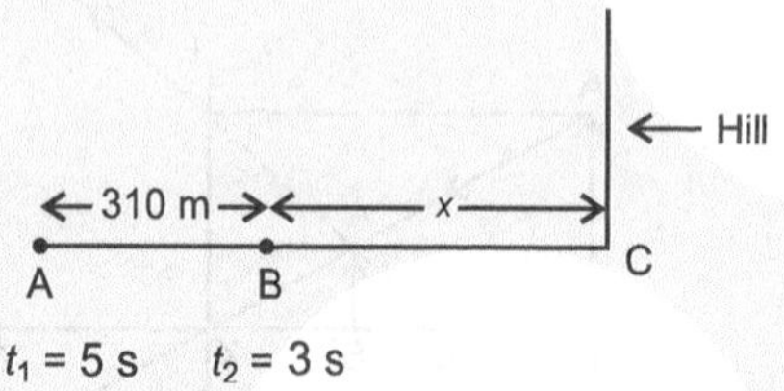

1st Case (Position A)

$$d_1 = AC = (x + 310) \text{ m}$$

$$t_1 = 5 \text{ s}$$

$$\therefore \quad v = \frac{2d_1}{t_1}$$

$$\text{or} \quad v = \frac{2(x+310)}{5} \quad \ldots\text{(i)}$$

2nd Case (Position B)

$$d_2 = BC = x \text{ m}$$

$$t_2 = 3 \text{ s}$$

$$\therefore \quad v = \frac{2d_2}{t_2}$$

$$\text{or} \quad v = \frac{2x}{3} \quad \ldots\text{(ii)}$$

From (i) and (ii),

$$\Rightarrow \quad \frac{2x}{3} = \frac{2(x+310)}{5}$$

$$\Rightarrow \quad 5x = 3x + 930$$

$$\Rightarrow \quad 2x = 930$$

$$x = 465 \text{ m}$$

Putting $x = 465$ in (ii),

$$v = \frac{2x}{3} = \frac{2 \times 465}{3} \text{ ms}^{-1} = 310 \text{ ms}^{-1} \quad \textbf{Ans.}$$

2. An observer X is on an island 5500 m from a vertical cliff on the shore. A ship Y is anchored between the island and the cliff. A blast on the ship's siren is heard twice by X, the interval being 5 s. [Velocity of sound = 330 ms^{-1}]

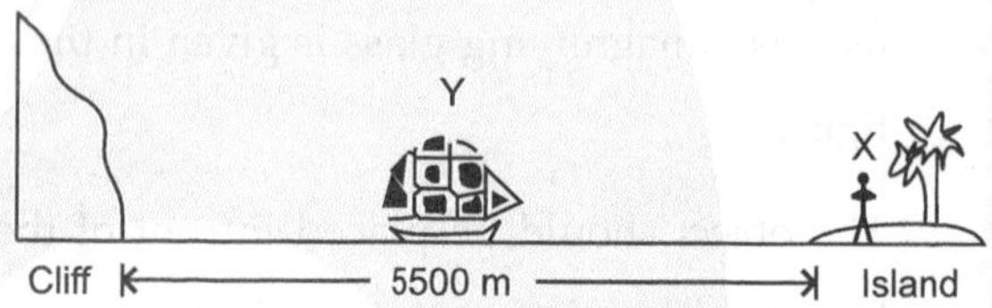

(i) Find the distance of the ship from the island.

(ii) Find the distance of the cliff from the ship.

Ans. (i) Let distance between the observer X on island and the ship Y = x m

$\therefore$ Distance between the ship Y and the cliff = $(5500 - x)$ m

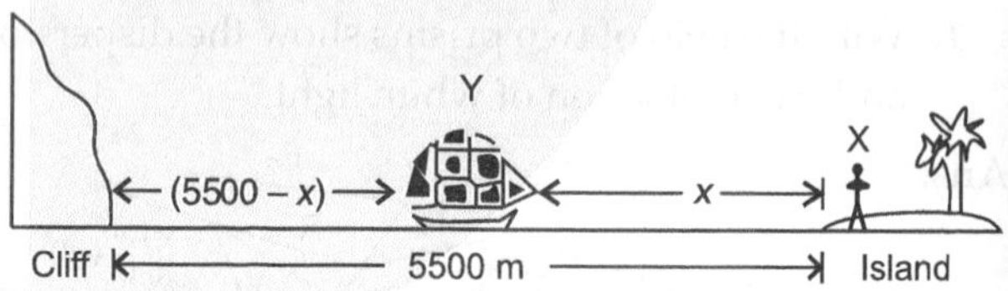

Let t_1 be the time at which the first blast is heard.

$$\therefore \quad t_1 = \frac{x}{v}$$

$$= \frac{x}{330} \quad \ldots\text{(i)}$$

$$\left(\because \text{ Velocity} = \frac{\text{Distance}}{\text{Time}}\right)$$

Let t_2 be the time at which the second blast is heard *i.e.,* time taken for the sound to travel from the ship to the cliff and then back to the observer on the island.

$$\therefore \quad t_2 = \frac{\text{Total distance travelled}}{\text{Velocity}}$$

Total distance = Y to cliff + Cliff to the island

Total distance = $(5500 - x) + 5500$

$= (11000 - x)$

$$\therefore \quad t_2 = \frac{11000 - x}{330}$$

But $t_2 - t_1 = 5$ s

∴ Subtracting equation (i) from equation (ii), we get

$$\frac{11000 - x}{330} - \frac{x}{330} = 5$$

$\therefore \quad 11000 - 2x = 5 \times 330$

$\therefore \quad 2x = 11000 - 1650$

$= 9350$

$x = 4675$ m.

∴ Distance between the ship Y and the island = 4675 m. Ans.

(ii) Distance between ship Y and the cliff

$= 5500 - x$

$= 5500 - 4675$

$\therefore \quad x = 825$ m Ans.

Distance of the ship from the island is 4675 m and the distance of the cliff from the ship is 825 m.

3. The diagram below shows the displacement-time graph for a vibrating body.

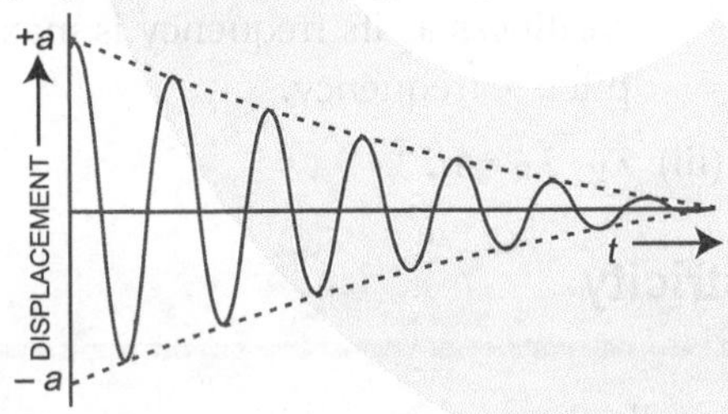

(i) Name the type of vibrations produced by the vibrating body.

(ii) Give one example of a body producing such vibrations.

(iii) Why is the amplitude of the wave gradually decreasing?

(iv) What will happen to the vibrations of the body after some time?

Ans. (i) The diagram shows damped vibrations.

(ii) A tuning fork vibrating in air.

(iii) The amplitude of the wave decreases due to frictional force which the surrounding medium exerts on the vibrating body.

(iv) After some time the amplitude gradually decreases and finally stops.

4. Explain free and forced vibrations.

Ans. When a system or a body vibrates without receiving impulses from another system or body, its vibrations are said to be free vibrations. However, when a system or body vibrates on account of impulses received from another system or body, the vibrations are called forced vibrations. In such a case, the body or the system vibrates with the frequency of the impulses received and not with its natural frequency.

5. Describe a simple experiment to illustrate the phenomenon of resonance and explain it.

Ans.

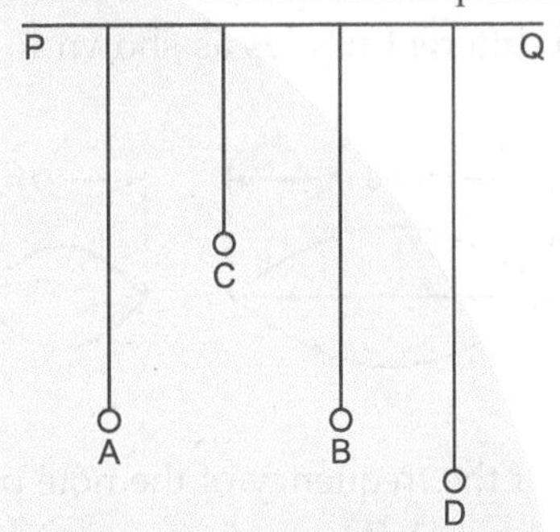

Four pendulums A, B, C and D are suspended from a rubber string PQ. The pendulums A and B are of equal lengths while C is shorter and D is longer. The pendulum A is set into vibration. It is observed that the pendulum B also starts vibrating and ultimately acquires the same amplitude as that of A. The pendulums C and D also vibrate, but with smaller amplitudes. The pendulum B is said to be in resonance with pendulum A.

Explanation: The natural frequency of pendulum B is equal to the frequency of vibration of pendulum A since both are of equal lengths. So the forced vibrations caused in pendulum B are in resonance with vibrations of pendulum A. On the other hand, the natural frequency of pendulum C is higher while of D is smaller than that of pendulum A. So these pendulums vibrate with the frequency of pendulum A with smaller amplitudes due to forced vibrations.

6. The diagram below shows three ways in which the string of an instrument can vibrate.

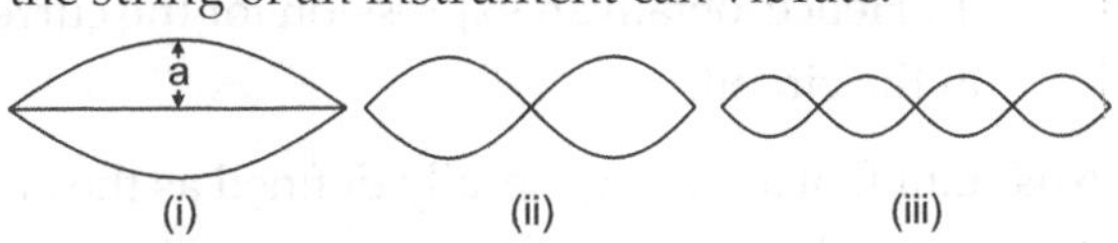

(i) Which of the above diagrams shows the fundamental note ?

(ii) Which has the frequency 2 times that of the fundamental ?

(iii) State the ratio between the frequency of the first and second.

Ans. (i) The diagram (i) represents the fundamental note because its frequency is lowest and amplitude maximum.

(ii) The diagram (ii) represents the sound note having frequency two times that of the (i) or fundamental.

(iii) The frequency of (ii) is two times of the frequency of (i) so the ratio between the frequencies of (i) and (ii) is 1 : 2.

7. A 0.6 m long stretched wire is made to vibrate in two different modes as shown in figure (a) and (b).

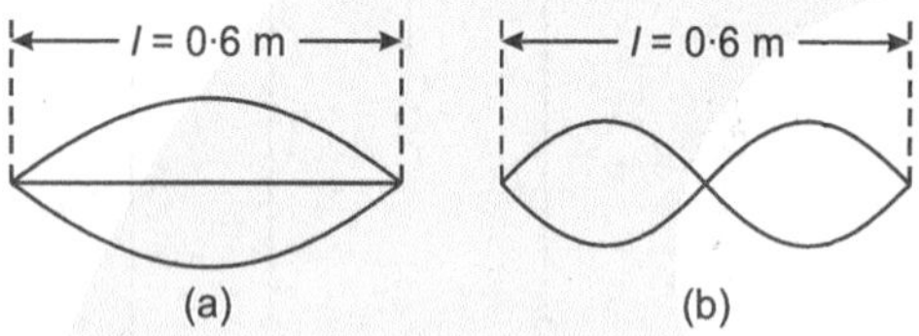

(i) If the frequency of the note produced in (b) is n, what is the frequency in case (a).

(ii) In which case (a) or (b), is the note louder? Give reasons.

(iii) In which case is the pitch higher? Give reason.

Ans. (i) The frequency of the note produced in (b) is 2 times that produced in (a). Since the frequency of the note in (b) is n, hence the frequency of note in (a) is $\frac{1}{2}n$ or $\frac{n}{2}$.

(ii) The note in (a) is louder, the reason being its amplitude is greater than the amplitude of note represented in (b), as the loudness being directly proportional to amplitude.

(iii) Pitch is directly proportional to the frequency of sound note. The frequency of note in figure (b) is more (double) than the frequency of note in (a). So the pitch of note (b) is higher than that of note (a).

8. The adjacent diagram shows three different modes of vibrations P, Q and R of the same string.

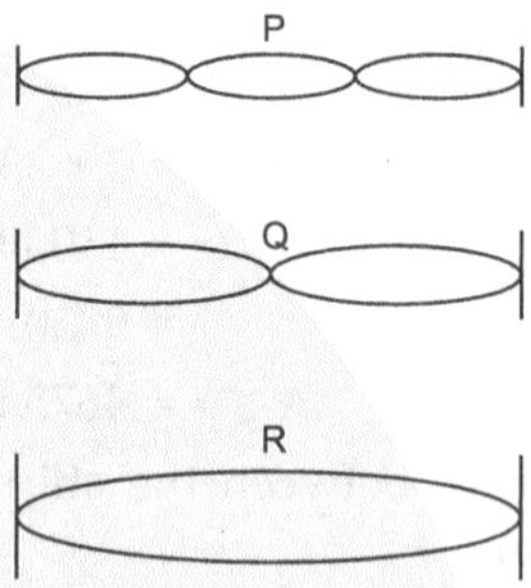

(i) Which vibration will produce a louder sound and why ?

(ii) The sound of which string will have maximum shrillness ?

(iii) State the ratio of wavelengths of P and R.

Ans. (i) Vibration R, as its amplitude is high and loudness $\propto A^2$.

(ii) Sound of string 'P' will have maximum shrillness as its frequency is maximum and pitch $\propto$ frequency.

(iii) $\lambda_P : \lambda_R = 1 : 3$.

Chapter 7. Electricity

1. Define the e.m.f. (E) and the terminal voltage (V) of a cell in terms of the work done in moving a unit charge. State the relation between these two works and the work done in moving a unit charge through a cell connected across the resistor. Take the internal resistance of the cell as 'r'. Hence, obtain an expression for the current (I) in the circuit.

Ans. E.m.f. of a cell is generally defined as the amount of work done (or the energy spent) in taking a unit positive charge around the complete circuit of the cell (*i.e.*, in the wire outside the cell and the electrolyte within the cell). It is also defined as the potential difference between the terminals of a cell when no current is drawn from it (or when the cell is in open circuit).

The terminal voltage of a cell is generally defined as the amount of work done in carrying a unit positive charge round the circuit connected across the terminals of the cell. If W is the amount of work done (in J) in moving a test charge q between the terminals of a cell through a resistor R, then the p.d. = $\frac{W}{q_0}$, *i.e.*, work done in moving a unit positive charge across the terminals of the cell.

Relation between E, V, external resistance R and the internal resistance r :

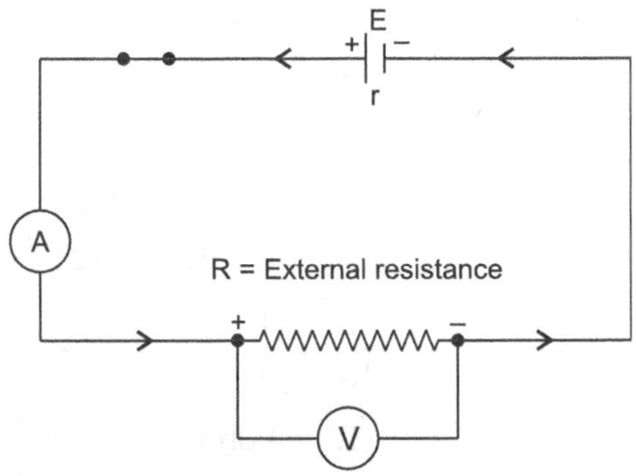

Applying Ohm's law to external resistance only

$$I = V/R$$

$$\Rightarrow \quad V = IR \quad \ldots(i)$$

Again applying Ohm's law to complete circuit

$$I = \frac{E}{R + r} \quad \ldots(ii)$$

$$\Rightarrow \quad E = I(R + r)$$

$$= IR + Ir$$

$$= V + v$$

$$\therefore \quad E = V + v \quad \ldots(iii)$$

v is the voltage drop in the cell.

2. A cell of e.m.f. 1·5 V and internal resistance 1 Ω is connected to the resistors of 4 Ω and 20 Ω in series. Draw a circuit diagram and calculate :

(i) current in the circuit,

(ii) p.d. across each resistor,

(iii) p.d. across the cell, and

(iv) the voltage drop when the current is flowing.

Ans.

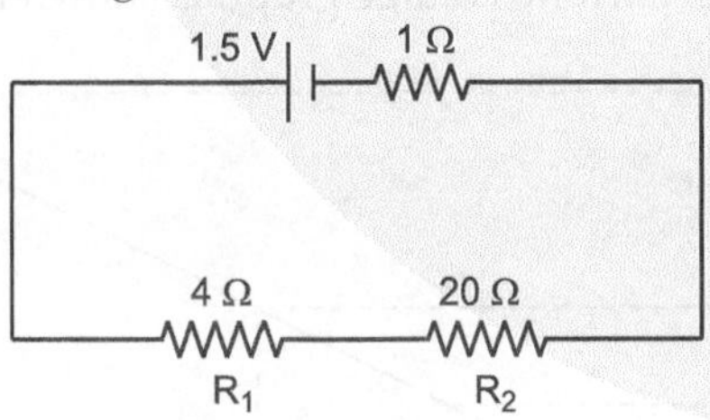

External resistance in circuit,

$$R = R_1 + R_2 = (4 + 20)\ \Omega = 24\ \Omega.$$

(i) Current in the circuit,

$$I = \frac{E}{R + r} = \frac{1·5}{24 + 1} A$$

$$= \frac{1·5}{25} A = 0·06\ A$$

(ii) p.d. across R_1,

$$V_1 = IR_1 = 0·06 \times 4$$

$$= 0·24\ V$$

p.d. across R_2,

$$V_2 = IR_2 = 0·06 \times 20$$

$$= 1·20\ V$$

(iii) p.d. across the cell

$$= \text{Terminal voltage}$$

$$= IR$$

$$= 0·06 \times 24$$

$$= 1·44\ V$$

(iv) Voltage drop (v)

$$= Ir$$

$$= 0·06 \times 1\ V$$

$$= 0·06\ V.$$

3. Four cells each of e.m.f. 2 V and internal resistance 0.1 Ω are connected in series to an ammeter of negligible resistance, a 1.6 Ω resistor and an unknown resistor R_1. The current in the circuit is 2 A. Draw a labelled diagram and calculate:

(i) Total resistance of the circuit, (ii) Total e.m.f. (iii) The value of R_1 and (iv) The p.d. across R_1.

Ans.

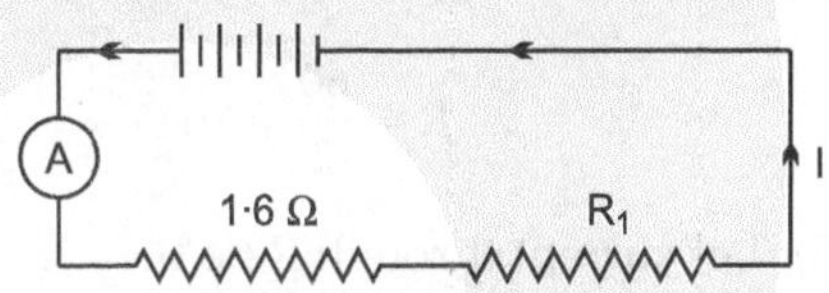

(i) The resistance in the circuit

$$= (1.6 + 4 \times 0.1 + R_1)\ \Omega$$

$$= (2 + R_1)\ \Omega \qquad \textbf{Ans.}$$

(ii) Total e.m.f. = No. of cell × e.m.f. of each cell

$$= 4 \times 2 = 8\ V. \qquad \textbf{Ans.}$$

(iii) Current in main circuit,

$$i = \frac{\text{Total e.m.f.}}{\text{Total resistance}}$$

or

$$2 = \frac{8}{2 + R_1}$$

$$\therefore \quad R_1 = 4/2 = 2\ \Omega. \qquad \textbf{Ans.}$$

∴ Total resistance = 2 + 2 = 4 Ω.

(iv) The p.d. across $R_1 = V = iR_1 = 2 \times 2 = 4$ V.

Ans.

4. In the figure below, the ammeter A reads 0·3 A. Calculate:

(i) the total resistance of the circuit

(ii) the value of R

(iii) the current flowing through R.

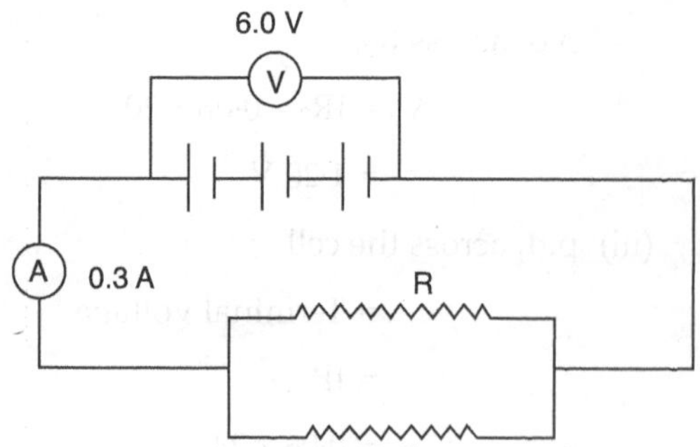

Ans. (i) Let the equivalent resistance of the circuit be R_p.

$\because \quad V = IR_p$

$\therefore \quad R_p = \frac{V}{I} = \frac{6 \cdot 0}{0 \cdot 3} = 20\ \Omega$ **Ans.**

(ii) Total resistance = 20 Ω

$\because$ Resistances are in parallel,

$\therefore \quad \frac{1}{20} = \frac{1}{R} + \frac{1}{60}$

$\frac{1}{R} = \frac{1}{20} - \frac{1}{60}$

$\frac{1}{R} = \frac{3-1}{60}$

$R = \frac{60}{2} = 30\ \Omega$ **Ans.**

(iii) Let current through R be I_1.

p.d. across resistance (R)

= p.d. across resistance (R_p)

$\Rightarrow \quad I_1 R = IR_p$

$\Rightarrow \quad I_1 \times 30 = 0 \cdot 3 \times 20$

$\therefore \quad I_1 = \frac{0 \cdot 3 \times 20}{30} A = 0 \cdot 2\ A$

$\therefore$ Current flowing through

R = 0·2 A. **Ans.**

5. Three resistors are connected to a 12 V battery as shown in the given figure:

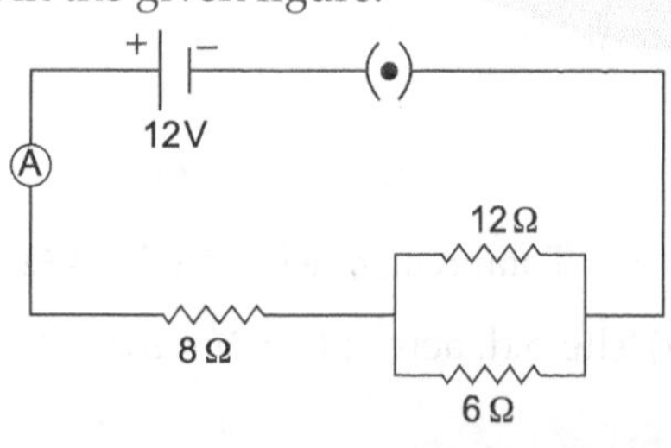

(i) What is the current through the 8 ohm resistor?

(ii) What is the potential difference across the parallel combination of 6 ohm and 12 ohm resistor?

(iii) What is the current through the 6 ohm resistor?

Ans. (i) 6 ohm and 12 ohm are connected in parallel, So

$$\frac{1}{R_1} = \frac{1}{6} + \frac{1}{12} = \frac{2+1}{12} = \frac{3}{12}$$

$$\Rightarrow \quad \frac{1}{R_1} = \frac{1}{4}$$

$R_1 = 4$ ohm.

$\therefore$ Total Resistance,

R = 8 + 4 = 12 ohm.

Now, V = IR

$\Rightarrow \quad 12 = I \times 12$

$\therefore \quad I = 1\ A$ **Ans.**

(ii) $V_1 = IR_1 = 1 \times 4 = 4$ volt **Ans.**

(iii) $V_1 = I_1 \times 6$

$\Rightarrow \quad 4 = I_1 \times 6$

$\therefore \quad I_1 = 4/6$

$= 0 \cdot 66\ A.$ **Ans.**

6. From the following observations taken while determining the resistance of a conductor, draw the current-voltage graph and calculate the resistance of conductor. Is the conductor ohmic?

Ammeter reading I (in ampere)	Voltmeter reading V (in volt)
0.2	0.4
0.3	0.6
0.5	1.0
0.8	1.6
1.5	3.0

Ans. The current-voltage graph is shown in figure:

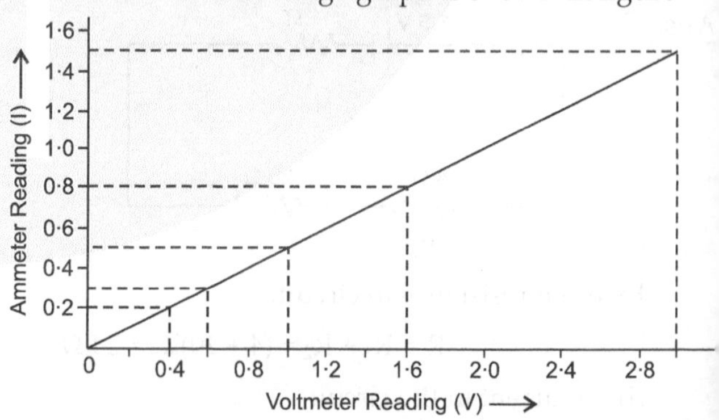

$$\text{Resistance, } R = \frac{\Delta V}{\Delta I} = \frac{(3.0 - 0.6)\ V}{(1.5 - 0.3)\ A}$$

$$= \frac{2.4\ V}{1.2\ A}$$

$= 2\ \Omega.$ **Ans.**

The conductor is ohmic.

7. Calculate equivalent resistance in the following cases:

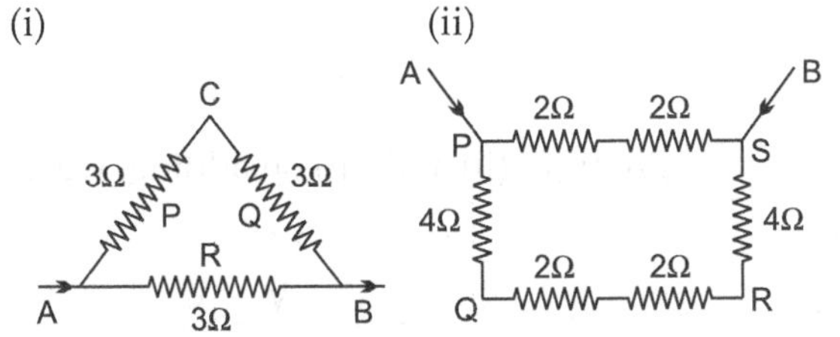

Ans. (i) Resistance P and Q are in series = 3 Ω + 3 Ω = 6 Ω.

Resistance of 6 Ω segment and 3 Ω are parallel.

$$\therefore \frac{1}{R} = \frac{1}{6} + \frac{1}{3} = \frac{1+2}{6} = \frac{3}{6} \text{ or } R = 2\,\Omega. \quad \textbf{Ans.}$$

(ii) Resistance of the arm PS = 2 Ω + 2 Ω = 4 Ω, (in series) and

Resistance of the arm QR = 2 Ω + 2 Ω = 4 Ω, (in series)

Resistance of arm PQ, QR and RS are in series

$$= (4 + 4 + 4)\,\Omega = 12\,\Omega$$

Now resistance of the arm PS and arms (PQ + QR + RS) are in parallel.

$$\frac{1}{R} = \frac{1}{4} + \frac{1}{12}$$

$$= \frac{3+1}{12} = \frac{4}{12}$$

or $R = 3\,\Omega$ **Ans.**

8. Three resistors are connected to a 6 V battery as shown in the figure given below:

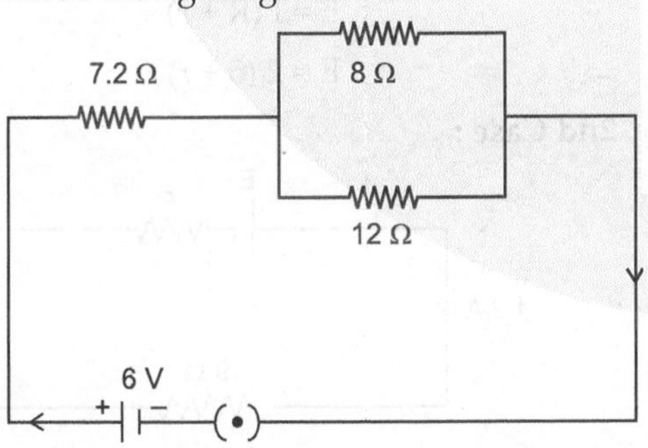

Calculate:

(i) the equivalent resistance of the circuit.

(ii) total current in the circuit.

(iii) potential difference across the 7.2 Ω resistor.

Ans. (i)

$$\frac{1}{R_1} = \frac{1}{8} + \frac{1}{12}$$

$$= \frac{3+2}{24} = \frac{5}{24}$$

$$R_1 = 4{\cdot}8 \text{ ohm.}$$

Total resistance = 4·8 + 7·2 = 12 ohm. **Ans.**

(ii)

$$V = IR$$

$$6 = I \times 12$$

$$I = 0{\cdot}5 \text{ A} \quad \textbf{Ans.}$$

(iii) Potential difference across the 7·2 Ω resistor,

$$= 0{\cdot}5 \times 7{\cdot}2$$

$$= 3{\cdot}6 \text{ V.} \quad \textbf{Ans.}$$

9. Five resistors of different resistances are connected together as shown in the figure. A 12 V battery is connected to the arrangement. Calculate :

(i) the total resistance in the circuit.

(ii) the total current flowing in the circuit

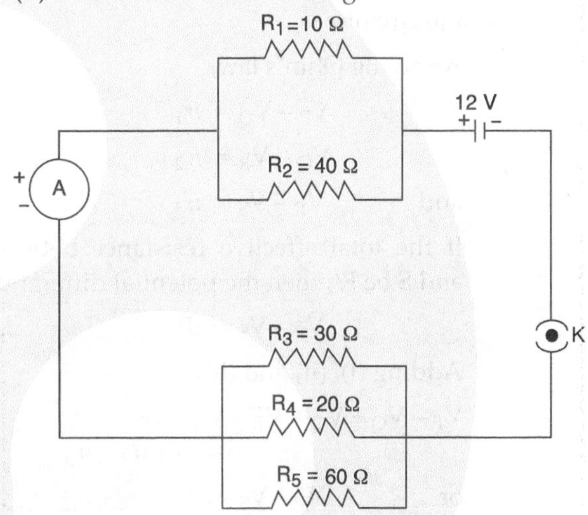

Ans. (i) From figure, R_1 and R_2 are in parallel.

$$\therefore \frac{1}{R'} = \frac{1}{R_1} + \frac{1}{R_2} = \frac{1}{10} + \frac{1}{40} = \frac{5}{40} = \frac{1}{8}$$

$$\therefore R' = 8\,\Omega$$

R_3, R_4 and R_5 are also in parallel,

$$\therefore \frac{1}{R''} = \frac{1}{R_3} + \frac{1}{R_4} + \frac{1}{R_5} = \frac{1}{30} + \frac{1}{20} + \frac{1}{60}$$

$$= \frac{6}{60} = \frac{1}{10}$$

$$\therefore R'' = 10\,\Omega$$

∴ Total resistance in the circuit

$$= R' + R''$$

$$= 8 + 10 = 18\,\Omega \quad \textbf{Ans.}$$

(ii) Total current flowing in the circuit,

$$I = \frac{V}{R} = \frac{12}{18}$$

$$= \frac{2}{3} = 0{\cdot}667 \text{ A.} \quad \textbf{Ans.}$$

10. Suppose there are three resistors A, B and C having resistances r_1, r_2 and r_3 respectively. If R represent their equivalent resistance, establish the following relations :

(i) $R = r_1 + r_2 + r_3$ when joined in series,

(ii) $\frac{1}{R} = \frac{1}{r_1} + \frac{1}{r_2} + \frac{1}{r_3}$, when joined in parallel.

Ans. (i)

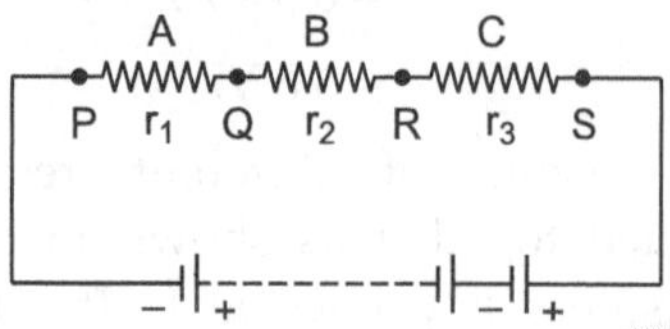

Let the three resistors be joined in series.

Let potential at P, Q, R and S are V_P, V_Q, V_R and V_S respectively and current i flows in the circuit.

Applying Ohm's law,

$$V_P - V_Q = ir_1 \quad ...(i)$$

$$V_Q - V_R = ir_2 \quad ...(ii)$$

and $$V_R - V_S = ir_3 \quad ...(iii)$$

If the total effective resistance between P and S be R′, then the potential difference :

$$V_P - V_S = iR' \quad ...(iv)$$

Adding (i), (ii) and (iii)

$$V_P - V_Q + V_Q - V_R + V_R - V_S = ir_1 + ir_2 + ir_3$$

or $$V_P - V_S = i(r_1 + r_2 + r_3) \quad ...(v)$$

Comparing (iv) and (v), we get

$$iR' = i(r_1 + r_2 + r_3)$$

or $$R' = r_1 + r_2 + r_3$$

(ii) When joined in parallel, let the main current be i which sub-divides at P into i_1 in r_1, i_2 in r_2 and i_3 in r_3 respectively.

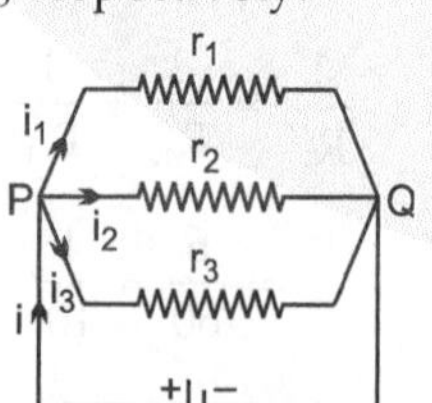

If potential difference across PQ be V then,

According to Ohm's law:

$$i_1 = \frac{V}{r_1} \quad ...(i)$$

$$i_2 = \frac{V}{r_2} \quad ...(ii)$$

and $$i_3 = \frac{V}{r_3} \quad ...(iii)$$

If R″ be the equivalent resistance of this combination, then

$$i = \frac{V}{R''} \quad ...(iv)$$

From (i), (ii) and (iii) on adding, we get

$$i = i_1 + i_2 + i_3$$

$$= \frac{V}{r_1} + \frac{V}{r_2} + \frac{V}{r_3}$$

$$= V\left(\frac{1}{r_1} + \frac{1}{r_2} + \frac{1}{r_3}\right) \quad ...(v)$$

Comparing (iv) and (v)

$$\frac{V}{R''} = V\left(\frac{1}{r_1} + \frac{1}{r_2} + \frac{1}{r_3}\right)$$

or $$\frac{1}{R''} = \frac{1}{r_1} + \frac{1}{r_2} + \frac{1}{r_3}$$

11. A cell supplies a current of 2 A when it is connected to a 5 Ω resistance and supplies a current of 1·2 A, if it is connected to a resistance of 9 Ω. Find the e.m.f. and internal resistance of the cell. Also, draw circuit diagram in both cases.

Ans. Let e.m.f. of the cell be E and internal resistance be r.

1st Case :

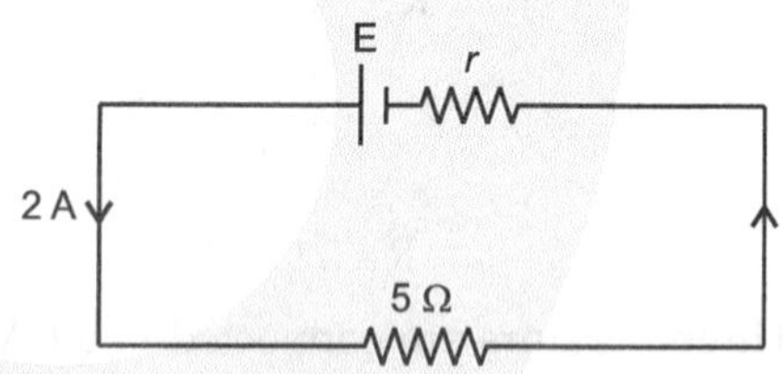

$$E = I(R + r)$$

$$\Rightarrow \quad E = 2(5 + r) \quad ...(i)$$

2nd Case :

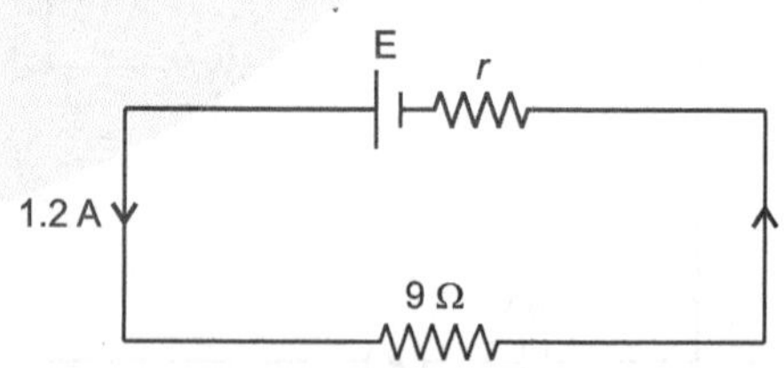

$$E = I(R + r)$$

$$\Rightarrow \quad E = 1{\cdot}2(9 + r) \quad ...(ii)$$

Dividing (i) by (ii), we get

$$\Rightarrow \quad 1 = \frac{2(5 + r)}{1{\cdot}2(9 + r)}$$

$$\Rightarrow \quad 12(9 + r) = 20(5 + r)$$

$$\Rightarrow \quad 108 + 12r = 100 + 20r$$

$$\Rightarrow \quad 20r - 12r = 108 - 100$$

$\Rightarrow$ $8r = 8$

$\therefore$ $r = 1\,\Omega$

From equation (i),

$E = 2\,(5 + r)$

$= 2\,(5 + 1)$ V

$= 2 \times 6$ V $= 12$ V. **Ans.**

12. Two resistors of 4 Ω and 6 Ω are connected in parallel to a cell to draw 0.5 A current from the cell.

(i) Draw a labelled circuit diagram showing the above arrangement.

(ii) Calculate the current in each resistor.

Ans. (i)

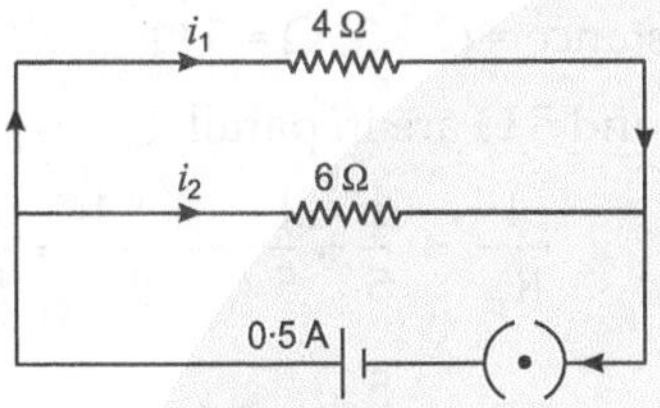

(ii) $\dfrac{1}{R} = \dfrac{1}{4} + \dfrac{1}{6} = \dfrac{3+2}{12} = \dfrac{5}{12}$

$\therefore$ $R = 2.4\,\Omega$

$\because$ $V = IR$

$\Rightarrow$ $V = 0.5 \times 2.4 = 1.2$ V

Now, $i_1 = \dfrac{V}{R_1}$

$= \dfrac{1.2}{4} = 0.3$ A

And $i_2 = \dfrac{1.2}{6}$

$= 0.2$ A **Ans.**

13. The figure shows a circuit.

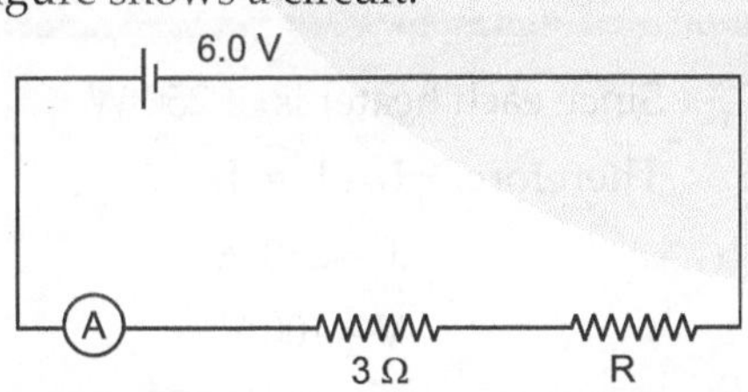

When the circuit is switched on, the ammeter reads 0·5 A.

(i) Calculate the value of the unknown resistor R.

(ii) Calculate the charge passing through the 3 Ω resistor in 120 s.

(iii) Calculate p.d. across 3 Ω resistor.

Ans. (i) We know, $V = IR'$

$\Rightarrow$ $6 = 0{\cdot}5 \times R'$

$\Rightarrow$ $R' = 12$ ohm

$\because$ $R' = 3 + R$

$\Rightarrow$ $12 = 3 + R$

$\therefore$ $R = 9$ ohm. **Ans.**

(ii) Charge, $q = it$

$\Rightarrow$ $q = 0{\cdot}5 \times 120 = 60$ coulomb **Ans.**

(iii) The p.d. across 3 Ω resistor

$= i \times 3$

$= 0{\cdot}5 \times 3$ V

$= 1{\cdot}5$ V **Ans.**

14. In the figure given alongside, A, B and C are three ammeters. The ammeter B reads 0.5 A. (All the ammeters have negligible resistance.)

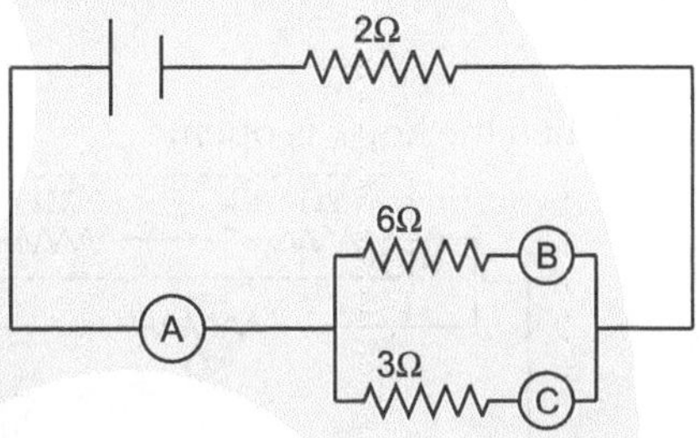

Calculate:

(i) the readings in the ammeters A and C.

(ii) the total resistance of the circuit.

Ans. (i) P.d. across 6 Ω = P.d. across 3 Ω

or $6 \times 0{\cdot}5 = 3 \times i_c$

or i_c *i.e.*, current passing through the ammeter C is:

$\dfrac{6 \times 0{\cdot}5}{3} = 1.0$ ampere **Ans.**

$\therefore$ Current through A

= Current passing through B

+ Current passing through C

$= 0{\cdot}5 + 1{\cdot}0 = 1{\cdot}5$ ampere **Ans.**

(ii) Here, $\dfrac{1}{R} = \dfrac{1}{6} + \dfrac{1}{3} = \dfrac{3}{6} = \dfrac{1}{2}$

or $R = 2\,\Omega$

$\therefore$ Total resistance of the circuit

$= 2\,\Omega + R$

$= 2\,\Omega + 2\,\Omega$

$= 4\,\Omega$ **Ans.**

Figure Based Long Answers

15.

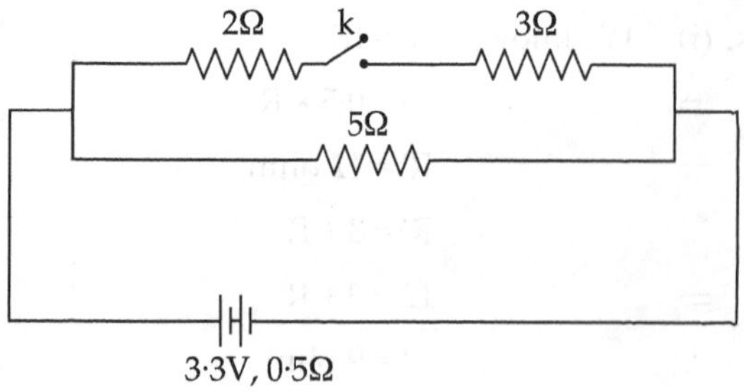

The diagram above shows a circuit with the key k open. Calculate:

(i) the resistance of the circuit when the key *k* is open.

(ii) the current drawn from the cell when the key *k* is open.

(iii) the resistance of the circuit when the key *k* is closed.

(iv) the current drawn from the cell when the key *k* is closed.

Ans. (i) When the key *k* is open:

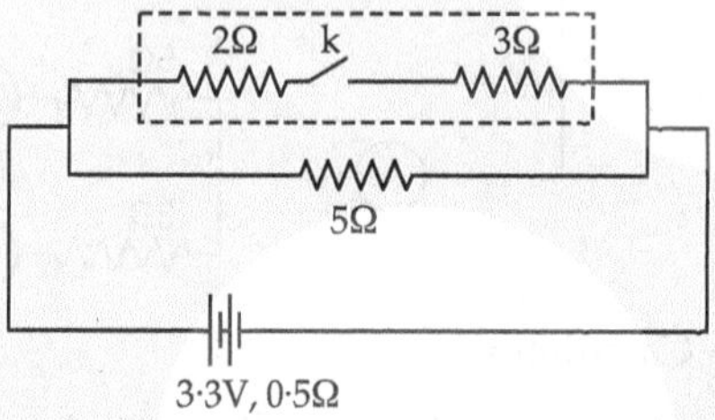

Resistance (R_1) of the circuit

$$= (5 + 0.5)\ \Omega$$

$$= 5.5\ \Omega$$

(ii) Current (I_1) drawn when key *k* is open:

$$I_1 = \frac{V}{R_1} = \frac{3.3}{5.5}\ \text{A}$$

$$= \frac{3}{5}\ \text{A} = 0.6\ \text{A}$$

(iii) When the key *k* is closed:

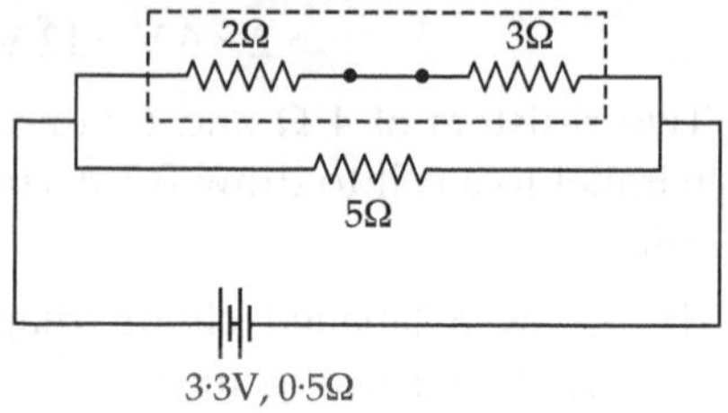

2 Ω and 3 Ω are in series and their equivalent resistance = (2 + 3) Ω = 5 Ω

5 Ω and 5 Ω are in parallel

$$\frac{1}{R_p} = \frac{1}{5} + \frac{1}{5} = \frac{1+1}{5} = \frac{2}{5}$$

$$R_p = \frac{5}{2}\ \Omega = 2.5\ \Omega$$

Resistance of circuit (R_2) when key k is closed

$$= (R_p + 0.5)\ \Omega$$

$$= (2.5 + 0.5)\ \Omega$$

$$= 3.0\ \Omega$$

(iv) Current (I_2) drawn when key k is closed

$$= \frac{V}{R_2}$$

$$= \frac{3.3}{3}\ \text{A} = 1.1\text{A}$$

Chapter 8. Electric Power and Household Circuits

1. Three 250 W heaters are connected in parallel to a 100 V supply, Calculate:

(i) Total current taken from the supply,

(ii) The resistance of each heater.

(iii) The energy supply in kWh to the three heaters in 5 hour.

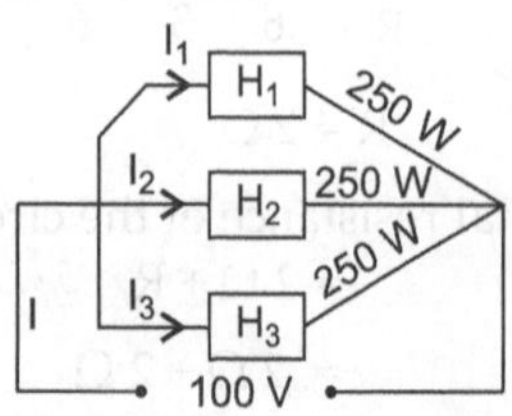

Ans. (i) Total current drawn from the supply

$$I = I_1 + I_2 + I_3$$

Since each heater is of 250 W.

Therefore, $I_1 = I_2 = I_3$

$$P = 250\ \text{W}$$

$$V = 100\ \text{V}$$

$$I_1 = \frac{P}{V} = \frac{250}{100}\ \text{A} = 2{\cdot}5\ \text{A}$$

∴ $I_2 = 2{\cdot}5$ A and $I_3 = 2{\cdot}5$ A

and $I = (2{\cdot}5 + 2{\cdot}5 + 2{\cdot}5)$ A

$= 7{\cdot}5$ A. **Ans.**

(ii) The resistance of each heater.

Since $P = \frac{V^2}{R}$

or $R = \frac{V^2}{P} = \frac{100 \times 100}{250} = 40\ \Omega$. **Ans.**

Therefore, 40 Ω is the resistance of each heater

(iii) The energy supplied to heaters:

Total power consumption

$= 250 + 250 + 250 = 750$ W.

∴ Energy supplied in kWh in 5 hours

$= \frac{\text{watt} \times \text{hour}}{1000} = \frac{750 \times 5}{1000}$ kWh

$= 3.75$ kWh. **Ans.**

2. An electric bulb is rated 240 V, 60 W and is working at 100% efficiency.

(i) Calculate the resistance of bulb.

(ii) (a) Draw the circuit diagram, if two such bulbs are connected in series.

(b) What is the rate of conversion of energy in each bulb?

(c) Total power used by the bulbs.

Ans. (i) We know that,

Power, $(P) = \frac{V^2}{R}$

Resistance of bulb (R)

$= \frac{V^2}{P} = \frac{240 \times 240}{60} = 960\ \Omega$

(ii) (a)

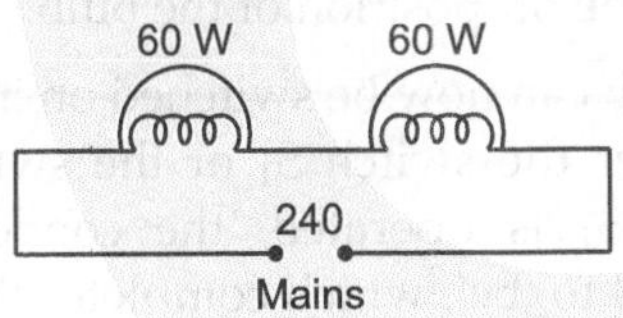

(b) Resistance of bulbs in series

$= (960 + 960)\ \Omega = 1920\ \Omega$.

∴ Current in series circuit (I)

$= \frac{V}{R} = \frac{240}{1920} = 0{\cdot}125$ A

∴ Rate of conversion of energy in any one bulb in 1 second

$= I^2Rt$

$= (0{\cdot}125)^2 \times 960 \times 1 = 15$ J

(c) Power of 1 bulb in series circuit

$= 15$ W

So power of 2 bulbs in series circuit

$= 2 \times 15\ \text{W} = 30$ W. **Ans.**

3. With reference to the diagram alongside calculate:

(i) The equivalent resistance between P and Q.

(ii) The reading of ammeter.

(iii) The electrical power between P and Q.

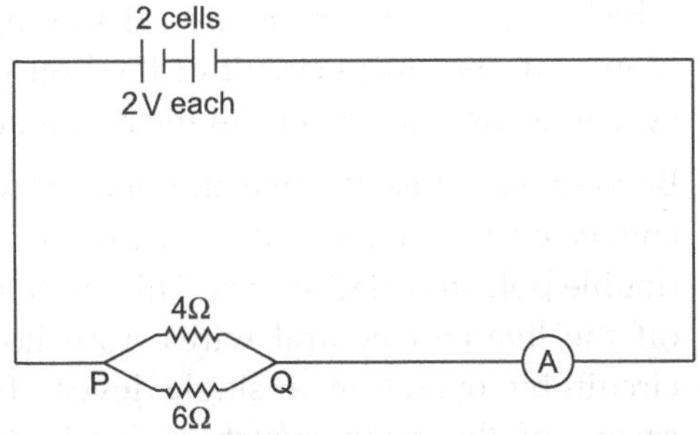

Ans. (i) $\frac{1}{R} = \frac{1}{4} + \frac{1}{6} = \frac{3+2}{12} = \frac{5}{12}$

So, equivalent resistance

$= \frac{12}{5} = 2{\cdot}4\ \Omega$. **Ans.**

(ii) Here, $R = 2{\cdot}4\ \Omega$

$E = 2 + 2 = 4$ V

From $E = IR$

$I = \frac{E}{R}$

or $I = \frac{4}{2{\cdot}4} = 1{\cdot}66$ A. **Ans.**

(iii) Power between P and Q

$= I^2R$

$= (1{\cdot}66)^2 \times (2{\cdot}4) = 6{\cdot}62$ W. **Ans.**

4. Explain the significance of kWh meter, main switch and main fuse in house-circuiting.

Ans. The electric cable coming to a house consists of three separate insulated wires, namely, live (phase) wire, neutral wire and earth wire. The neutral and earth wire on the supply end are connected together so that both of them are at zero potential.

Inside the house, the live and neutral wires are connected to the input terminals of kWh meter, while the earth wire is connected to the body of the kWh meter. The kWh meter reads the units of electrical energy (in kWh), consumed in a house. Without the kWh meter, it is not possible to correctly note the electrical energy consumed in a house or by an electric appliance.

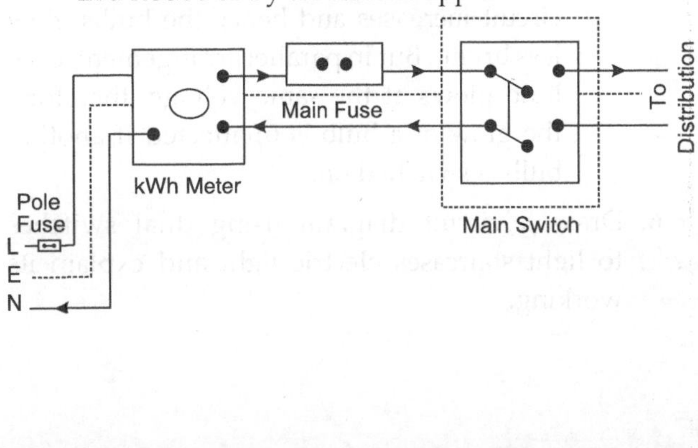

The live wire coming out from the output terminal of kWh meter has another fuse in it which is called the main fuse. If the main fuse blows off, the entire circuit of the house wiring system is disconnected from the mains supply.

Beyond main fuse the live and neutral wires are connected to a main switch which consists of a double pole iron clad switch. This switch can cut off the live and neutral wires from household circuit by operating a single lever. The iron casing of the main switch is locally earthed. Then the power flowing out of the main switch is distributed in house by the common ring or three wire system. The diagram illustrates the working of the kWh meter, the main fuse and the main switch.

5. The diagram below shows two ways of connecting three bulbs A, B, C to 220 V mains in a room. Name the two arrangements. Which of them would you prefer? Give two reasons for your answer.

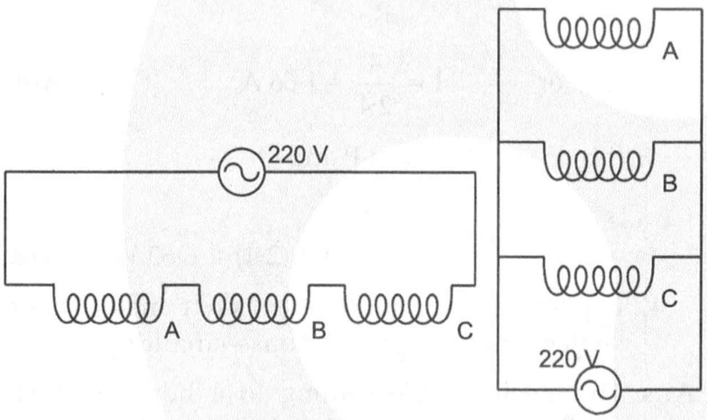

(a) (b)

Ans. The arrangement shown in the figure (a) is a series arrangement.The arrangement in figure (b) is the parallel arrangement.

We prefer the parallel arrangement. The reasons are:

(i) In series arrangement, if one of the bulb is fused, the other bulbs also cease to glow. But in parallel arrangement, if one bulb fuses, the other bulbs continue to glow.

(ii) In series arrangement, the resistance of the circuit increases and hence the bulbs glow less bright. But in parallel arrangement, each bulb glows at the same voltage, therefore, the glow of a bulb is unaffected if another bulb is switched on.

6. Draw a circuit diagram using dual switches to light staircases electric light and explain its working.

Ans.

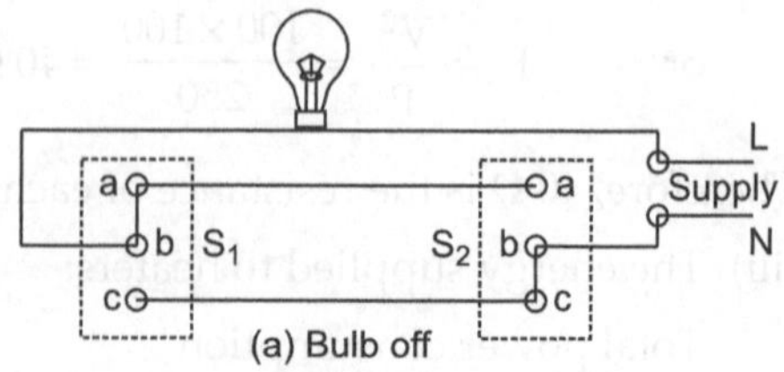

(a) Bulb off

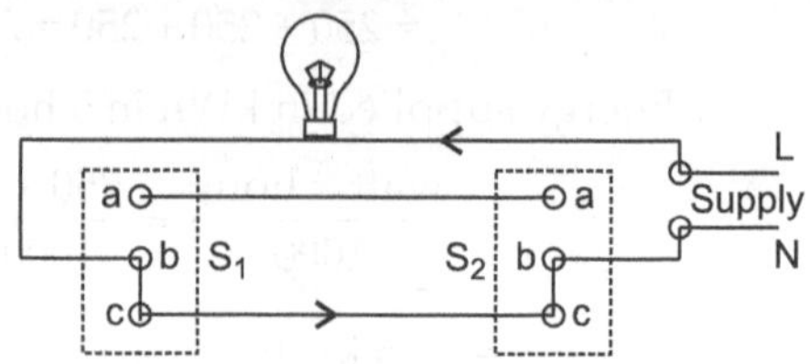

(b) Bulb on through switch S_1

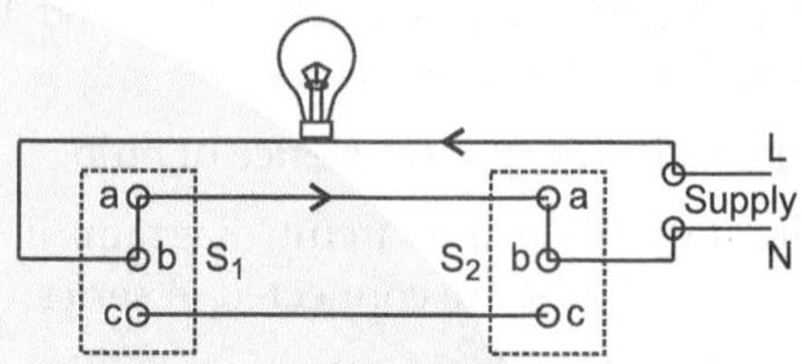

(c) Bulb on through switch S_2

Dual control switches are the double pole type switches which are generally used at the top and bottom of a staircase, or at the opposite ends of a long corridor, etc. With such switches, the appliance (say, a bulb) can be switched on or off from two different places.

The working and circuits of a dual control switch are illustrated in Fig. (a), (b) and (c). Let a switch S_1 be fitted at the top of the staircase. Fig. (a) shows the off position of the bulb.

The bulb can now be switched on independently by either the switch S_1 or the switch S_2. If the switch S_1 is operated, the connection 'ab' is changed to 'bc', which completes the circuit and the bulb lights up [Fig. (b)]. Similarly, on operating the switch S_2, the connection 'bc' changes to 'ba', which again completes the circuit [Fig. (c)].

Similarly, if the bulb is in on position as shown in Fig. (b), one can switch off the bulb by changing the connections 'ab' to 'bc' either by the switch S_1 or the switch S_2.

In a staircase, while going up a person puts on the light by operating the switch S_1 so that the connection 'ab' changes to 'bc' and makes the current to flow in the circuit. On reaching at the top he operates the switch S_2 to put off the light so that the connection 'cb' changes to 'ba' and the flow of current stops.

The same system is operative when a person puts on light by operating the switch S_2 to change the connection 'cb' to 'ba' and switch off the light by operating the switch S1 to change the connection 'ab' to 'bc'.

Chapter 9. Magnetic Effect of Current

1. Describe an experiment to show that there is a magnetic field around a current carrying conductor.

Ans.

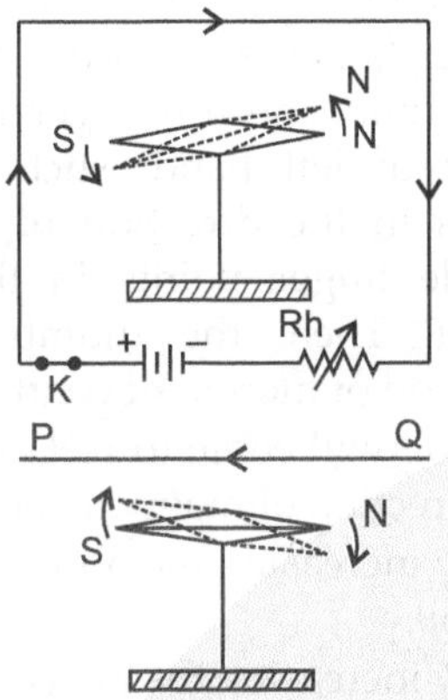

In 1820, Oersted discovered that when an electric current passes through a conductor, a magnetic field is produced around the conductor, *i.e.*, so long the current continues to flow through the conductor, it behaves like a magnet. This can be illustrated by a simple experiment described below.

Take a magnetic needle pivoted at its centre and free to move in horizontal plane.

Take a thick and insulated copper wire PQ and connect it to the battery circuit through a key K and a variable resistance Rh.

Hold the wire PQ parallel and just above the axle of the magnetic needle and press the key. It is observed that N-pole of the needle points towards the west so long the current continues to flow through the wire. It is also observed that if the current is increased or decreased, the deflection of the magnetic needle also increases or decreases.

If the direction of the current through PQ is reversed, *i.e.*, if the current flows from Q to P, then N-pole of the needle is deflected towards East, this experiment shows that a magnetic field is set up around a current carrying conductor.

2. Describe a set up for plotting the magnetic lines of force in:

(i) A straight wire, (ii) A circular coil and (iii) A straight solenoid.

Ans. (i) A set-up for plotting the magnetic lines of force in a straight wire is shown in the diagram. A smooth cardboard with a hole at the centre is placed horizontally and a wire passes vertically through the hole. Some iron filings are sprinkled on the cardboard and an electric current is passed through the wire. On slightly tapping the cardboard the iron filing arrange themselves in concentric circle around the wire, as shown in the diagram. The direction of magnetic lines of force is anticlockwise since the current in the straight wire is flowing upwards. If the current flow is downwards then the direction of lines of force is clockwise. [See diagram (a) and (b)]

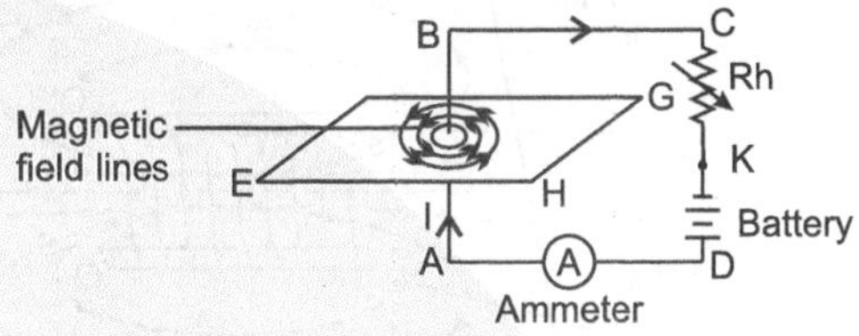

(a)

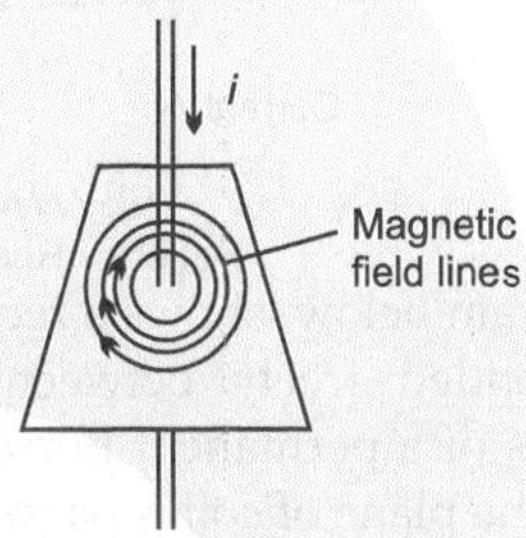

(b)

(ii) Arrangement of the apparatus is the same as in (a), only the wire is in the form of a circular coil and is passing through two holes, A and B in the cardboard. On passing the current through the circular coil, iron filings sprinkled on the horizontal surface of the smooth cardboard make a pattern of the lines of force, as shown in the diagram.

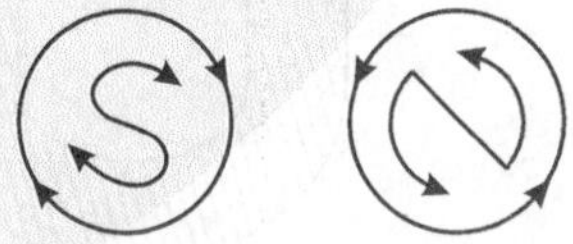

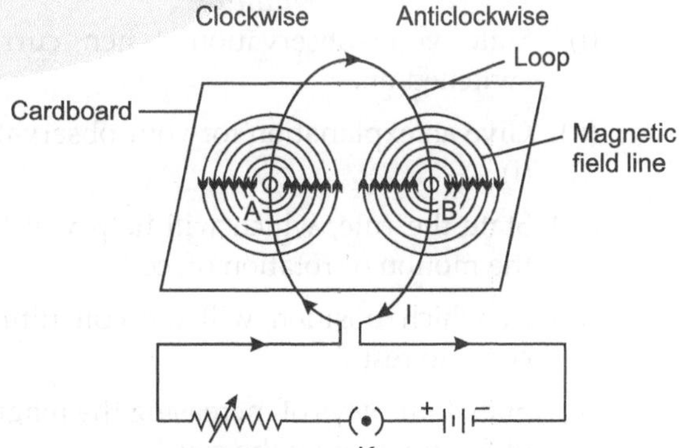

One face of the coil has North polarity while the other face shows South polarity, in accordance with the following rule : Looking at one face of the coil, if the direction of the current is anticlockwise, that face has North polarity and the opposite face has South polarity.

(iii) The accompanying diagram illustrates the pattern of magnetic lines of force in a straight solenoid. As shown in the diagram, the solenoid behaves like a cylindrical magnet having North and South poles at free ends. If the current flows in the reverse direction, the polarity at the free ends is also reversed.

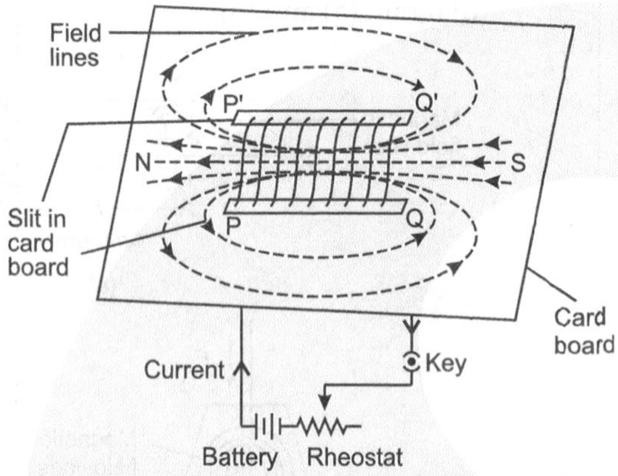

3. Diagram below, shows a rectangular coil ABCD, suspended freely between the concave pole pieces of a permanent horse shoe magnet, such that the plane of coil is parallel to magnetic field.

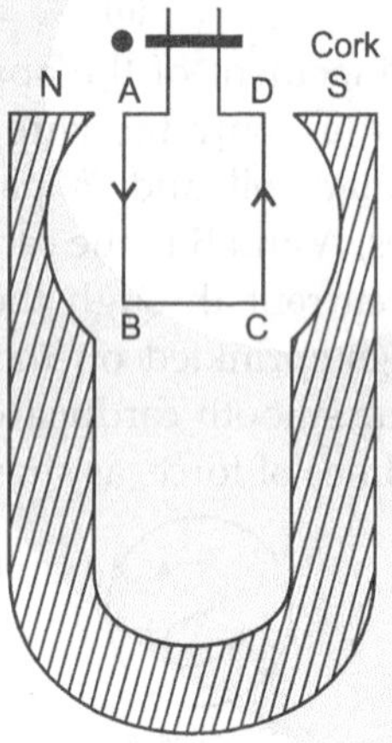

(i) State your observation, when current is switched on.

(ii) Give an explanation for your observation in (i).

(iii) State the rule, which will help you to find the motion of rotation of coil.

(iv) In which position will the coil ultimately come to rest ?

(v) State four ways of increasing the magnitude of force acting on the coil.

Ans. (i) The coil ABCD will turn. The arm AB of the coil will move out of the plane of paper and arm CD into the plane of paper.

Thus, coil will turn in anticlockwise direction.

(ii) A magnetic field is set up by the coil due to the passage of electric current. The magnetic field of the coil is at right angles to the magnetic field of permanent magnet. Thus, a magnetic couple acts, which turns the coil.

(iii) Fleming's Left Hand Rule: It states: Stretch the thumb, the forefinger and the middle finger of left hand, such that forefinger points in the direction of magnetic field; middle finger points in the direction of curent. Then, the thumb points in the direction of motion of conductor.

(iv) The coil will come to rest at right angles to the direction of magnetic field.

(v) (a) By increasing the number of turns in the coil.

(b) By increasing the area of cross section of the coil.

(c) By placing laminated soft iron core within the coil.

(d) By increasing the magnitude of current in the coil.

4. Figures (a) to (c) show the working of a simple A.C. generator. Study the diagrams and give the answers of the following questions.

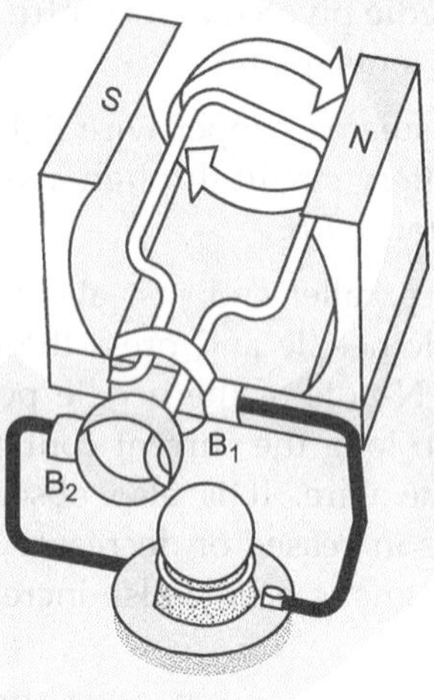

(a)

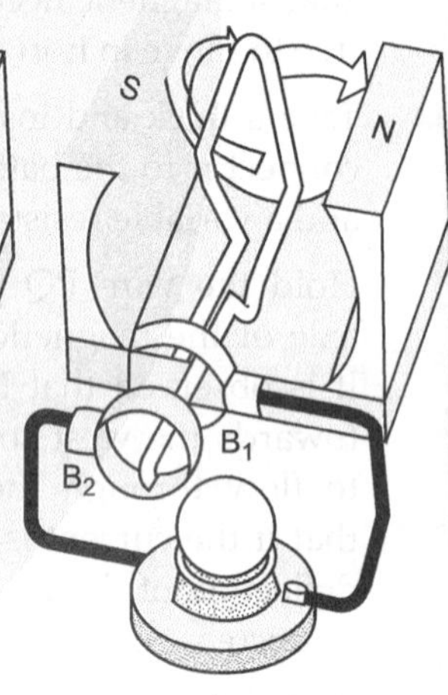

(b)

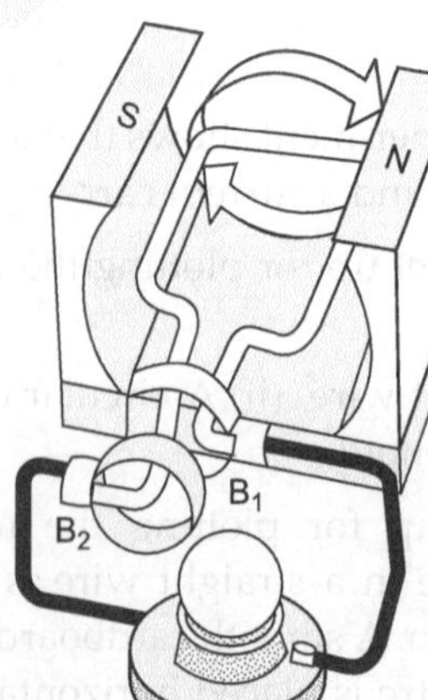

(c)

(i) State and explain the principle underlying the working of a simple A.C. generator.

(ii) Where is the loop of wire placed?

(iii) What happens when the loop is rotated?

(iv) Indicate the direction of the current flowing through the wire for the first half of the turn (Figure a). Name and state the rule used in finding the direction of the current.

(v) Indicate the direction of the current for the case shown in figure b.

(vi) Indicate the direction of current in the outer circuit (*i.e.*, electric bulb) in figure 1 and in figure c.

(vii) What type of current is shown in the above diagrams? Explain.

Ans. (i) The generator works on the principle of Electromagnetic Induction. Whenever the flux passing through a closed circuit changes, e.m.f. is induced in it and a current begins to flow. Its direction is given by Fleming's Right Hand Rule or Lenz Law. As long as the loop is moving within the magnetic field, an electric current will flow.

(ii) The loop of wire is placed between the two poles of a U-shaped magnet.

(iii) When the loop of wire is rotated between the two poles of a U-shaped magnet, an electric current flows through the wire.

(iv) As rotation of the coil is in the clockwise direction and the direction of magnetic field is from right to left, the current will flow in the anticlockwise direction, *i.e.*, from B_2 and B_1.The law used is Lenz's law.

Lenz's law: "The direction of the induced e.m.f. is such that it opposes the change responsible for its production."

(v) No electric current flows when the loop of wire is halfway through the turn, *i.e.*, at an angle of 90°. [as sin 90° = 0]

(vi) **Figure 1:** From brush B_2 to B_1

Figure 3: From brush B_1 to B_2

(vii) The type of current shown is an alternating current or A.C. current. When the loop of wire makes the second half of the turn, the direction of the electric current through the bulb is reversed and it flows back in the opposite direction.

5. Copy the given diagram of a transformer and complete it. Name the parts A and B. Name the part you have drawn to complete the diagram. What is the material of this part? Is this transformer a step-up or step-down ? Give reason.

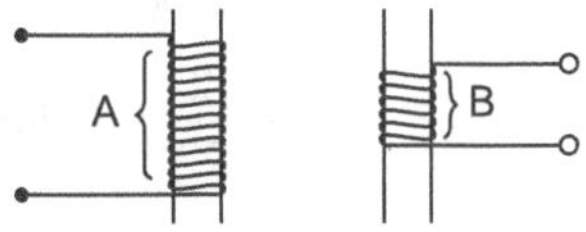

Ans. The completed diagram is given below:

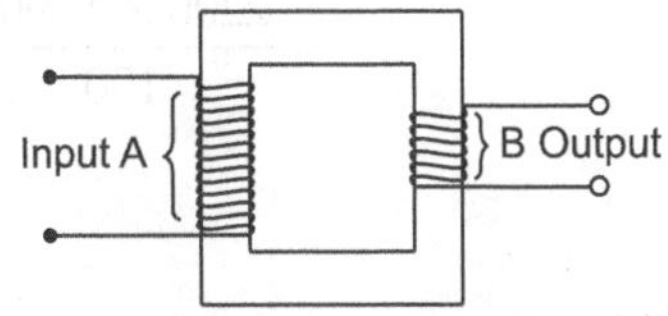

The part A is called primary coil. The part B is called secondary coil.

The part drawn is called core.

The material of core is soft iron.

This transformer is step-down since it has more turns in primary coil than in secondary coil.

6. The power supply to the primary coil of a transformer is 200 W. Find:

(i) Current in primary coil if the e.m.f. supplied to it is equal to 220 V.

(ii) The number of turns in the primary coil is equal to 80 and that in secondary is 800. What is the transformation ratio ?

(iii) Name the type of transformer.

(iv) What will be the output voltage?

(v) What is the current in the secondary coil for an ideal transformer?

(vi) What is the output power?

(vii) Is output and input power equal?

(viii) Compare the current flowing in a secondary coil and in a primary coil.

Ans. (i) P (Power) = 200 W = VI = $E_p \times I_p$

$\therefore$ P = 200 = 220 × I_p

$\therefore$ Current in the primary coil

$$= \frac{200}{220}$$

$$I_p = 0.909 \text{ A.}$$ **Ans.**

(ii) Transformation ratio,

$$\frac{E_s}{E_p} = \frac{N_s}{N_p} = \frac{800}{80} = 10.$$ **Ans.**

(iii) It is a step up transformer.

(iv) Using $\frac{E_s}{E_p} = \frac{N_s}{N_p} = 10$

$$\therefore \quad \frac{E_s}{220} = 10$$

$$\therefore \quad E_s = 2200 \text{ V.}$$ **Ans.**

(v) Using $E_p \times I_p = 2200 \times I_s$

$\therefore$ $200 = 2200 \times I_s$

$\therefore$ $I_s = \dfrac{2}{22} = 0.90909$ **Ans.**

(vi) Output power $= E_s \times I_s$

$= \dfrac{2200 \times 0.0909}{100} = 200$ W.

Ans.

(vii) Yes. It is equal, because it is an ideal transformer and for an ideal transformer,

$$\text{Efficiency} = \eta = \frac{\text{Output power}}{\text{Input power}} = 1 \text{ or } 100\%$$

i.e., there is no energy loss due to any cause.

(viii) $I_s = 0.0909$ A and $I_p = 0.909$ A

Less current flows through the secondary coil than in the primary coil because it is a step-up transformer.

Chapter 10. Specific Heat Capacity and Latent Heat

1. A piece of ice is heated at a constant rate. The variation of temperature with heat input is shown in the graph below:

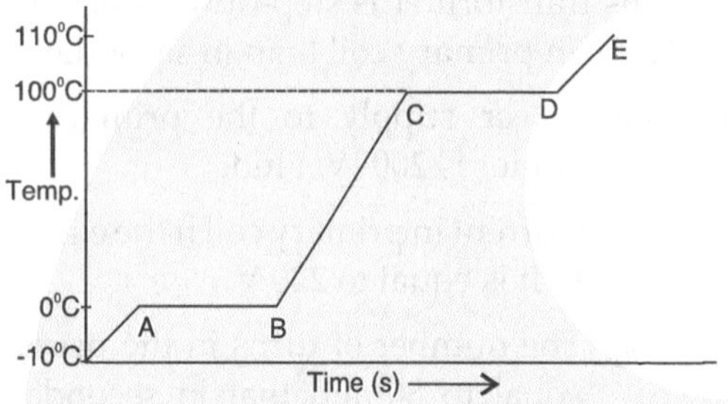

(i) What is being represented by AB and CD?

(ii) What conclusion can you draw regarding the nature of ice from the above graph?

Ans. (i) AB part of the graph represents time in seconds taken by ice at 0°C to convert into water at 0°C.

CD part of the graph represents time in seconds taken by water at 100°C to convert into steam at 100°C.

(ii) From the graph, we conclude that ice takes less time in heating from −10°C to 0°C. But in the process of melting, it takes comparatively longer time. It shows that ice has a high specific latent heat of fusion.

2. What observation will you record and how will you determine the specific latent heat of fusion of ice?

Ans.

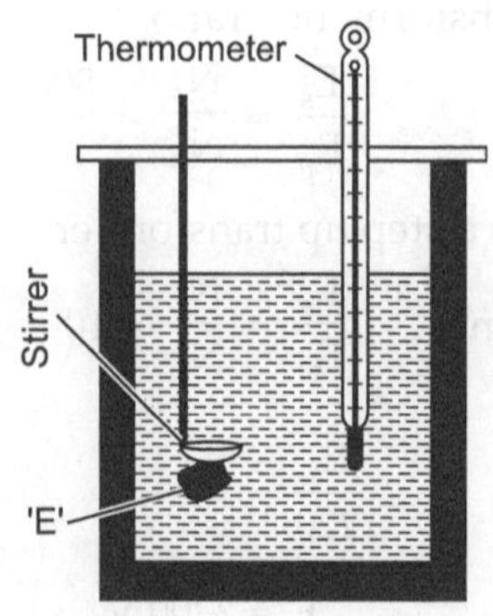

A clean and dry copper calorimeter with a stirrer, is weighed when empty, again weighed when filled nearly with water. Then initial temperature is noted. Some pieces of dry ice are transferred into the calorimeter. The calorimeter is placed inside the box and is continuously stirred till the whole of ice melts. The final temperature is carefully noted. Then the mass of calorimeter with water and water from ice is again weighed. The observations recorded are :

Mass of empty calorimeter with stirrer $= m_1$ g

Mass of calorimeter + water $= m_2$ g

Initial temperature of water $= t_1$°C

Final temperature of mixture $= t_2$°C

Mass of calorimeter + Water + Ice (in the form of water) $= m_3$ g

Temperature of ice = 0°C

If L J/g be the specific latent heat of fusion of ice and s_c be the specific heat capacity of copper, then :

Mass of ice melted $= (m_3 - m_2)$ g

Heat required by ice water at 0°C to rise to t_2°C

$$= (m_3 - m_2) \times 4.2 \times (t_2 - 0)$$

Heat lost by calorimeter

$$= m_1 \times s_c \times (t_1 - t_2)$$

Heat lost by water $= (m_2 - m_1) \times 4.2 \times (t_1 - t_2)$

From the heat equation,

Heat gained = Heat lost

$$(m_3 - m_2)\,L + (m_3 - m_2) \times 4.2 \times t_2 = m_1 \times s_c \times (t_1 - t_2) + (m_2 - m_1) \times 4{\cdot}2 \times (t_1 - t_2)$$

Therefore,

$$L = \left[\frac{(m_1 s_c)(t_1 - t_2) + (m_2 - m_1) \times 4.2 \times (t_1 - t_2) - (m_3 - m_2) \times 4.2 \times t_2}{(m_3 - m_2)}\right]$$

Chapter 11. Radioactivity and Nuclear Energy

1. Radioactive radiations were allowed to fall on a photographic plate, placed in an electric field for few seconds. The photographic plate was developed and three distinct spots (A, B and C) were visible.

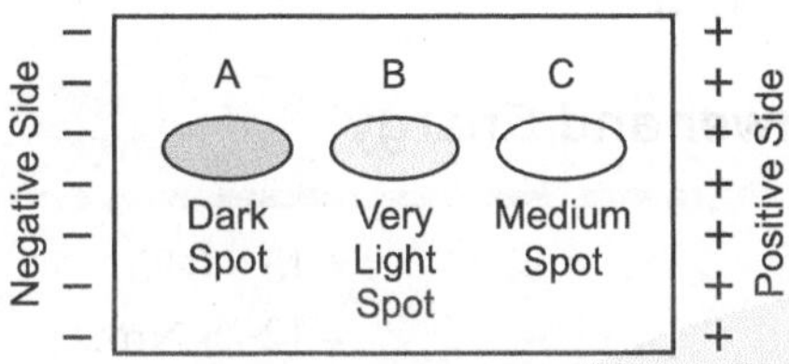

Developed Photographic Plate

Identify the radiations, responsible for each spot with reasons.

Ans. Spot A is on the negative side of the plate, hence it is due to α-radiations.

Spot B is in the middle (*i.e.*, undeviated), hence it is due to γ-radiations.

Spot C is on the positive side of the plate, hence it is due to β-radiations.

2. The diagram shows a thick leads cube having a cavity in the middle. In the cavity, is placed some radioactive substance. Copy the diagram and trace the paths of α-particles, β-particles and γ-radiations as they pass through a powerful electric field.

Ans.

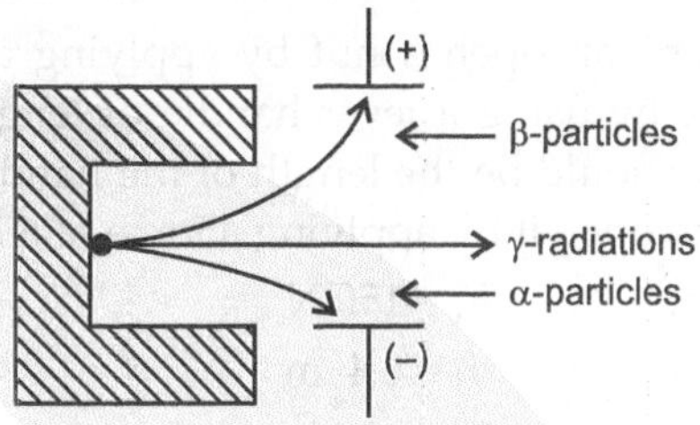

Short Numericals

Set 5

Chapter 1. Force, Work, Power and Energy

1. A man can open a nut by applying a force of 150 N by using a lever handle of length 0·4 m. What should be the length of the handle if he is able to open it by applying a force of 60 N?

Ans. Given, $F_1 = 150$ N

$l_1 = 0{\cdot}4$ m

$F_2 = 60$ N

$l_2 = ?$

$\because$ Force × Perpendicular distance

= Torque required to open the nut in both cases.

$$F_1 \times l_1 = F_2 \times l_2$$

$$150 \times 0{\cdot}4 = 60 \times l_2$$

$$l_2 = 1 \text{ m}$$ **Ans.**

2. The perpendicular distance between the point of application of force and turning point is 1·75 m, when a force of 80 N acts on a rigid body. Calculate the moment of force.

Ans. Force (F) = 80 N

Perpendicular distance

= 1·75 m

$\therefore$ Moment of force = F × Perpendicular distance

= 80 × 1·75 = 140 Nm. **Ans.**

3. Calculate the minimum distance between point of application of force and axis of rotation of a rigid body, if a force 125 N produces a moment of force of 25·75 Nm.

Ans. Moment of force = Force × Perpendicular distance

25·75 = 125 × Perpendicular distance

Perpendicular distance

$$= \frac{25{\cdot}75}{125} = 0{\cdot}206 \text{ m.}$$ **Ans.**

4. A couple of 15 N force acts on a rigid body, such that arm of couple is 85 cm. Calculate the moment of couple in S.I. system.

Ans. Moment of couple = Force × Arm of couple

= 15 × 85 Ncm.

$$= 15 \times 85 \times \frac{1}{100} \text{ Nm.}$$

= 15 × 0·85

= 12·75 Nm. **Ans.**

5. Calculate the magnitude of the force of a couple, when arm of couple is 63 cm and moment of couple is 21 Nm.

Ans. Moment of couple = F × Arm of couple

21 Nm = F × 63 cm

21 Nm = F × 0·63 m

$$F = \frac{21}{0{\cdot}63} = 33{\cdot}33 \text{ N.}$$ **Ans.**

6. Calculate the length of arm of couple, if a force of 13 N produces a moment of couple of 14·3 Nm.

Ans. Moment of couple = Force × Arm of couple

14·3 = 13 × Arm of couple

$$\text{Arm of couple} = \frac{14{\cdot}3}{13}$$

= 1·1 m. **Ans.**

7. A uniform meter scale is in equilibrium as shown in the diagram :*

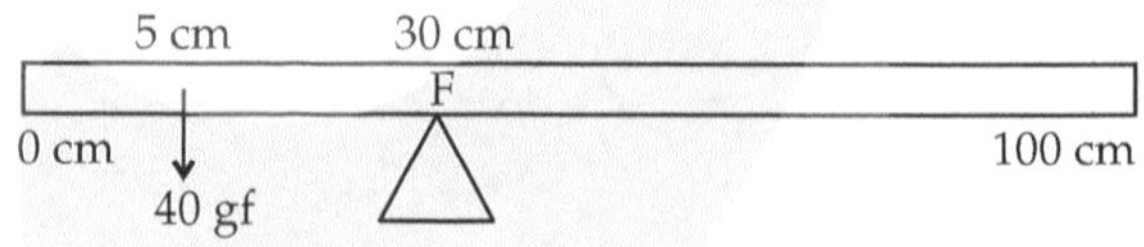

(i) Calculate the weight of the meter scale.

(ii) Which of the following options is correct to keep the ruler in equilibrium when 40 gf wt is shifted to 0 cm mark ?

F is shifted towards 0 cm.

Or

F is shifted towards 100 cm.

Ans. (i) Let the weight of the meter scale be x gf and it acts at the centre of gravity (*i.e.*, 50 cm mark)

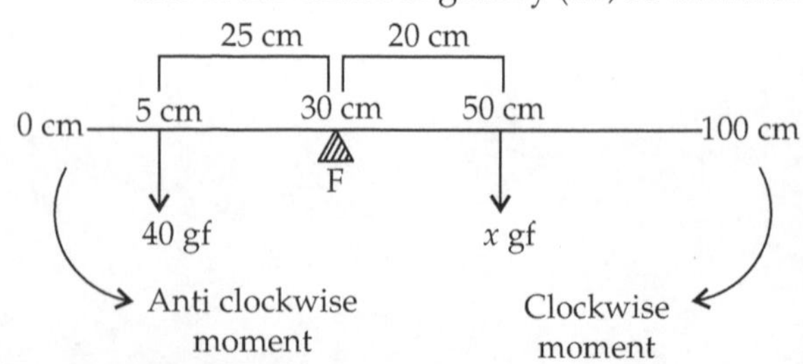

Anticlockwise moment = 40 ×25 gf cm

* **Frequently asked previous years Board Exam Questions.**

Clockwise moment = $x \times 20$ gf cm

when the meter scale is balanced,

Clockwise moment = Anticlockwise moment

$$\Rightarrow \quad x \times 20 = 40 \times 25$$

$$\Rightarrow \quad x = \frac{40 \times 25}{20} \text{ gf} = 50 \text{ gf}$$

∴ Weight of meter scale is 50 gf.

(ii) F is shifted towards 0 cm.

8. (i) With reference to the direction of action, how does a centripetal force differ from a centrifugal force during uniform circular motion ?

(ii) Is centrifugal force the force of reaction of centripetal force ?

(iii) Compare the magnitudes of centripetal and centrifugal force.*

Ans. (i) Centripetal force acts in a direction towards the centre of circular path whereas centrifugal force acts in a direction away from the centre of circular path.

(ii) No, centrifugal force is not the force of reaction of centripetal force because action and reaction do not act on the same body.

(iii) Magnitudes of centripetal and centrifugal forces are in the ratio 1 : 1.

9. An energy of 4000 J causes a displacement of 80 m in 4 s. Calculate (i) Force (ii) Power.

Ans. (i) Energy = Work = 4000 J

$$W = F \times S$$

$$4000 = F \times 80$$

$$\therefore \quad F = \frac{4000}{80} \text{ N} = 50 \text{ N} \qquad \textbf{Ans.}$$

(ii) $$\text{Power} = \frac{\text{Work}}{\text{Time}} = \frac{4000}{4} \text{ W} = 1000 \text{ W}$$

Ans.

10. Calculate the work done when:

(i) A 5 kg weight is lifted 10 m vertically. ($g = 9.8 \text{ m/s}^2$)

(ii) A car is moved on a rough road through 30 m against a frictional resistance of 75 N.

Ans. (i) Force = $m \times g = 5 \times 9.8 = 49$ N; Distance = 10 m

∴ Work done = F × S = 49 × 10 = 490 J. **Ans.**

(ii) Resisting force = 75 N; Distance moved = 30 m

∴ Work done = 75 × 30 J = 2250 J. **Ans.**

11. A coolie carrying a load on his head and moving on a frictionless horizontal platform does no work. Explain the reason why.

Ans. We know $W = F S \cos\theta$.

Since force is normal to the displacement.

So, $\theta = 90°$

Hence, Work done,

$$W = FS \cos\theta = FS \cos 90° = 0. \qquad \textbf{Ans.}$$

12. A force is applied on a body of mass 20 kg moving with a velocity of 40 ms^{-1}. The body attains a velocity of 50 ms^{-1} in 2 seconds. Calculate the work done by the body.

Ans. $$\text{Work done, } W = \frac{1}{2} m (v^2 - u^2) = \frac{1}{2} \times 20 \times (50^2 - 40^2) = 9000 \text{ J}$$

Ans.

13. How long should an electric motor of power 2 H.P. operate, so as to pump 5 m^3 of water from a depth of 15 m. [Take $g = 10 \text{ N kg}^{-1}$]

Ans. Mass of water = Volume × Density = 5 × 1000 = 5000 kg.

Height = 15 m, t = ?, $g = 10 \text{ N kg}^{-1}$

P = 2 H.P. = 2 × 750 = 1500 W

Work done in operating electric motor

$= P \times t = 1500 \times t$

Work done in raising water

$= mgh = 5000 \times 10 \times 15 = 750000$ J

∴ $1500 \times t = 750000$

$t = 500$ second. **Ans.**

14. State the C.G.S. unit of work. How it is related to its S.I. unit?

Ans. The C.G.S. unit of work is erg.

1 erg = 1 dyne × 1 cm.

The S.I. unit of work is joule.

1 joule = 1 newton × 1 metre

Relationship: 1 J = 1 N × 1 m

$= 10^5$ dyne × 100 cm = 10^7 erg.

Ans.

15. If the power of a motor is 40 kW, at what speed can it raise a load of 20,000 N?

Ans. Given : Power = 40 kW = 40,000 W; F = 20,000 N

∴ Power = Force × Average speed

or 40,000 = 20,000 × Average speed

$$\therefore \quad \text{Average speed} = \frac{40{,}000}{20{,}000} = 2 \text{ m/s.} \qquad \textbf{Ans.}$$

16. (i) The world record for weight lifting was held by Sergei Didyk of USSR. He lifted 261 kg to a height of 2.3 m in 4 s; find: (a) weight lifted by him, (b) work done by him, (c) power developed by him. ($g = 10 \text{ m/s}^2$)

* Frequently asked previous years Board Exam Questions.

(ii) The work done by the heart is 1 joule per beat. Calculate the power of the heart if it beats 72 times in one minute.

Ans. (i) (a) Weight lifted by him

$= m \times g = 261 \times 10 = 2610$ N. **Ans.**

(b) Work done by him

$= 2610 \times 2{\cdot}3 = 6003$ J. **Ans.**

(c) Power developed by him

$$= \frac{\text{Work (in J)}}{\text{Time (in s)}} = \frac{6003}{4} \text{ W}$$

$= 1500.75$ W. **Ans.**

(ii) Total work done by the heart in one minute (*i.e.*, 60 s) is 72 beats or 72 joule.

∴ Power of heart

$$= \frac{\text{W (in joules)}}{\text{Time (in s)}} = \frac{72}{60} = 1{\cdot}2 \text{ Watt.}$$

Ans.

17. A machine raises a load of 750 N through a height of 16 m in 5 seconds. Calculate the power at which the machine works.

Ans. We have, $\text{Power} = \frac{\text{Force} \times \text{Distance}}{\text{Time}}$

$$= \frac{750 \times 16}{5} = 2400 \text{ W}$$

$= 2{\cdot}4$ kW. **Ans.**

18. Calculate the change in the kinetic energy of a moving body if its velocity is reduced to 1/3rd of the initial velocity.

Ans.

$$\text{K.E}_1 = \frac{1}{2} mv^2$$

$$\text{K.E}_2 = \frac{1}{2} m\left(\frac{1}{3} v\right)^2$$

$$\frac{\text{K.E.}_1}{\text{K.E.}_2} = \frac{\frac{1}{2} mv^2}{\frac{1}{2} m\left(\frac{1}{3} v\right)^2}$$

$$= \frac{1}{2} mv^2 \times \frac{2 \times 9}{mv^2}$$

$$= \frac{9}{1}$$

$$\text{K.E.}_2 = \frac{1}{9} \text{K.E}_1$$

So, K.E. becomes $\frac{1}{9}$th of its initial K.E. **Ans.**

19. A girl of mass 35 kg climbs up from the first floor of a building at a height 4 m above the ground to the third floor at a height 12 m above the ground. What will be the increase in her gravitational potential energy? [$g = 10 \text{ ms}^{-2}$]

Ans. Increase in gravitational potential energy

$= mg\,(h_2 - h_1)$

$= 35 \times 10 \times (12 - 4) = 350 \times 8 = 2{,}800$ J

Ans.

20. A man having a box on his head, climbs up a slope and another man having an identical box walks the same distance on a levelled road. Who does more work against the force of gravity and why?

Ans. Man having a box on his head who climbs up a slope does more work against the force of gravity because he has more potential energy by virtue of his position *i.e.*, height.

As, P. E. = Work done = F × S

$= mg \times h$

21. A body of mass 0.2 kg falls from a height of 10 m to a height of 6 m above the ground. Find the loss in potential energy taking place in the body. [$g = 10 \text{ ms}^{-2}$]

Ans. Given, Mass = 0·2 kg

Height, h = 10 m to 6 m

Loss in potential energy

$= mg\,(h_1 - h_2)$

$= 0{\cdot}2 \times 10 \times (10 - 6) = 8$ J **Ans.**

22. A ball of mass 200 g falls from a height of 5 m. What will be its kinetic energy when it just reaches the ground? [$g = 9{\cdot}8 \text{ ms}^{-2}$]

Ans. Given, $m = 200$ g

$h = 5$ m

When the ball reaches the ground then

P.E. = K.E.

$$\therefore \quad \text{K.E.} = mgh = \frac{200}{1000} \times 9{\cdot}8 \times 5 = 9{\cdot}8 \text{ joule}$$

Ans.

23. Two bodies, A and B of equal mass are kept at heights 20 m and 30 m respectively. Calculate the ratio of their potential energies.

Ans. Let the equal mass of two bodies A and B be m.

Given: height, $h_1 = 20$ m

and $h_2 = 30$ m

Potential energy of body A

$= mgh_1$

$= m \times g \times 20$

Potential energy of body B

$= mgh_2 = m \times g \times 30$

$$\frac{U_A}{U_B} = \frac{m \times g \times 20}{m \times g \times 30} = \frac{2}{3}$$ **Ans.**

24. A ball of mass 0·20 kg is thrown vertically upward with an initial velocity of 20 m/s. Calculate the maximum potential energy it gains as it goes up.

Short Numericals

Ans. Maximum kinetic energy

$$= \frac{1}{2} mv^2 = \frac{1}{2} \times 0{\cdot}20 \times 20 \times 20 = 40 \text{ J}$$

According to the principle of conservation of energy,

Maximum potential energy

= Maximum kinetic energy = 40 J. **Ans.**

25. A body is thrown vertically upwards. Its velocity keeps on decreasing. What happens to its kinetic energy as its velocity becomes zero?

Ans. When $v = 0$ then, kinetic energy $\left(\frac{1}{2}mv^2\right)$ also becomes zero.

The kinetic energy completely changes to potential energy.

26. A moving body weighing 400 N possesses 500 J of kinetic energy. Calculate the velocity with which the body is moving. [$g = 10 \text{ ms}^{-2}$]

Ans. Given : Weight = 400 N,

$$W = mg$$

$$400 = m \times 10$$

$$m = 40 \text{ kg}$$

Now, Kinetic Energy

$$= \frac{1}{2} mv^2$$

$$500 = \frac{1}{2} \times 40 \times v^2$$

$$v = 5 \text{ m/s}$$ **Ans.**

Chapter 2. Simple Machines

1. The length of a nut-cracker is 12 cm. A nut when kept at a distance of 4 cm from its fulcrum requires an effort of 100 gf to crack it. What force will be required to crack the nut without using the nut-cracker?

Ans. Given : Effort arm = 12 cm, Load arm = 4 cm, Effort (E) = 100 gf, Load (L) = ?

(Nut cracker is the lever of second order)

Since, $$\frac{\text{Load}}{\text{Effort}} = \frac{\text{Effort arm}}{\text{Load arm}}$$

$$\therefore \quad \frac{L}{100} = \frac{12}{4}$$

or $$L = 300 \text{ gf.}$$

Thus, a force of 300 gf is required to crack the nut without using the nut cracker. **Ans.**

2. A crow bar of length 100 cm is used to lift a load of 5 kgf. It has its fulcrum at a distance of 20 cm from the load. Calculate: (i) the mechanical advantage of crow bar, and (ii) the effort applied at the other end.

Ans. Given: L = 5 kgf, Load arm = 20 cm, Effort arm = 100 – 20 = 80 cm.

(Since the crow bar is the lever of first order with fulcrum in between the load and the effort).

(i) Mechanical advantage

$$= \frac{\text{Effort arm}}{\text{Load arm}} = \frac{80}{20} = 4$$

(ii) $$\text{M.A.} = \frac{\text{Load (L)}}{\text{Effort (E)}}$$

$$\therefore \quad \text{Effort (E)} = \frac{\text{Load (L)}}{\text{M.A.}} = \frac{5}{4}$$

$$= 1{\cdot}25 \text{ kgf.}$$

3. In a single movable pulley system, a load of 125 kgf is lifted by an effort of 75 kgf. Find the percentage efficiency of system.

Ans. V.R. of single movable pulley

= Number of supporting segments of string = 2

M.A. of single movable pulley

$$= \frac{L}{E} = \frac{125}{75} = \frac{5}{3}$$

$$\% \text{ efficiency} = \frac{\text{M.A.}}{\text{V.R.}} \times 100$$

$$= \frac{5 \times 100}{3 \times 2} = 83{\cdot}3\%.$$

4. A woman draws water from a well using a fixed pulley. The mass of the bucket and water together is 6·0 kg. The force applied by the woman is 70 N. Calculate the mechanical advantage. [Take $g = 10 \text{ m/s}^2$]

Ans. $$\text{Load} = 6 \times 10 \text{ N}$$

$$= 60 \text{ N}$$

$$\text{M.A.} = \frac{\text{Load}}{\text{Effort}}$$

$$= \frac{6 \times 10}{70} = \frac{60}{70} = \frac{6}{7} = 0{\cdot}85$$ **Ans.**

5. A man raises a load of 360 kgf using a block and tackle system of 3 pulleys. If the efficiency of the pulley system is 60%, what effort is applied by the man?

Ans. Given: L = 360 kgf, E = ?, Efficiency = 60% = 0.6

Velocity ratio =Number of pulleys in the system = 3.

Mechanical advantage (M.A.)

$= \text{V.R.} \times \text{Efficiency} = 3 \times 0{\cdot}6 = 1{\cdot}8$

But, $\quad \text{M.A.} = \dfrac{\text{Load (L)}}{\text{Effort (E)}}$

$\therefore \quad \text{Effort (E)} = \dfrac{\text{Load (L)}}{\text{M.A.}}$

$= \dfrac{360}{1.8} = 200 \text{ kgf.}$ **Ans.**

Chapter 3. Refraction of Light

1. A diver in water looks obliquely at an object AB in air.

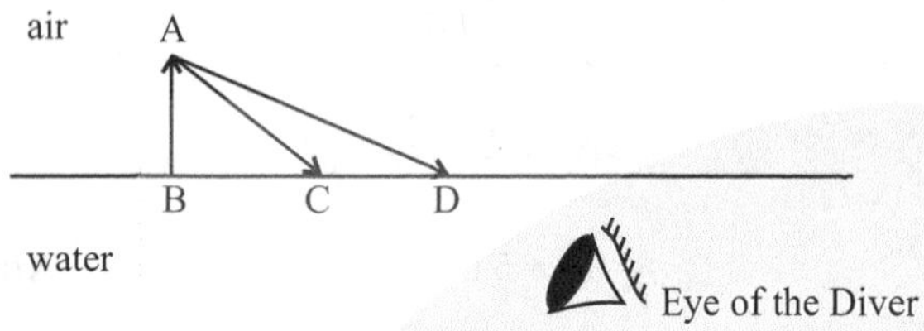

(i) Does the object appear taller, shorter or of the same size to the diver ?

(ii) Show the path of two rays AC and AD starting from the tip of the object as it travels towards the diver in water and hence obtain the image of the object.*

Ans. (i) The object will appear taller.

(ii) A′B is the image formed of the object AB.

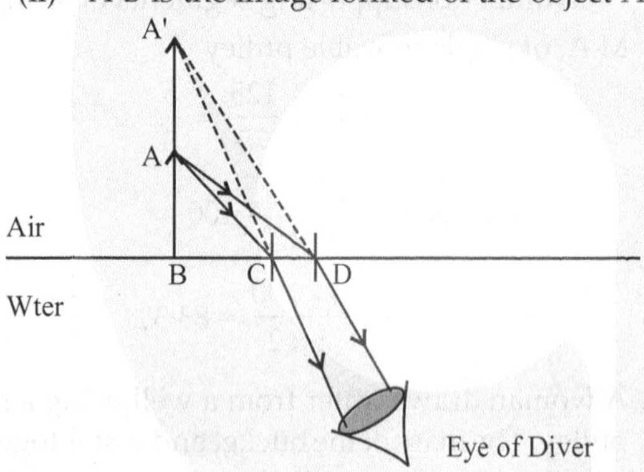

2. Complete the path of the ray AB through the glass prism in PQR till it emerges out of the prism. Given the critical angle of the glass as 42°.*

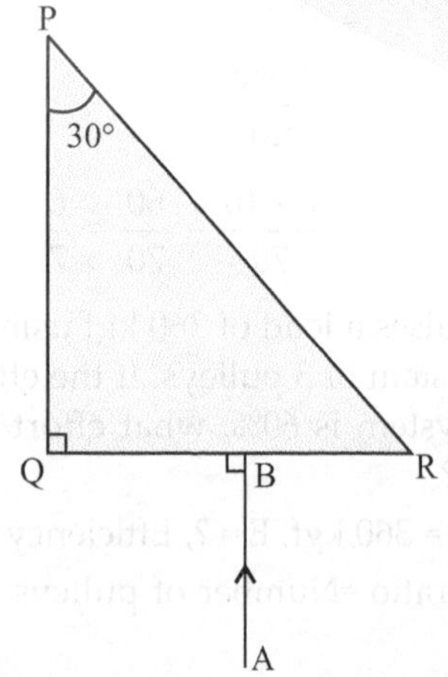

Ans. The path of the ray till it emerges out of the prism is ABCDE.

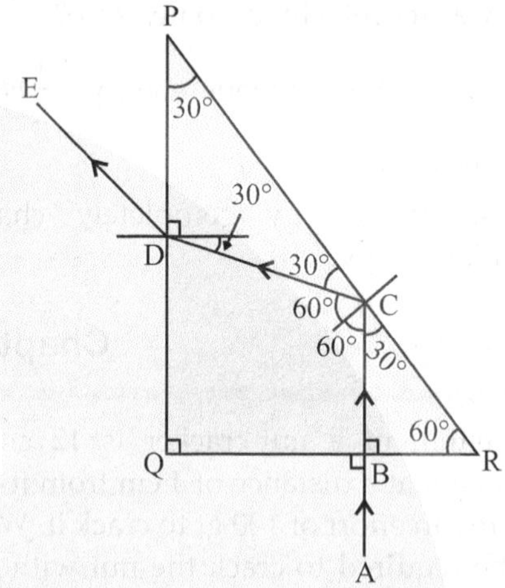

3. How does the angle of deviation formed by a prism change with the increase in the angle of incidence ?

Draw a graph showing the variation in the angle of deviation with the angle of incidence at a prism surface.*

Ans. Experimentally it has been observed that as the angle of incidence increases, the angle of deviation first decreases, reaches to a minimum value for a certain angle of incidence and then an further increasing the angle of incidence, the angle of deviation begins to increase.

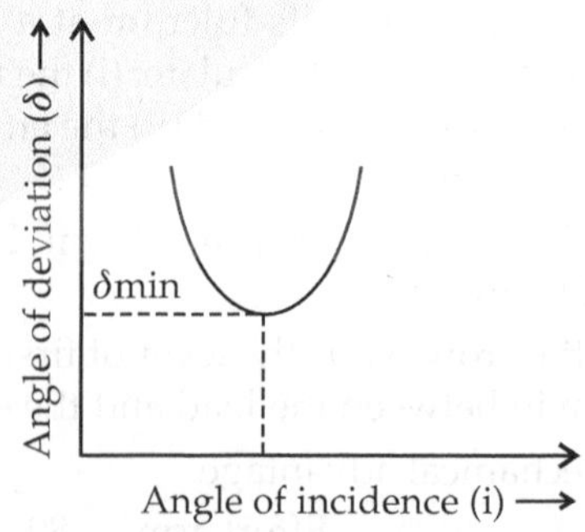

4. The speed of light in air is 3×10^8 ms^{-1}. Calculate the speed of light in water. The refractive index of water is 4/3.

Ans. Refractive index of water

$$= \frac{\text{Speed of light in air}}{\text{Speed of light in water}}$$

* **Frequently asked previous years Board Exam Questions.**

$\therefore$ Speed of light in water

$$= \frac{\text{Speed of light in air}}{\text{Refractive index of water}}$$

$$= \frac{3 \times 10^8}{4/3} = 2.25 \times 10^8 \text{ ms}^{-1}. \quad \textbf{Ans.}$$

5. The velocity of light in diamond is 121,000 kms^{-1}. What is its refractive index?

Ans. Speed of light in diamond

$$= 121{,}000 \text{ km/s}$$

$$= 121 \times 10^6 \text{ m/s}$$

Refractive index of diamond

$$= \frac{\text{Speed of light in air}}{\text{Speed of light in diamond}}$$

$$= \frac{3 \times 10^8}{121 \times 10^6} = 2{\cdot}48 \quad \textbf{Ans.}$$

6. Write the approximate values of speed of light in (i) air and (ii) glass. Use these values to calculate the refractive index of glass with respect to air.

Ans. (i) Speed of light in air = 3×10^8 m/s

(ii) Speed of light in glass = 2×10^8 m/s

Refractive index of glass with respect to air

$$= \frac{\text{Speed of light in air}}{\text{Speed of light in glass}}$$

$$= \frac{3 \times 10^8}{2 \times 10^8} = \frac{3}{2} = 1.5 \quad \textbf{Ans.}$$

7. The refractive index of glass with respect to air is 3/2. What is the refractive index of air with respect to glass?

Ans. Given: $\quad {}_a\mu_g = \frac{3}{2}$

$${}_g\mu_a = \frac{1}{{}_a\mu_g}$$

$$= \frac{1}{3/2} = \frac{2}{3} = 0.67 \quad \textbf{Ans.}$$

8. A glass block is having a refractive index of 3/2, light ray is incident at an angle of 45°. Find the sine of angle of refraction inside the glass block.

Ans. According to snell's law of refraction:

$$\mu = \frac{\sin i}{\sin r}$$

$$\frac{3}{2} = \frac{\sin 45°}{\sin r}$$

$$\Rightarrow \quad \frac{3}{2} = \frac{1/\sqrt{2}}{\sin r}$$

$$\sin r = \frac{2}{3\sqrt{2}} = \frac{\sqrt{2}}{3} \quad \textbf{Ans.}$$

9. A coin placed at the bottom of a beaker appears to be raised by 4·0 cm. If the refractive index of water is 4/3, find the depth of the water in the beaker.

Ans. Let $\quad$ Real depth = x

Refractive index,

$$\mu = \frac{\text{Real depth}}{\text{Apparent depth}}$$

$$\Rightarrow \quad \frac{4}{3} = \frac{x}{x-4}$$

$$\Rightarrow \quad 4x - 16 = 3x$$

$$\therefore \quad x = 16 \text{ cm.} \quad \textbf{Ans.}$$

10. A ray of light is incident on a glass surface at an angle of 50° with the corresponding angle of refraction of 30°. Find the value of the refractive index of glass.

Ans. Refractive index = $\frac{\sin i}{\sin r}$

$${}_a\mu_g = \frac{\sin i}{\sin r} = \frac{\sin 50°}{\sin 30°}$$

$$(\because \angle i = 50° \text{ and } \angle r = 30°)$$

$$= \frac{0.7660}{0.5000}$$

$${}_a\mu_g = 1.532. \quad \textbf{Ans.}$$

11. (i) The refractive index of glass with respect to air is 1·5. What is the value of the refractive index of air with respect to glass?

(ii) A ray of light is incident as a normal ray on the surface of separation of two different mediums. What is the value of the angle of incidence in this case?

Ans. (i) $\quad {}_a\mu_g = 1{\cdot}5$

$${}_g\mu_a = \frac{1}{{}_a\mu_g}$$

$$= \frac{1}{1{\cdot}5} = 0{\cdot}666$$

$$= 0{\cdot}67 \quad \textbf{Ans.}$$

(ii) Value of the angle of incidence,

$$\angle i = 0° \quad \textbf{Ans.}$$

Chapter 4. Refraction through lenses and Optical Instruments

1. A convex lens is of focal length 20 cm. Find its power.

Ans. Given : Focal length, f = 20 cm = 0.2 m.

$$\text{Power} = \frac{1}{\text{Focal length (in m)}}$$

$$= \frac{1}{0.2} = 5\text{ D}$$ **Ans.**

2. If lens is of focal length 25 cm. Calculate the power of the lens.

Ans. Focal length = 25 cm = 0.25 m

$$P = \frac{1}{0.25} = 4\text{ D}$$ **Ans.**

Chapter 5. Spectrum

1. What is the wavelength of the wave whose frequency is 10^{12} Hz? Name the electromagnetic wave.

Ans. Frequency of the wave = 10^{12} Hz.

Velocity of electromagnetic wave

$= 3 \times 10^8$ m/sec.

$\therefore$ Wavelength of the wave $= \frac{v}{f} = \frac{3 \times 10^8}{10^{12}}$

$= 3 \times 10^{-4}$ m

$= 3 \times 10^{-2}$ cm. **Ans.**

These waves are heat waves or infrared waves whose wavelength varies from 10^{-2} to 10^{-4} cm.

Chapter 6. Echoes and Vibrations of Sound

1. A boy standing in front of a wall at a distance of 17 m produce 10 claps per second. He notices that the sound of his clapping coincides with the echo. Echo is heard only once when clapping is stopped. Calculate the speed of sound.

Ans. Let d be the distance of wall from the boy.

To hear the echo, sound has to travel a distance

$2d = 2 \times 17 = 34$ m.

Since 10 claps are produced in one second, therefore each clap is produced after 1/10 s which is equal to the time taken for the echo to be heard

i.e., $t = \frac{1}{10}$ s = 0.1 s

Now, speed of sound,

$$v = \frac{\text{Distance travelled}}{\text{Time taken}} = \frac{34}{0.1}$$

$= 340\text{ ms}^{-1}$. **Ans.**

2. An observer standing in front of a vertical wall claps 10 times in one second. He adjusts his distance from the wall in such a way that his second clapping coincides with the echo. This happens when his distance from the wall is 17.5 m. What is the velocity of sound heard by him?

Ans. Given: $d = 17.5$ m

An observer claps 10 times in one second, *i.e.,* he claps at an interval of 1/10th of a second or 0.1 s. As the sound of the echo coincides with the clap this means that the echo is heard after 0.1 s.

Therefore, $t = 0.1$ s

$$v = \frac{2d}{t} = \frac{2 \times 17{\cdot}5}{0.1} = 350\text{ m/s.}$$ **Ans.**

3. Two waves of the same pitch have their amplitudes in the ratio 2 : 3.

(i) What will be the ratio of their loudness?

(ii) What will be the ratio of their frequencies?

Ans. (i) Given, Amplitude ratio, $a_1 : a_2 = 2 : 3$

We know that,

$$I \propto a^2$$

$$\therefore \quad \frac{I_1}{I_2} = \frac{a_1^2}{a_2^2} = \frac{2^2}{3^2} = \frac{4}{9}$$

$\Rightarrow$ $I_1 : I_2 = 4 : 9$

(ii) We know that, Frequency $\propto$ Pitch

$\therefore$ Ratio of frequencies = 1 : 1.

4. An observer stands at a certain distance away from a cliff and produces a loud sound. He hears the echo of the sound after 1·8 s. Calculate the distance between the cliff and the observer if the velocity of sound in air is 340 ms^{-1}.

Ans. We know, $v = \frac{2d}{t}$

$\Rightarrow$ $d = \frac{vt}{2}$

$\Rightarrow$ $d = 340 \times \frac{1{\cdot}8}{2}$

$\therefore$ $d = 306$ m. **Ans.**

5. A man standing between two cliffs produces a sound and hears two successive echoes at intervals of 3 s and 4 s respectively. Calculate the distance between the two cliffs. The speed of the sound in the air is 330 ms^{-1}.

Ans. The first echo is heard from the nearest cliff, so let its distance be d_1.

$$\therefore \quad d_1 = \frac{v \times t}{2} = 330 \times \frac{3}{2} = 495 \text{ m}$$

The second echo is heard from farther cliff, so let its distance be d_2.

$$\therefore \quad d_2 = \frac{v \times t}{2} = 330 \times \frac{4}{2} = 660 \text{ m}$$

$\therefore$ Total distance = 660 + 495 = 1155 m. **Ans.**

6. An instrument is able to detect the reflected waves from an enemy aeroplane, after a time interval of 0.02 milliseconds. If the velocity of the waves is 3×10^8 ms^{-1}, calculate the distance of the plane from the radar.

Ans. The distance of the plane from the radar is

$$\text{Distance } (d) = \frac{v \times t}{2}$$

$$= \frac{3 \times 10^8 \times 0{\cdot}02 \times 10^{-3}}{2} \text{ m}$$

$$= 3 \times 10^3 \text{ m}$$

$$= 3 \text{ km.}$$ **Ans.**

7. The smoke from the gun barrel is seen 2 second before the explosion is heard. If the speed of sound in air is 340 ms^{-1}, calculate the distance of the observer from the gun. State the approximation used.

Ans. The light has a much larger speed (= 3×10^8 ms^{-1}) in comparison to the sound (= 340 ms^{-1}), therefore we can assume that the light takes negligible time and the sound takes 2 second to reach the observer.

$$\text{Now,} \quad \text{Speed} = \frac{\text{Distance travelled}}{\text{Time taken}}$$

$$\therefore \quad \text{Distance} = \text{Speed} \times \text{Time}$$

$$= 340 \times 2 = 680 \text{ m.}$$ **Ans.**

8. A person fires a gun while standing at a distance of 850 m from a building. If the speed of sound is 350 ms^{-1}, calculate the time in which the person hears an echo of the gun fire.

Ans. Speed of sound in air = 350 ms^{-1}

$\therefore$ Time in which echo is heard,

$$t = \frac{2d}{v} = \frac{2 \times 850}{350} = 4{\cdot}86 \text{ s.}$$ **Ans.**

9. A radio signal sent towards the surface of the moon is received in 2.5 seconds. If the speed of the radio waves is 3×10^8 ms^{-1}, find the distance of the moon from the earth.

Ans. Let 'd' be the required distance. The radio waves then travel a total distance $2d$ (to and from moon) in 2.5 second.

$$\text{Hence,} \quad \frac{2d}{2.5} = \text{speed} = 3 \times 10^8$$

$$d = 3 \times 10^8 \times \frac{2.5}{2} \text{ m}$$

$$= 3.75 \times 10^8 \text{ m.}$$ **Ans.**

10. A sound made on the surface of a lake takes 3 s to reach a boatman. How much time will it take to reach a diver inside the water at the same depth?

Velocity of sound in air = 330 ms^{-1}

Velocity of sound in water = 1450 ms^{-1}

Ans. We have, $$v = \frac{d}{t}$$

$$\Rightarrow \quad 330 = \frac{d}{3}$$

$$\Rightarrow \quad d = 990 \text{ m}$$

Now for the diver inside the water,

$$v' = \frac{d}{t'}$$

$$t' = \frac{990}{1,450}$$

$$\therefore \quad t' = 0{\cdot}682 \text{ s.}$$ **Ans.**

11. A boy stands 66.4 m in front of a high wall and then blows a whistle. Calculate the time interval when he hears an echo.

[speed of sound = 332 m/s].

Ans. Given: d = 66.4 m

v = 332 m/s

For an echo the distance travelled by the sound is $2d$.

$$\therefore \quad \text{Time} = \frac{2d}{v} = \frac{2 \times 66.4}{332} = 0.4 \text{ second.}$$ **Ans.**

12. A ship which is stationary at a distance of 680 m from the shore, sends a signal to the coast. Its echo is heard in 4 s. Find the velocity of sound.

Ans. Given: d = 680 m

t = 4 s

Sound takes 4 s to travel to and fro *i.e.*, in 4 s the distance travelled is 2×680 m.

$$\therefore \quad v = \frac{2d}{t} = \frac{2 \times 680}{4} = 340 \text{ m/s.}$$ **Ans.**

13. An observer stands at a distance of 850 m from a cliff and fires a gun. After what time gap will he hear the echo? If sound travels at a speed of 350 m/s in air.

Ans. Given: d = 850 m, t = ?, speed, v = 350 m/s

$$\therefore \quad t = \frac{2d}{v}$$

$$\Rightarrow \quad t = \frac{2 \times 850}{350}$$

$$\Rightarrow \quad t = \frac{2 \times 17}{7} = 4{\cdot}86 \text{ s.}$$ **Ans.**

14. The wavelength and frequency of sound wave in a certain medium is 40 cm and 835 Hz respectively. In the same medium if another wave has wavelength equal to 32 cm, calculate its frequency.

Ans. Given: λ = 40 cm = 0.40 m, f = 835 Hz

$$v = f \times \lambda = 835 \times 0.40 = 334 \text{ ms}^{-1}$$

In the second case, since medium is same so the velocity of sound will remain the same.

Now, $v = 334 \text{ ms}^{-1}$, λ = 32 cm = 0.32 m

From $\quad v = f\lambda$

$$\therefore \quad f = \frac{v}{\lambda} = \frac{334}{0.32} = 1043.75 \text{ Hz.}$$ **Ans.**

15. A pendulum has a frequency of 5 vibration per second. An observer starts the pendulum and fires simultaneously and he hears the echo from a cliff after 8 vibrations of the pendulum. If the velocity of sound in air is 340 m/sec, what is the distance between the cliff and the observer?

Ans. Given:

Frequency of pendulum = 5 vibration/s

$$\therefore \quad \text{Time taken} = \frac{8}{5} = 1{\cdot}6 \text{ s}$$

and $\quad$ Velocity of sound = 340 m/s

$$\text{But} \quad d = \frac{v \times t}{2} = \frac{340 \times 1{\cdot}6}{2}$$

$$\therefore \quad d = 272{\cdot}0 \text{ m/s}$$ **Ans.**

16. A radar sends a signal to an aircraft at a distance of 30 km away and receives it back after 2×10^{-4} second. What is the speed of the signal?

Ans. We know that,

$$\text{Speed, } v = \frac{2d}{t}$$

$$\Rightarrow \quad v = \frac{2 \times 30 \times 10^3}{2 \times 10^{-4}}$$

$$\Rightarrow \quad v = \frac{30 \times 10^3}{10^{-4}}$$

$$\Rightarrow \quad v = 3 \times 10^4 \times 10^4$$

$\therefore$ Speed of the signal,

$$v = 3 \times 10^8 \text{ m/s}$$ **Ans.**

17. A sound wave travelling in water has wavelength 0.4 m.

Is this wave audible in air? (The speed of sound in water = 1400 ms^{-1})*

Ans.

$$\lambda = 0.4 \text{ m}$$

$$V = 1400 \text{ ms}^{-1}$$

$$\because \quad V = f\lambda$$

$$\Rightarrow \quad f = \frac{V}{\lambda} = \frac{1400}{0.4} \text{ Hz}$$

$$= 3500 \text{ Hz}$$

$\because$ Frequency remains unchanged in air, the wave is audible in air because 3500 Hz falls in the audible range of frequency.

18. Two waves of the same pitch have amplitudes in the ratio 1 : 3. What will be the ratio of their:*

(i) intensities and

(ii) frequencies?

Ans. (i) Same pitch

$$\frac{A_1}{A_2} = \frac{1}{3}$$

Intensities, $I \propto A^2$

$$\frac{I_1}{I_2} = \left(\frac{A_1}{A_2}\right)^2 = \frac{1}{9}$$

(ii) 1:1, Pitch is same $\Rightarrow$ frequency is same.

19. A pendulum has a frequency of 4 vibrations per second. An observer starts the pendulum and fires a gun simultaneously. He hears the echo from the cliff after 6 vibrations of the pendulum. If the velocity of sound in air is 340 m/s, find the distance between the cliff and the observer.*

Ans. Time taken to complete 4 vibrations = 1 second

Time taken to complete 1 vibration = $\frac{1}{4}$ second

$\therefore$ Time taken to complete 6 vibrations

$$= \frac{1}{4} \times 6 \text{ second}$$

$$= 1.5 \text{ second}$$

Time (t) = 1.5 s

Velocity (v) = 340 m/s

$\therefore$ Distance between the cliff and observer

$$d = \frac{v \times t}{2}$$

$$= \frac{340 \times 1.5}{2} \text{ m}$$

$$= 255 \text{ m}$$

* Frequently asked previous years Board Exam Questions.

Short Numericals

Chapter 7. Electricity

1. A cell of e.m.f 2·0 V and internal resistance 1 Ω is connected to the resistors of 3 Ω and 6 Ω in series. Calculate: (i) the current drawn from the cell, (ii) the p.d. across each resistor, (iii) the terminal voltage of the cell and (iv) the voltage drop.

Ans. Total resistance of the circuit, R = 1 + 3 + 6 = 10 Ω

(i) The current drawn from the cell

$$= \frac{E}{R} = \frac{2·0}{10} = 0·2 \text{ A.}$$ **Ans.**

(ii) The p.d. across 3 Ω resistor,

$$V_1 = IR_1 = 0·2 \times 3 = 0·6 \text{ V}$$

The p.d. across 6 Ω resistor,

$$V_2 = IR_2 = 0·2 \times 6 = 1·2 \text{ V.}$$ **Ans.**

(iii) The terminal voltage of the cell,

$$V = V_1 + V_2 = 0·6 + 1·2 = 1·8 \text{ V.}$$ **Ans.**

(iv) The voltage drop

$$= E - V = 2·0 - 1·8 = 0·2 \text{ V.}$$ **Ans.**

2. A current of 100 mA flows through a wire. The charge on an electron is $1·6 \times 10^{-19}$ C. Find the number of electrons passing per second through the cross section of the conductor.

Ans. Given: i = 100 mA = 100×10^{-3} A = 0·1 A, $e = 1·6 \times 10^{-19}$ C

Let n electrons are passing per second.

Charge on one electron

$= e$

Total charge passed in one second

$= ne$

Thus, $i = ne$

or $$n = \frac{i}{e} = \frac{0·1}{1·6 \times 10^{-19}}$$

$$= 6·25 \times 10^{17}.$$ **Ans.**

3. A current of 0.2 A flows through a conducting wire for 5 minutes. How much charge will flow?

Ans. Given: I = 0·2 A, t = 5 minute = 5 × 60 s = 300 s.

Charge, Q = I × t = 0·2 × 300 = 60 C. **Ans.**

4. An ammeter placed in series with an electric radiator reads 0·5 A and a voltmeter placed across it reads 230 volts. What is the resistance of the radiator?

Ans. Using Ohm's law,

$$R = \frac{V}{I}$$

$$= \frac{230}{0·5} \quad \text{(Given V = 230 V, and I = 0·5 A)}$$

$$R = 460\ \Omega$$ **Ans.**

5. An electric bulb draws 0·5 A current at 3·0 V. Calculate the resistance of the filament of the bulb.

Ans. Given : I = 0·5 A, V = 3·0 V

Resistance of filament of the bulb,

$$R = \frac{V}{I} = \frac{3·0}{0·5} = 6\ \Omega.$$ **Ans.**

6. Two bulbs are marked 100 W, 220 V and 60 W, 110 V. Calculate the ratio of their resistances.

Ans. Given:

I Bulb	II Bulb
$P_1 = 100$ W	$P_2 = 60$ W
$V_1 = 220$ V	$V_2 = 110$ V.

We know, $$P = \frac{V^2}{R}$$

So, $$R = \frac{V^2}{P}$$

So, $$\frac{R_1}{R_2} = \frac{\frac{V_1^2}{P_1}}{\frac{V_2^2}{P_2}} = \frac{V_1^2 \times P_2}{P_1 \times V_2^2}$$

$$= \frac{220^2 \times 60}{100 \times 110 \times 110} = \frac{12}{5}$$

∴ $R_1 : R_2 = 12 : 5$ **Ans.**

7. Calculate the value of the resistance which must be connected to a 15 Ω resistance to provide an effective resistance of 6 Ω.

Ans. Resistance decreases in parallel combination. Let a resistance R be connected to 15 Ω resistance in parallel to make resultant 6 Ω.

Now, $$\frac{1}{6} = \frac{1}{15} + \frac{1}{R}$$

or $$\frac{1}{R} = \frac{1}{6} - \frac{1}{15}$$

$$= \frac{5-2}{30} = \frac{3}{30} = \frac{1}{10}$$

∴ R = 10 Ω **Ans.**

8. A metal wire of resistance 6 Ω is stretched so that its length is increased to twice its original length. Calculate its new resistance.

Ans. When the wire is stretched to twice its original length, its area of cross section becomes half.

∵ $R \propto l$ and $R \propto \frac{1}{a}$

∴ $R' = n^2R$

$= 2^2 \times 6$ (n = 2)

$= 24\ \Omega.$ **Ans.**

9. Two resistors having 2 Ω and 3 Ω resistance are connected (i) in series, and (ii) in parallel. Find the equivalent resistance in each case.

Ans. Given : $R_1 = 2\ \Omega$, $R_2 = 3\ \Omega$

(i) In series : Equivalent resistance,

$$R = R_1 + R_2$$

or $$R = 2 + 3 = 5\ \Omega$$

(ii) In parallel : Equivalent resistance,

$$R = \frac{R_1R_2}{R_1 + R_2}$$

or $$R = \frac{2 \times 3}{2+3} = \frac{6}{5} = 1.2\ \Omega.$$ **Ans.**

10. A wire of uniform thickness with a resistance of 27 Ω is cut into three equal pieces and they are joined in parallel. Find the resistance of the parallel combination.

Ans. Resistance of given wire = 27 Ω

∴ Resistance of each small wire = $\frac{27}{3} = 9\ \Omega$

Let R_P be the equivalent resistance for parallel combination.

Then $$\frac{1}{R_P} = \frac{1}{R_1} + \frac{1}{R_2} + \frac{1}{R_3}$$

$$= \frac{1}{9} + \frac{1}{9} + \frac{1}{9}$$

$$\frac{1}{R_P} = \frac{3}{9}$$

$$R_P = \frac{9}{3}$$

$$R_P = 3\ \Omega$$ **Ans.**

11. The resistance of two resistors joined in series is 8 Ω and in parallel is 1.5 Ω. Find the value of the two resistances.

Ans. In series, $R_1 + R_2 = 8\ \Omega$ ...(i)

In parallel,

$$\frac{R_1R_2}{R_1 + R_2} = 1.5\ \Omega \quad ...(ii)$$

∴ $R_1R_2 = 8 \times 1.5 = 12\ \Omega$

Now $(R_1 - R_2)^2 = (R_1 + R_2)^2 - 4R_1R_2$

∴ $(R_1 - R_2)^2 = (8)^2 - 4 \times 12$

or $(R_1 - R_2)^2 = 64 - 48 = 16$

or $R_1 - R_2 = 4\ \Omega$ ...(iii)

On solving equations (i) and (iii), we get

$R_1 = 6\ \Omega$ and $R_2 = 2\ \Omega$ **Ans.**

12. A bulb when connected to a 12 V battery draws 1·5 A current when glowing. What is the resistance of the filament of the bulb? Will the resistance be more, same or less when the bulb is not glowing?

Ans. Given : V = 12 V, I = 1·5 A

Using ohm's law;

$$V = IR$$

$$R = \frac{V}{I} = \frac{12}{1 \cdot 5}\ \Omega$$

$$= 8\ \Omega$$

Resistance will be less when the bulb is not glowing.

13. A cell supplies a current of 0.6 A through a 2 Ω coil and a current of 0.3 A through a 8 Ω coil. Calculate the e.m.f and internal resistance of the cell.

Ans. Given: $I_1 = 0.6$ A, $R_1 = 2\ \Omega$ and $I_2 = 0.3$ A, $R_2 = 8\ \Omega$

Let the e.m.f. of the cell is E and its internal resistance is r.

We have, $$I = \frac{E}{R + r}$$

$$0.6 = \frac{E}{2 + r} \text{ and } 0.3 = \frac{E}{8 + r}$$

∴ $E = 0.6\,(2 + r) = 1.2 + 0.6r$

and $E = 0.3\,(8 + r) = 2.4 + 0.3r$

Thus, $1.2 + 0.6r = 2.4 + 0.3r$

or $0.6r - 0.3r = 2.4 - 1.2$

or $0.3r = 1.2$

or $r = \frac{1.2}{0.3} = 4\ \Omega$. **Ans.**

and $E = 1.2 + 0.6 \times 4 = 1.2 + 2.4 = 3.6$ V. **Ans.**

14. A charge of 750 C flows through a conductor for 2 minutes 30 second. Find the magnitude of current flowing through the conductor.

Ans. Given:

Charge (Q) = 750 C

Time (t) = (2 × 60 + 30) s = 150 s

Current (I) = $\frac{Q}{t}$

$$= \frac{750}{150}\text{ A}$$

$$= 5\text{ A.}$$ **Ans.**

15. Two wires of same material and same lengths have radii in the ratio of 2 : 3. Compare their resistances.

Ans. Given : $r_1 : r_2 = 2 : 3$

Area of cross-section of wire,

$$a = \pi r^2$$

∴ $$a_1 : a_2 = r_1^2 : r_2^2 = 4 : 9$$

Resistance of wire,

$$R \propto \frac{1}{a}$$

$$\therefore \quad \frac{R_1}{R_2} = \frac{a_2}{a_1} = \frac{9}{4}. \quad \textbf{Ans.}$$

16. The resistance of a wire of length 100 cm and of uniform area of cross-section 0·030 cm^2, is found to be 1·5 Ω. Calculate the resistivity of the wire.

Ans. Given: l = 100 cm, a = 0·030 cm^2, R = 1·5 Ω

$$\text{Resistivity } (\rho) = \frac{Ra}{l}$$

$$= \frac{1{\cdot}5 \times 0{\cdot}030}{100}\ \Omega\ \text{cm}$$

$$= 0{\cdot}00045\ \Omega\ \text{cm}. \quad \textbf{Ans.}$$

17. What should be the length of a nickel wire of area of cross-section 3 mm^2 used for making a rheostat of 750 Ω? [r of nickel = 6·9 × 10^{-8} Ω m].

Ans. Given: Specific resistance,

$$\rho = 6{\cdot}9 \times 10^{-8}\ \Omega\text{m}$$

$$R = 750\ \Omega$$

$$A = 3\ \text{mm}^2 = 3 \times 10^{-6}\ \text{m}^2$$

As
$$l = \frac{R \times A}{\rho} = \frac{750 \times 3 \times 10^{-6}}{6{\cdot}9 \times 10^{-8}}$$

∴ Length, l = 32608.7 m

The length of a nickel wire should be 32608.7 m. **Ans.**

18. (i) What are superconductors?*

(ii) Calculate the current drawn by an appliance rated 110 W, 220 V when connected across 220 V supply.

(iii) Name a substance whose resistance decreases with the increase in temperature.

Ans. (i) Superconductors are substances of zero resistance at very temperatures.

(ii)
$$P = 110\ \text{W}$$
$$V = 220\ \text{volt}$$
$$\therefore \quad \text{Current I} = \frac{P}{V} = \frac{110}{220}\ \text{A}$$
$$= 0.5\ \text{A}$$

(iii) For semiconductors such as silicon, germanium, resistance decreases with the increase in temperature.

Chapter 8. Electric Power and Household Circuits

1. State (i) the S. I. unit, (ii) the household unit of electrical energy. How are these units related ?

Ans. (i) The S.I. unit of electrical energy is joule (J).

(ii) The household unit of electrical energy is kilo-watt hour (kWh).

Relationship:

$$1\ \text{kWh} = 1\ \text{kW} \times 1\ \text{h}$$
$$= 1000\ \text{W} \times 3600\ \text{s}$$
$$= 3.6 \times 10^6\ \text{J}.$$

2. An electric kettle is rated 3 kW, 250 V. What should be the current capacity of the fuse with it?

Ans. The maximum current capacity of kettle,

$$I = \frac{P}{V} = \frac{3000}{250} = 12\ \text{A}.$$

Thus, the fuse should be of current capacity slightly less than 12 A.

3. How can electrical energy consumed by an electric appliance be calculated in kilowatt hour (kWh)?

Ans. The electrical energy in kWh can be calculated as follows:

Energy in kWh

$$= \frac{\text{Power (in watt)} \times \text{Time (in hour)}}{1000}$$

$$\text{or} \quad \text{kWh} = \frac{\text{V (volt)} \times i \text{ (ampere)} \times t \text{ (hour)}}{1000}$$

$$\therefore \text{Energy (in kWh)} = \frac{Vit}{1000}.$$

4. In what unit does the domestic electric meter measure the electrical energy consumed? State the value of this unit in S.I. unit.

Ans. The domestic electric meter measures the electrical consumption in kilowatt-hour (kWh).

$$1\ \text{kWh} = 1\ \text{kilowatt} \times 1\ \text{hour}$$
$$= 1000\ \text{js}^{-1} \times (60 \times 60)\ \text{s}$$
$$= 3{\cdot}6 \times 10^6\ \text{J in S.I. unit.} \quad \textbf{Ans.}$$

5. An electric bulb is marked 100 W, 230 V. What is the energy consumed by the bulb in one hour.

Ans. Energy consumed in 1s is 100 J.

∴ Energy consumed in hour

$$= 100 \times 60 \times 60\ \text{J} = 360000\ \text{J} \quad \textbf{Ans.}$$

6. State other bigger units of electric power and express them in watt.

Ans. The other bigger units of power commonly used are:

1 Kilowatt = 1000 watt

1 Horse power (H. P.) = 746 watt

Megawatt = 10^6 watt

* **Frequently asked previous years Board Exam Questions.**

7. An I ampere of current is passed through a resistor of resistance R ohm for time t second. Obtain expression for (i) the electrical energy, and (ii) the electrical power spent.

Ans. Potential difference across the resistor,

$$V = IR \text{ volt}$$

Charge passed in resistor,

$$Q = It \text{ coulomb}$$

(i) Work done (or the electrical energy spend) in passing charge Q coulomb through the resistor at potential difference V volt = QV joule

$$= (It) \times (IR) = I^2 Rt \text{ joule.}$$

(ii) Electrical power spent $= \dfrac{\text{Energy}}{\text{Time}} = \dfrac{I^2Rt}{t}$

$= I^2R$ watt.

8. What is the resistance of an appliance rated at V volt, P watt?

Ans. Let I be the current drawn by the appliance.

Then Power, $P = V \times I$

or $I = P/V$

Now, Resistance, $R = V/I$

Hence, $R = V \times \dfrac{V}{P} = \dfrac{V^2}{P}.$

9. Establish the relation that heat produced in a metallic conductor due to the flow of current is $\dfrac{V^2t}{R}$ joules, where V is the potential difference across the ends of the conductor, R is its resistance and t is the time in second for which current flows.

Ans. We know that,

$$W = Vit \text{ joule}$$

Applying Ohm's law,

$$V = iR$$

or $i = V/R$

Substituting the value of i, we get

$$W = V.\frac{V}{R}t = \frac{V^2t}{R} \text{ joule.}$$

10. An electric lamp A of 40 W and another electric lamp B of 100 W are connected to 220 V supply. Calculate the ratio of their filament resistances?

Ans. We know that,

$$P = V.I. = \frac{V.V}{R} = \frac{V^2}{R}$$

$\therefore$ $40 = \dfrac{(220)^2}{R_A}$ and $100 = \dfrac{(220)^2}{R_B}$

$R_A = \dfrac{220 \times 220}{40}$ and $R_B = \dfrac{220 \times 220}{100}$

$R_A = 110 \times 11$ and $R_B = 22 \times 22$

$\therefore$ $\dfrac{R_A}{R_B} = \dfrac{110 \times 11}{22 \times 22} = \dfrac{5}{2}$

$\therefore$ $R_A : R_B = 5 : 2$ **Ans.**

11. Find the energy released by a current of 0.25 A flowing through a heater for 5 minutes. The p.d. is 230 V.

Ans. Given: V = 230 V; t = 5 minutes = 60 × 5 = 300 s, I = 0.25 A

$$W = VIt = 230 \times \frac{25}{100} \times 300$$

$\therefore$ W = 17250 J. **Ans.**

12. An electric iron is rated at 230 V, 750 W. What is its resistance? What maximum current can be passed through it?

Ans. Given : V = 230 V, Power = 750 W

We know that, $P = V \times I$

$\therefore$ $750 = 230 \times I$

$\therefore$ $I = 750/230$

$\therefore$ $I = 3.26$ A

By Ohm's law, $\dfrac{V}{I} = R$

$\therefore$ $R = \dfrac{230}{3.26}$

$R = 70.53\ \Omega$. **Ans.**

13. An electric bulb is marked 100 W, 230 V. What current does it take?

Ans. Current flowing through the bulb,

$i = \dfrac{P}{V} = \dfrac{100}{230} = \dfrac{10}{23}$ A = 0.435 A **Ans.**

14. An electric bulb is marked 100 W, 230 V. How long would this lamp take to consume one unit of electricity?

Ans. Energy consumed in kWh

$$= \frac{\text{Power (in watt)} \times \text{Time (in hour)}}{1000}$$

$$= \frac{10 \times t}{1000}$$

If energy consumed is 1 kWh, then

$$1 = \frac{100 \times t}{1000}$$

$\therefore$ Time required (t) = 10 hour. **Ans.**

15. The equation I^2R seems to suggest that the rate of heating in a resistor is reduced if resistance

Short Numericals

decreases, whereas equation $P = \frac{V^2}{R}$ suggests just the opposite. How do you reconcile this problem?

Ans. There is no contradiction in the relation. If V is constant, on decreasing resistance, the current increases and so as $P = I^2R$ the same result we gets from equation $P = \frac{V^2}{R}$.

16. Find the heat produced in joule when a current of 10 A flows through a connecting wire of resistance $2 \times 10^{-2}\ \Omega$ and heating element of resistance 250 Ω for 2 second.

Ans. Heat produced when current flows in the wire for 2 second,

$$Q_1 = I^2Rt$$

$$Q_1 = 10 \times 10 \times 2 \times 10^{-2} \times 2$$

$$Q_1 = 4\text{ J}$$

Heat produced in heating element in 2 second,

$$Q_2 = 10 \times 10 \times 250 \times 2$$

$$Q_2 = 50{,}000\text{ J}$$ **Ans.**

17. A family uses a light bulb of 100 W, a fan of 100 W, and a heater of 1000 W, each for 8 hours a day. If the cost of electricity is ₹ 2 per unit, what is the expenditure of the family per day, on electricity?

Ans. Power consumed in watt hours

$$= 100 \times 8 + 100 \times 8 + 1000 \times 8$$

$$= (800 + 800 + 8000)\text{ Wh}$$

$$= 9600\text{ Wh}$$

i.e., Electric energy consumed in kWh (units)

$$= \frac{9600}{1000} = 9.6\text{ kWh.}$$

Cost at ₹ 2 per unit or per kWh

$$= ₹\ 2 \times 9.6 = ₹\ 19.20.$$ **Ans.**

18. An electric bulb is rated as 100 W, 250 V. What information does it convey? Calculate: (i) the resistance of its filament while glowing, and (ii) the current through the filament of the bulb.

Ans. The rating 100 W, 250 V on the bulb means that if the bulb is lighted on a 250 V supply, it consumes 100 W of electrical power (*i.e.,* 100 J electrical energy in 1 second).

Given : P = 100 W, V = 250 V.

(i) The resistance of filament,

$$R = \frac{V^2}{P} = \frac{(250)^2}{100} = 625\ \Omega$$

(ii) The current through the filament,

$$I = \frac{P}{V} = \frac{100}{250} = 0.4\text{ A.}$$ **Ans.**

19. Out of the three fuses with current ratings 5 A, 10 A, 15 A, which one is to be connected in a geyser circuit marked 3 kW, 220 V? Give reason for your answer.

Ans. The current flowing through geyser

$$= \frac{3 \times 1000}{220} = 13.63\text{ A} = 14\text{ A.}$$ **Ans.**

The rating of fuse should be just matching with the maximum current carried by the circuit. So the fuse rating should be of 15 A.

Chapter 9. Magnetic Effect of Current

1. The primary coil of a transformer has 200 turns while the secondary coil has 1000 turns. What type of transformer is this ? If the input voltage is 10 V, what will be the output voltage?

Ans. The transformer is a step-up transformer.

Given : $N_p = 200$, $N_s = 1000$, $E_p = 10$ V, E_s = ?

By relation,

$$\frac{E_s}{E_p} = \frac{N_s}{N_p}$$

$$\therefore \quad \frac{E_s}{10} = \frac{1000}{200}$$

or $$E_s = \frac{10 \times 1000}{200} = 50\text{ V.}$$ **Ans.**

2. The voltage supply to primary coil is 210 V. If the number of turns in primary coil is 200 turns and that of in secondary coil is 20 turns, then find out the output voltage. Name the type of the transformer.

Ans. Using, $$\frac{E_p}{E_s} = \frac{N_p}{N_s}$$

$$\frac{210}{E_s} = \frac{200}{20}$$

or $$E_s = \frac{210 \times 20}{200}$$

Output voltage,

$$E_s = 21\text{ V.}$$

It is a step down transformer. **Ans.**

3. A transformer lowers e.m.f. 220 V to 12 V. If the number of turns in primary are 8800, how many turns are in secondary coil ?

Ans. Using, $\frac{E_p}{E_s} = \frac{N_p}{N_s}$

$\therefore \quad \frac{220}{12} = \frac{8800}{N_s}$

$\therefore \quad N_s = \frac{8800 \times 12}{220} = 40 \times 12 = 480$

∴ Number of turns in secondary coil = 480. **Ans.**

4. A transformer lowers e.m.f. from 220 V to 15 V. If 400 W power is given in primary, calculate (i) the current in primary coil and (ii) the current in secondary coil.

Ans. (i) We know that,

$$\text{Power} = I_p \times E_p$$

or $\quad 400 = I_p \times 220$

$\therefore \quad I_p = 1.8 \text{ A.}$ Ans.

(ii) Using,

$$I_p \times E_p = I_s \times E_s$$

$\therefore \quad I_s = \frac{I_p \times E_p}{E_s}$

$I_s = \frac{1.8 \times 220}{15} = 26.4 \text{ A.}$ **Ans.**

5. A transformer lowers e.m.f. from 220 V to 15 V. If the number of turns in primary are 3520, how many turns are in the secondary coil?

Ans. We know that,

$$\frac{E_s}{E_p} = \frac{N_s}{N_p}$$

or $\quad \frac{15}{220} = \frac{N_s}{3520}$

$N_s = \frac{3520 \times 15}{220} = 240$

Number of turns in secondary coil are 240. **Ans.**

Chapter 10. Specific Heat Capacity and Latent Heat

1. Water falls from a height of 50 m. Calculate the rise in the temperature of water when it strikes the bottom.
[g = 10 ms^{-2}; Specific heat capacity of water = 4200 J/kg°C]

Ans. If the whole of the energy due to fall of water changes into heat energy, and θ°C be the rise in the temperature of water, then

$$mgh = ms\,\theta$$

or $\quad m \times 10 \times 50 = m \times 4200 \times \theta$

or Rise in temperature of water,

$$\theta = \frac{m \times 10 \times 50}{m \times 4200} = \frac{5}{42}\,°\text{C}$$

$= 0{\cdot}119°\text{C} = 0{\cdot}12°\text{C}.$ **Ans.**

2. A solid of mass 25 g at 120°C is placed in 100 g of water at 20°C. Calculate the final temperature of the mixture. [Specific heat capacity of solid = 0·8 Jg^{-1}°C^{-1}, Specific heat capacity of water = 4·2 Jg^{-1}°C^{-1}].

Ans. Let final temperature of mixture = x°C

Heat lost by 25 g solid at 120°C

= Heat gained by 100 g water at 20°C

$$m_1 c_1 \Delta\theta_1 = m_2 c_2 \Delta\theta_2$$

$$25 \times 0{\cdot}8 \times (120 - x) = 100 \times 4{\cdot}2 \times (x - 20)$$

$$2400 - 20x = 420x - 8400$$

$$440x = 10800$$

$\therefore \quad x = \frac{10800}{440} = 24{\cdot}54°\text{C}.$ **Ans.**

3. 200 g of hot water at 80°C is added to 300 g of cold water at 10°C. Calculate the final temperature of the mixture of water. Consider the heat taken by the container to be negligible. [Specific heat capacity of water is 4200 J kg^{-1} °C^{-1}]

Ans. Given: For hot water:

m_1 = 200 g, temperature, t_1 = 80°C

For cold water:

m_2 = 300 g, temperature, t_2 = 10°C

Let final temperature of mixture = θ.

By the principle of calorimetry,

$$\text{Heat given} = \text{Heat taken}$$

$$200 \times c \times (80 - \theta) = 300 \times c \times (\theta - 10)$$

$$200 \times 80 - 200\,\theta = 300\,\theta - 300 \times 10$$

$$16000 - 200\,\theta = 300\,\theta - 3000$$

$$19000 = 500\,\theta$$

$\theta = 38°\text{C}$ **Ans.**

4. An electric immersion heater is rated 1250 W. Calculate the time in which it will heat 20 kg of water at 5°C to 65°C.

Ans. Let the time taken be 't' second.

Energy supplied by heater = P × t

= 1250 × t J

Heat required by water to raise its temperature

= $mc\,\Delta\theta$

= 20 × 4200 × (65 – 5) J

= 5040000 J

Energy supplied by heater

= Heat gained by water

$1250 \times t = 5040000$

Short Numericals

$$t = \frac{5040000}{1250} \text{ s}$$

$= 4032$ s

$=$ 1 hour 7 minutes 12 second. **Ans.**

5. Some hot water was added to three times its mass of cold water at 10°C and the resulting temperature was found to be 20°C. What was the temperature of the hot water?

Ans. Let the temperature of hot water be t°C and mass of hot water = m gram

$\therefore$ Mass of cold water = $3m$ gram

Temperature of cold water = 10°C

Final temperature of mixture = 20°C

Heat taken by cold water = $3m \times c \times (20° - 10°)$

Heat given by hot water = $m \times c \times (t - 20°)$

Heat given = Heat taken

$m \times c \times (t - 20°) = 3m \times c \times 10°$

$t - 20° = 30°$

$\therefore$ $t = 30° + 20° = 50°C$ **Ans.**

6. A piece of iron of mass 2·0 kg has a thermal capacity of 966 J/°C. How much heat is needed to warm it by 15°C?

Ans. Given : m = 2·0 kg, Thermal capacity = 966 J/°C

Rise in temperature = 15°C

$$\text{Thermal capacity, } c' = \frac{\text{Heat required}}{\text{Rise in temperature}}$$

$\therefore$ Heat needed = 966 × 15 = 14490 J **Ans.**

7. Two metallic blocks P and Q having masses in ratio 2 : 1 are supplied with the same amount of heat. If their temperatures rise by same degree, compare their specific heat capacities.*

Ans. Let specific heat capacity of block P and block Q be C_p and C_q respectively.

Metallic block P	Metallic block B
$m_1 = 2m$	$m_2 = m$
$c_1 = C_p$	$c_2 = C_q$
$\Delta T = \Delta T$	$\Delta T = \Delta T$
$H_1 = m_1 c_1 \Delta T$	$H_2 = m_2 c_2 \Delta T$
$= 2m\, C_P \Delta T$	$= m\, C_Q \Delta T$

$\because$ $H_1 = H_2$

$\Rightarrow$ $2m\, C_P \Delta T = m C_Q \Delta T$

$$\therefore \quad \frac{C_P}{C_Q} = \frac{m \times \Delta T}{2m \times \Delta T} = \frac{1}{2}$$

$\therefore$ Ratio of specific heat capacities of block P and block Q is 1 : 2.

8. The temperature of a lead piece of mass 400 g rises from 20°C to 50°C when 1560 J of heat is supplied to it. Calculate :

(i) Heat capacity of lead piece, (ii) Specific heat capacity of lead.

Ans. (i) Heat capacity of lead piece

$$= \frac{\text{Heat supplied}}{\text{Rise in temperature}} = \frac{1560}{(50° - 20°)}$$

$= 52$ J/°C. **Ans.**

(ii) Specific heat capacity of lead

$$= \frac{\text{Heat supplied}}{\text{Mass} \times \text{Rise in temperature}}$$

$$= \frac{1560}{(400) \times (50° - 20°)} = 0{\cdot}13 \text{ J/g°C.}$$ **Ans.**

9. 63.2 g of copper at 50°C can just melt 3.8 g of ice. If the specific latent heat of ice is 336 J/g, find the specific heat capacity of copper.

Ans. Let the specific heat capacity of copper be s.

$\therefore$ Heat lost by copper

$= 63.2 \times s \times (50 - 0) = 3160s$ J

Heat gained by ice = 3.8 × 336 J.

Since Heat gained = Heat lost

We have $3160\, s = 3.8 \times 336 = 1276.8$

or $s = (1276.8/3160)$ J/g°C

$= 0.404$ J/g°C. **Ans.**

10. 50 g of metal piece at 27°C requires 2400 J of heat energy so as to attain a temperature of 327°C. Calculate the specific heat capacity of the metal.

Ans. Given: $m = 50 \text{ g or } \frac{50}{1000} \text{ kg}$

$H = 2400$ J

$\theta_1 = 27°C$

$\theta_2 = 327°C$

$H = mc\,\theta$

$$c = \frac{H}{m\theta} = \frac{2400}{\frac{50}{1000} \times (327 - 27)}$$

$$= \frac{2400}{\frac{5}{100} \times 300}$$

$$= \frac{2400}{15} = 160 \text{ J kg}^{-1}{}^{\circ}\text{C}^{-1}$$ **Ans.**

11. A hot solid of mass 60 g at 100°C is placed in 150 g of water at 20°C. The final steady temperature recorded is 25°C. Calculate the specific heat capacity of the solid. [Specific heat capacity of water = 4200 J kg^{-1}°C^{-1}]

Ans. Given: Mass of hot solid,

$m_1 = 60$ g at 100°C

Mass of water, $m_2 = 150$ g at 20°C

Final temperature = 25°C.

* **Frequently asked previous years Board Exam Questions.**

By the principle of calorimetry,

Heat given = Heat taken

$\Rightarrow \quad m_1c_1\Delta\theta_1 = m_2c_2\Delta\theta_2$

$\Rightarrow \; 60 \times c \times (100 - 25) = 150 \times 4{\cdot}2 \times (25 - 20)$

∴ Required specific heat capacity,

$c_1 = 0{\cdot}7$ J/g °C

12. Find the time taken by a 500 W heater to raise the temperature of 50 kg of material of specific heat capacity 960 J/kg°C from 18°C to 38°C. Assume that all the heat from the heater is given to the material.

Ans. Power (P) = 500 W = 500 J/sec.

Mass of the material

(m) = 50 kg.

Rise in temperature (t)

= 38 – 18 = 20°C

Specific heat capacity of the material

(s) = 960 J/kg°C

∴ Quantity of heat required (Q)

$= mst = 50 \times 20 \times 960 = 960{,}000$ J.

∴ Time required $= \dfrac{960000}{500} = 1920$ s = 32 min.

13. A piece of iron of mass 2·0 kg has a thermal capacity of 966 J/°C. What is its specific heat capacity in S.I. units?

Ans. Thermal capacity = mass × specific heat capacity

∴ Specific heat capacity

$= \dfrac{\text{Thermal capacity}}{\text{mass}}$

$= \dfrac{966}{2}$ J kg^{-1}°C^{-1}

= 483 J kg^{-1}°C^{-1}. **Ans.**

14. How much heat energy is gained when 0·5 kg of water at 20°C is brought to its boiling point? [Specific heat capacity of water = 4200 Jkg^{-1}°C^{-1}]

Ans. $Q = mc\,\Delta\theta$

$= 0{\cdot}5 \times 4200 \times (100 - 20)$ J

$= 0{\cdot}5 \times 4200 \times 80$ J

= 168000 J **Ans.**

15. 600 calorie of heat is required to raise the temperature of 0·060 kg of a substance from 0°C to 100°C. Calculate its specific heat capacity: (i) in calorie kg^{-1}°C^{-1}, (ii) in J kg^{-1}°C^{-1}.

Ans. (i) Rise in temperature (Δθ)

= (100 – 0)°C = 100°C

$Q = mc\,\Delta\theta$

$600 = 0{\cdot}060 \times c \times 100$

$c = \dfrac{600}{0{\cdot}060 \times 100}$ calorie kg^{-1}°C^{-1}

= 100 calorie kg^{-1}°C^{-1}

(ii) $c = 100 \times 4{\cdot}2$ Jkg^{-1}°C^{-1}

= 420 J kg^{-1}°C^{-1}. **Ans.**

16. How much heat energy is released when 5 g of water at 20°C changes to ice at 0°C?
[Specific heat capacity of water = 4·2 J g^{-1}°C^{-1}
Specific latent heat of fusion of ice = 336 J g^{-1}]

Ans. $Q = mc\,\Delta\theta + mL$

$= 5 \times 4{\cdot}2 \times 20 + 5 \times 336$

= 420 + 1,680

= 2,100 J **Ans.**

Chapter 11. Radioactivity and Nuclear Energy

1. An element P disintegrates by alpha-emission and the new element suffers two further disintegration, both by beta-emission, to form an element Q. Show that P and Q are isotopes.

Ans. Let mass number of P = A

Let atomic number of P = Z

Nuclear reaction for α-emission,

$_{Z}P^{A} \rightarrow {}_{Z-2}X^{A-4} + {}_{2}He^{4}$

α-particle

Nuclear reactions for two further β-emission,

$_{Z-2}X^{A-4} \rightarrow {}_{Z-1}Y^{A-4} + {}_{-1}e^{0}$

β-particle

$_{Z-1}Y^{A-4} \rightarrow {}_{Z}Q^{A-4} + {}_{-1}e^{0}$

β-particle

The atomic numbers (Z) of P and Q are the same but their mass numbers [A and (A–4)] are different. So, P and Q are isotopes. **Ans.**

2. An element X has the symbol $_{84}X^{202}$. It emits an alpha-particle and then a beta-particle. The final nucleus is $_{b}Y^{a}$. Find a and b.

Ans. $_{84}X^{202} \xrightarrow{\alpha} {}_{82}X^{198}_{1} + {}_{2}He^{4}$

α-particle

$_{82}X^{198} \xrightarrow{\beta} {}_{83}X^{198} + {}_{-1}e^{0}$

β-particle

$_{b}Y^{a} = {}_{83}Y^{198}$

∴ $a = 198$ and $b = 83$. **Ans.**

3. Complete the following fission reaction :
$${}_{92}U^{235} + {}_{0}n^{1} \longrightarrow {}_{a}Sb^{133} + {}_{41}Co^{b} + 4\,{}_{0}n^{1} + \text{energy}$$
Write down the values of a and b.

Ans. The given equation is:
$${}_{92}U^{235} + {}_{0}n^{1} \longrightarrow {}_{a}Sb^{133} + {}_{41}Co^{b} + 4\,{}_{0}n^{1}$$
Hence, $235 + 1 = 133 + b + 4$
or $236 = b + 137$
$\therefore$ $b = 236 - 137 = 99$
Similarly, $92 = a + 41$
$\therefore$ $a = 92 - 41 = 51$
Hence,
$${}_{92}U^{235} + {}_{0}n^{1} \rightarrow {}_{51}Sb^{133} + {}_{41}Co^{99} + 4\,{}_{0}n^{1}.$$ **Ans.**

4. Calculate the amount of energy released in MeV due to a loss of mass of 2 kg.

Ans. Given : $\Delta m = 2$ kg, $c = 3 \times 10^8$ m/s
According to Einstein's mass-energy equivalence,
$$E = (\Delta m)\,c^2$$
$$= 2 \times (3 \times 10^8)^2 = 1{\cdot}8 \times 10^{17}\text{ J}$$
$$= \frac{1{\cdot}8 \times 10^{17}}{1{\cdot}6 \times 10^{-19} \times 10^{6}}\text{ MeV}$$
$$= 1{\cdot}125 \times 10^{30}\text{ MeV.}$$ **Ans.**

5. In a nuclear fusion reaction, the loss in mass is 0·5%. How much energy is released in the fusion of 2 kg mass.

Ans. Given: Mass lost, $\Delta m = 0{\cdot}5\%$
$\therefore$ Loss in mass in nuclear fusion of 2 kg mass
$$= \frac{0{\cdot}5}{100} \times 2\text{ kg} = 0{\cdot}01\text{ kg}$$
Energy released, $E = (\Delta m)\,c^2$
$$= 0{\cdot}01 \times (3 \times 10^8)^2$$
$$= 9 \times 10^{14}\text{ J.}$$ **Ans.**

Long Numericals

| Set 6 |

Chapter 1. Force, Work, Power and Energy

1. On a see-saw, two children of masses 30 kgf and 50 kgf are sitting on one side of it at distance 2 m and 2·5 m respectively, from its middle. Where should a man of mass 74 kgf sit to balance it?

Ans. Let the two children be sitting on the left arm. They will produce anticlockwise moment due to their weights about the middle point.

Total anticlockwise moment

$= 30 \text{ kgf} \times 2 \text{ m} + 50 \text{ kgf} \times 2{\cdot}5 \text{ m}$

$= 60 \text{ kgf m} + 125 \text{ kgf m} = 185 \text{ kgf m}$

To balance it, the man should sit on the right arm (to produce clockwise moment). Let his distance from the middle be x m. Then

Clockwise moment

$= 74 \text{ kgf} \times x \text{ m} = 74x \text{ kgf m}$

By the principle of moments,

Anticlockwise moment

= Clockwise moment

$185 = 74x$

or $x = \dfrac{185}{74}$ m

= 2·5 m (on the other side).

Ans.

2. A boy of mass 40 kg runs up a height of 30 steps each 20 cm high. Find:

(i) The force of gravity acting on the boy.

(ii) The work done by the boy against gravity. (Take $g = 9{\cdot}8 \text{ ms}^{-2}$)

Ans. Given: $m = 40$ kg, $g = 9{\cdot}8 \text{ ms}^{-2}$

Distance moved, $s = 30 \times 20$ cm

$= 600$ cm $= 6$ m

(i) The force of gravity,

$F = mg = 40 \times 9{\cdot}8 = 392$ N **Ans.**

(ii) The work done by the boy,

$W = F \times s = 392 \times 6 = 2352$ J **Ans.**

3. Calculate the work done when:

(i) A 5 kg weight is lifted 10 m vertically upward [$g = 9{\cdot}8 \text{ ms}^{-2}$].

(ii) A car is moved on a rough road through 30 m against a frictional resistance of 75 N.

Ans. (i) Given: Force $= m \times g = 5 \times 9{\cdot}8 = 49$ N; Distance = 10 m

$\therefore$ Work done $= F \times s = 49 \times 10 = 490$ J **Ans.**

(ii) Given: Resisting force = 75 N; Distance moved = 30 m

$\therefore$ Work done $= 75 \times 30$ J $= 2250$ J. **Ans.**

4. An electric motor of power 100 W is used to drive the stirrer in a water bath. If 50% of the energy supplied to the motor is spent in stirring the water. Calculate the work done on the water in one minute.

Ans. Power (P) = 100 W

Time (t) = 1 minute = 60 seconds

$\therefore$ Work done by motor

= Produced energy

$= P \times t = 100 \times 60 = 6000$ J

$\therefore$ Work done on water by motor

= 50% of total produced energy

$= \dfrac{50}{100} \times 6000 = 3000$ J. **Ans.**

5. The weights of two bodies are 2·0 N and 2·0 kgf respectively. What is the mass of each body? Calculate the ratio of work done on both bodies in lifting them through a height of 5 m. [$g = 10 \text{ m/sec}^2$].

Ans. For first body:

Let Weight, $W_1 = 2{\cdot}0$ N

Then Weight, $W_1 = m_1 g$

Substituting the values, we get

$2 = m_1 \times 10$

$\Rightarrow \quad m_1 = \dfrac{2}{10} = 0{\cdot}2$ kg

For second body:

Let Weight, $W_2 = 2{\cdot}0 \text{ kgf} = 2{\cdot}0 \times 10$ N

Then Weight, $W_2 = m_2 g$

Substituting the values, we get

$2 \times 10 = m_2 \times 10$

$\Rightarrow \quad m_2 = 2$ kg.

$\therefore$ Ratio of work done $= \frac{m_1gh}{m_2gh} = \frac{0{\cdot}2}{2}$

$= \frac{1}{10} = 1 : 10$ **Ans.**

6. A truck weighing 1000 kgf changes its speed from 36 km/h to 72 km/h in 2 minutes. Calculate the work done by the engine and its power.

Ans. Weight of truck = 1000 kgf

$\therefore$ Mass (m) = 1000 kg

Here, u = 36 kmh^{-1} = 10 ms^{-1}, v = 72 kmh^{-1} = 20 ms^{-1}, t = 2 minutes = 2 × 60 s = 120 s

Work done = Increase in kinetic energy

$= \frac{1}{2} m (v^2 - u^2)$

$= \frac{1}{2} \times 1000 \times (20^2 - 10^2)$ J

= 500 (400 – 100) J

= 500 × 300 J = 150000 J

= $1{\cdot}5 \times 10^5$ J

Power of the engine $= \frac{W}{t}$

$= \frac{150000}{120}$

$= 1{\cdot}25 \times 10^3$ W **Ans.**

7. A force of 1 kgf displaces a body by a distance of 10 cm (i) in direction of force, (ii) normal to the force (iii) at an angle of 60° to the direction of force. Calculate the amount of work done in each case. [Take: $g = 9{\cdot}8$ ms^{-2} and cos 60° $= \frac{1}{2}$]

Ans. Given: F = 1 kgf = 1 × g N = 9·8 N, $s = \frac{10}{100}$ m = 0·1 m.

(i) When displacement is in the direction of force,

θ = 0°, cos 0° = 1

W = F × s = 9·8 × 0·1 = 0·98 J **Ans.**

(ii) When displacement is normal to the force,

θ = 90°, cos 90° = 0, *i.e.*, W = 0. **Ans.**

(iii) When displacement is at an angle of 60° to the direction of force,

θ = 60°, cos 60° $= \frac{1}{2}$

W = F.s cos θ = 9·8 × 0·1 × $\frac{1}{2}$ = 0·49 J

Ans.

8. Calculate the horse power of an engine, which lifts 5000 m^3 of water from a depth of 40 m in 20 minutes. [g = 10 ms^{-2}, density of water = 1000 kgm^{-3}, 1 H.P. = 750 W].

Ans. Mass (m) of water = Volume × Density

= 5000 × 1000

= 5 × 10^6 kg

Time = 20 × 60 s = 1200 s

Work done in lifting water

= mgh

= 5 × 10^6 × 10 × 40 J

= 2 × 10^9 J

Power of engine $= \frac{W}{t} = \frac{2 \times 10^9}{1200}$ W

Power (in horse power)

$= \frac{2 \times 10^9}{1200 \times 750}$ H.P.

= 2222·2 H.P. **Ans.**

9. An electric motor of power 1000 W is switched on for 1 minute and 40 second. If 80% of the energy of the motor is useful, calculate: (i) useful work done by the motor, (ii) load lifted by it through a vertical height of 5 m. [g = 10 ms^{-2}]

Ans. (i) Time (t) = 1 minute 40 second

= (60 + 40) s = 100 s

Energy supplied to the motor

= P × t = 1000 × 100 J = 10^5 J

Useful work done by the motor

= 80% of 10^5 J

$= \frac{80}{100} \times 100000$ J = 80000 J **Ans.**

(ii) Useful work done by the motor in lifting the load = mgh

By the law of conservation of energy,

mgh = 80000

m × 10 × 5 = 80000

$\therefore \quad m = \frac{80000}{10 \times 5}$ kg = 1600 kg

$\therefore$ Load lifted = 1600 kgf. **Ans.**

10. A coolie carries a load of 30 kgf through a distance of 500 m in 5 minutes while another coolie B carries the same load through the same distance in 10 minutes. Compare the (i) work done, and (ii) power developed. [Take: g = 10 ms^{-2}]

Ans. Given: F = 30 kgf = 30 × g N = 30 × 10 N = 300 N, s = 500 m.

For coolie A, work done (W_1)

= F × s = 300 × 500 = 150,000 J

Time (t_1) = 5 minutes = 5 × 60 s = 300 s

$\therefore$ Power developed (P_1)

Long Numericals

$$= \frac{W_1}{t_1} = \frac{150,000}{300} = 500 \text{ Js}^{-1}$$

For coolie B, work done (W_2)

$$= F \times s = 300 \times 500 = 150,000 \text{ J}$$

Time (t_2) = 10 minutes = 10 × 60 s = 600 s

∴ Power developed (P_2)

$$= \frac{W_2}{t_2} = \frac{150,000}{600} = 250 \text{ Js}^{-1}$$

(i) $\frac{W_1}{W_2} = \frac{150,000}{1,50,000} = \frac{1}{1}$. **Ans.**

(ii) $\frac{P_1}{P_2} = \frac{500}{250} = \frac{2}{1}$. **Ans.**

11. A girl of mass 40 kg runs a height of 80 stairs, each 25 cm high with a load of 20 kg on her head in 25 s. If g is 10 m/s^2, find:

(i) Gravitational force acting on the girl.

(ii) Work done by her.

(iii) Useful work done by her.

(iv) Her power in watt.

Ans. (i) Gravitational force acting on the girl

$$= \text{Mass (in kg)} \times g$$

$$= 40 \times 10 = 400 \text{ N.}$$ **Ans.**

(ii) Total gravitational force

$$= (\text{Mass of girl + load}) \times g$$

$$= (40 + 20) \times 10 = 600 \text{ N.}$$

Distance or height

$$= \frac{80 \times 25}{100} = 20 \text{ m.}$$

∴ Work done by the girl

$$= 600 \times \text{distance}$$

∴ Total work done

$$= 600 \times 20 = 12000 \text{ J.}$$ **Ans.**

(iii) Useful work done

$$= \text{Load} \times g \times \text{distance covered}$$

$$= 20 \times 10 \times 20 = 4000 \text{ J.}$$ **Ans.**

(iv) $$\text{Power} = \frac{\text{Total work done}}{\text{Time}}$$

$$= \frac{12000}{25} \text{ W} = 480 \text{ W.}$$ **Ans.**

12. A man weighing 700 N runs up a flight of 15 steps, each 20 cm high, in 5 seconds. Calculate the power of the man.

Ans. Total height covered

$$= 15 \times 20 \text{ cm} = 300 \text{ cm} = 3 \text{ m}$$

Work done = force × distance = 700 × 3

$$= 2100 \text{ J}$$

Since time taken is 5 s, the power of the man is

Power = (2100/5) Js^{-1} = 420 W. **Ans.**

13. A water pump raises 50 kg of water through a height of 25 m in 5 s. Calculate the power supplied by the pump. (Take g = 10 N kg^{-1}).

Ans. Given: m = 50 kg, g = 10 N kg^{-1}, s = 25 m, t = 5 s.

Force of gravity, F = mg = 50 × 10 = 500 N

Work done, W = F × s = 500 × 25 = 12,500 J

$$\text{Power, P} = \frac{W}{t} = \frac{12,500}{5} = 2500 \text{ W.}$$

Ans.

14. An electric motor pump is 60% efficient and is rated 2·5 H.P. Calculate the maximum load it can lift through a height of 10 m in 8 s. [Take 1 H.P. = 750 W and g = 10 ms^{-2}].

Ans. Useful power supplied by the motor

$$= \frac{2.5 \times 750 \times 60}{100} \text{ W}$$

$$= 1125 \text{ W}$$

If maximum load lifted is W kg, then

Work done in lifting up the load

$$= mgh = W \times 10 \times 10 \text{ J}$$

and Work done by the pump

$$= 1125 \times 8 \text{ J}$$

∴ W × 10 × 10 = 1125 × 8

Maximum amount of load lifted,

$$W = \frac{1125 \times 8}{10 \times 10} \text{ kg} = 90 \text{ kg.}$$ **Ans.**

15. In a hydroelectric power station, 1000 kg of water is allowed to drop through a height of 100 m in 1 s. If the conversion of potential energy to electric energy is 60%. Calculate the power output. [Take : g = 10 m/sec^2]

Ans. In a hydroelectric power station:

Mass of water (m) = 1000 kg.

Height (h) = 100 m.

Time (t) = 1 s

Produced potential energy (P.E.)

$$= mgh = 1000 \times 10 \times 100 \text{ J}$$

$$= 1000000 \text{ J}$$

Electrical energy = 60% of potential energy

$$= \frac{60}{100} \times 1000000 = 600000 \text{ J}$$

$$\therefore \text{Power output} = \frac{\text{Energy}}{\text{Time}}$$

$$= \frac{600000}{1} \text{ W} = 600000 \text{ W}$$

$$= \frac{600000}{1000} \text{ kW}$$

$$= 600 \text{ kW.}$$ **Ans.**

16. For the same kinetic energy of a body, what should be the change in its velocity if its mass is increased nine times?

Ans. Let initial mass (m_1) = m, initial velocity (v_1) = v

Final mass (m_2) = $9m$, final velocity (v_2) = ?

For the same kinetic energy,

$$\frac{1}{2}m_1v_1^2 = \frac{1}{2}m_2v_2^2$$

or $$\frac{1}{2}mv^2 = \frac{1}{2} \times 9m \times v_2^2$$

$\therefore$ $$v_2 = \sqrt{\frac{1}{9}v^2} = \frac{1}{3}v$$ **Ans.**

17. Calculate the kinetic energy of a body of mass 0·1 kg and momentum 40 kg m/s.

Ans. We know that, K.E. = $\frac{1}{2}mv^2$ and momentum,

$p = mv$

On eliminating v, we get

$$\text{K.E.} = \frac{1}{2}m \times \left(\frac{p}{m}\right)^2 = \frac{p^2}{2m}$$

Given: p = 40 kg m/s, m = 0·1 kg.

$$\text{K.E.} = \frac{(40)^2}{2 \times 0{\cdot}1} = 8000 \text{ J} = 8 \times 10^3 \text{ J}$$ **Ans.**

18. A body of mass 50 kg has a momentum of 3000 kg ms^{-1}. Calculate:

(i) the kinetic energy of the body.

(ii) the velocity of the body.

Ans. (i) Given: m = 50 kg, p = 3000 kg ms^{-1}

Momentum, $p = mv$

$\therefore$ $$v = \frac{p}{m} = \frac{3000}{50} = 60 \text{ ms}^{-1}$$

Kinetic energy of the body

$$= \frac{1}{2}mv^2$$

$$= \frac{1}{2} \times 50 \times (60)^2 = 25 \times 3600$$

$$= 90000 \text{ J}$$ **Ans.**

(ii) Velocity of the body,

v = 60 ms^{-1}. **Ans.**

19. A body of mass m falls from a height h_1 to a height h_2, above the ground ($h_1 > h_2$). What is the loss in potential energy?

Ans. At height h_1, initial potential energy of body

$= mgh_1$

At height h_2, final potential energy of body

$= mgh_2$

$\therefore$ Loss in potential energy

= Initial potential energy

– Final potential energy

$= mgh_1 - mgh_2 = mg(h_1 - h_2)$

Ans.

20. Obtain an expression for the kinetic energy of a body of mass m moving with a velocity v?

Ans. Let the body travelling with a velocity 'v' be brought to rest by a retarding force 'F' after travelling a distance 's'.

From relation, $v^2 = u^2 - 2as$

Retardation, $$a = \frac{v^2}{2s}$$

$\therefore$ Retarding force,

$$F = ma = m \times \frac{v^2}{2s} = \frac{1}{2}\frac{mv^2}{s}$$

$$\text{Work done, W} = F \times s = \frac{1}{2}\frac{mv^2}{s} \times s = \frac{1}{2}mv^2$$

Hence, Kinetic energy = $\frac{1}{2}mv^2$.

21. A body of mass m is taken from a height h to $2h$. What is the increase in its potential energy?

Ans. Initial potential energy = mgh

Final potential energy = $mg \times 2h = 2mgh$

Increase in potential energy

= Final potential energy

– Initial potential energy

$= 2mgh - mgh = mgh.$

22. A ball of mass 50 g falls from a height of 2 m and rebounds from the ground to 1.6 m. Find:

(i) the potential energy possessed by the ball when initially at rest.

(ii) the kinetic energy of the ball before it hits the ground.

(iii) the final potential energy of the ball.

(iv) the loss in kinetic energy of the ball on collision.

[Take: g = 10 N kg^{-1}]

Ans. Given: $m = 50 \text{ g} = \frac{50}{1000} \text{ kg} = 0.05 \text{ kg}$, g = 10 N kg^{-1}, h_1 = 2 m, h_2 = 1.6 m.

(i) Initial potential energy possessed by the ball = $mgh_1 = 0.05 \times 10 \times 2 = 1.0$ J **Ans.**

(ii) The K.E. of the ball before it hits the ground

= Initial potential energy possessed by the ball

= 1.0 J **Ans.**

(iii) The final potential energy of the ball = mgh_2

$= 0.05 \times 10 \times 1.6 = 0.8$ J **Ans.**

(iv) The kinetic energy of the ball after striking the ground
= Final potential energy of the ball = 0.8 J

∴ The loss in kinetic energy of the ball on collision = (1·0 – 0·8) J = 0·2 J. **Ans.**

Chapter 2. Simple Machines

1. A block and tackle system of pulleys has velocity ratio 4.

(i) Draw a neat, labelled diagram of the system indicating clearly the points of application and direction of load and effort.

(ii) What will be its V. R. if the weight of the movable block is doubled ?*

Ans. (i) Diagram of a block and tackle system having velocity ratio = 4

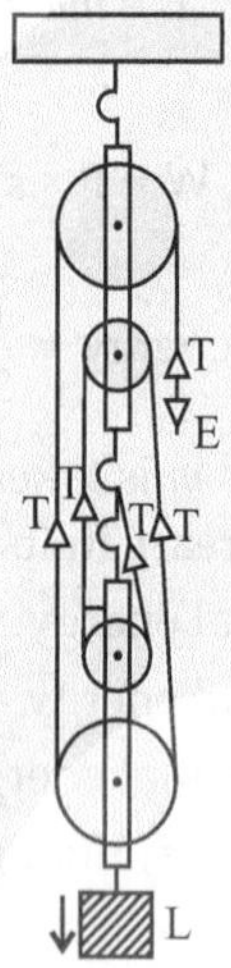

(ii) If weight of movable pulley is doubled there will be no change in the velocity ratio.

2. The diagram below shows a pulley arrangement :*

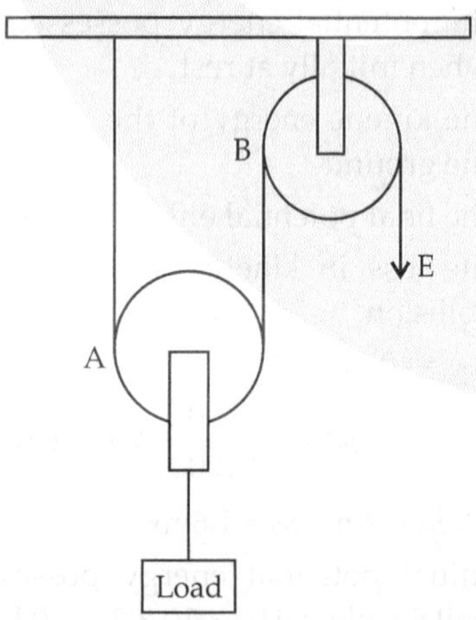

(i) Copy the diagram and mark the direction of tension on each strand of the string.

(ii) What is the velocity ratio of the arrangement ?

(iii) If the tension acting on the string is T, then what is the relationship between T and effort E ?

(iv) If the free end of the string moves through a distance x, find the distance by which the load is raised.

Ans. (i) The diagram with direction of tension on each strand is shown below.

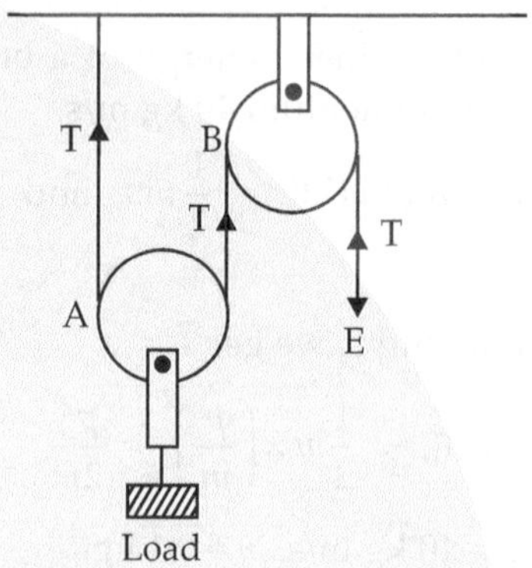

(ii) Velocity ratio = 2

(iii) E = T

(iv) Load is raised by a distance $\frac{x}{2}$.

3. A machine is driven by a 50 kg mass falling at a rate of 10·0 m in 5 s. It lifts a load of 250 kgf. Taking the force of gravity on 1 kg mass as 10 N. Calculate the power input to the machine. If the efficiency of the machine is 60%, find the height to which the load is raised in 5 s.*

Ans. Given: Effort, E = 50 kgf = 50 × 10 N = 500 N, d_E = 10·0 m, t = 5 s.

Load, L = 250 kgf = 250 × 10 N = 2500 N, Efficiency = 60% = 0.6

Power input to the machine

$$= \frac{\text{Effort} \times \text{Displacement of effort}}{\text{Time}}$$

$$= \frac{500 \times 10{\cdot}0}{5} = 1000 \text{ W.}$$ **Ans.**

Let the load is raised to a height d_L in 5 s, then

$$\text{Efficiency} = \frac{\text{Power output}}{\text{Power input}} = \frac{(L \times d_L)/t}{(E \times d_E)/t}$$

or $$0{\cdot}6 = \frac{2500 \times d_L}{500 \times 10}$$

or $$d_L = \frac{500 \times 10 \times 0{\cdot}6}{2500} = 1.2 \text{ m.}$$ **Ans.**

* **Frequently asked previous years Board Exam Questions.**

4. The mechanical advantage of a machine is 5 and its efficiency is 80%. It is used to lift a load of 200 kgf to a height of 20 m. Calculate:

(i) The effort required, and

(ii) The work done on the machine [$g = 10\ ms^{-2}$].

Ans. Given: M.A. = 5, Efficiency = 80% = 0.8

Load, L = 200 kgf = 200 × 10 N = 2,000 N, $d_L = 20$ m.

(i) $$\text{M.A.} = \frac{L}{E}$$

$$\therefore \quad \text{Effort, E} = \frac{L}{\text{M.A.}} = \frac{200}{5}\ \text{kgf} = 40\ \text{kgf}.$$ **Ans.**

(ii) Work output = Load × Displacement of load = 2,000 N × 20 m = 40,000 J

$$\text{Efficiency} = \frac{\text{Work output}}{\text{Work input}}$$

$$\therefore \quad \text{Work input} = \frac{\text{Work output}}{\text{Efficiency}} = \frac{40000}{0{\cdot}8}$$

= 50,000 J. **Ans.**

5. An effort of 10 kgf is applied on a machine through a distance of 40 cm, when a load of 50 kgf moves through a distance of 4 cm. Calculate:

(i) Velocity ratio.

(ii) Mechanical advantage.

(iii) Percentage efficiency of the machine.

Ans. (i) Velocity ratio (V.R.)

$$= \frac{\text{Displacement of effort } (d_E)}{\text{Displacement of load } (d_L)}$$

$$= \frac{40\ \text{cm}}{4\ \text{cm}} = 10$$

(ii) Mechanical advantage (M.A.)

$$= \frac{\text{Load}}{\text{Effort}} = \frac{50\ \text{kgf}}{10\ \text{kgf}} = 5$$

(iii) $$\text{Efficiency } (\eta) = \frac{\text{M.A.}}{\text{V.R.}} \times 100\% = \frac{5}{10} \times 100\%$$

= 50%. **Ans.**

6. A cook used a fire tong pair of length 32 cm to lift a piece of burning coal of mass 500 g. If he applies his effort at a distance of 8 cm from the fulcrum, what is his effort? Assume friction etc. to be absent. Also obtain values of the M.A. and the V.R. of this machine.

Ans. The fire tong pair is a lever of class III in which the point of application of the effort is in between the load and the fulcrum. Hence, when friction, etc., are assumed to be absent, we have :

Effort × 8 cm = 0.5 kgf × 32 cm

$$\therefore \quad \text{Effort} = 0.5 \times \frac{32}{8} = 2.0\ \text{kgf}.$$ **Ans.**

Thus, $$\text{M.A.} = \frac{\text{Load}}{\text{Effort}} = \frac{0.5}{2.0} = \frac{1}{4} = 0.25.$$ **Ans.**

Also $$\text{V.R.} = \frac{\text{Effort arm}}{\text{Load arm}} = \frac{8}{32} = 0.25.$$ **Ans.**

7. If weight of the movable pulley is W, obtain the expression for M.A., V.R. and efficiency of a single movable pulley.

Ans. If weight of the movable pulley with its frame is W, in equilibrium

$$E = T$$
$$L + W = 2T$$

or $$L + W = 2E$$

$\Rightarrow$ $$L = 2E - W$$

$$\text{M.A.} = \frac{L}{E} = \frac{2E - W}{E} = 2 - \frac{W}{E}$$

V.R. will remain same.

Now, $$\eta = \frac{\text{M.A.}}{\text{V.R.}} = \frac{2 - \frac{W}{E}}{2} = 1 - \frac{W}{2E}$$

If W << 2E, then

$$2E = L$$

or $$\eta = \left(1 - \frac{W}{L}\right).$$ **Ans.**

8. A simple machine enables us to lift a load of 10,000 N by the application of an effort of 500 N. However, when the point of application of the effort moves through 2.5 m, the load gets raised through 10 cm only. What are the values of the : (i) M.A. (ii) V.R. and (iii) Efficiency of the machine?

Ans. (i) $$\text{M.A.} = \frac{\text{Load}}{\text{Effort}} = \frac{10{,}000}{500} = 20$$ **Ans.**

(ii) $$\text{V.R.} = \frac{\text{Distance moved by the effort}}{\text{Distance moved by the load}} = \frac{2.5}{(10/100)} = 25$$ **Ans.**

(iii) $$\text{Efficiency} = \frac{\text{M.A.}}{\text{V.R.}} \times 100\% = \frac{20}{25} \times 100\% = \frac{4}{5} \times 100\% = 80\%$$ **Ans.**

9. A pulley system has a velocity ratio of 4 and an efficiency of 90%. Calculate :

(i) the mechanical advantage of the system.

(ii) the effort required to raise a load of 300 N by the system.

Ans. Given: V.R. = 4, η = 90%

(i) $$\text{M.A.} = \text{V.R.} \times \eta\%$$
$$= \frac{4 \times 90}{100} = 3{\cdot}6$$ **Ans.**

(ii) $$\text{M.A.} = \frac{\text{Load}}{\text{Effort}}$$
$$\therefore \quad \text{Effort} = \frac{\text{Load}}{\text{M.A.}}$$
$$= \frac{300}{3{\cdot}6}$$
$$= 83{\cdot}33 \text{ N}$$ **Ans.**

10. A 'block and tackle' system used 3 pulleys in the lower block and 4 pulleys in the upper block. What is the velocity ratio of this system ? If the load is to be lifted by a person capable of applying a maximum effort of 1000 N, what is the maximum load that can be lifted under ideal conditions?

The actual maximum load that gets lifted turns out to be 6300 N. What are the values of the actual M.A. and efficiency of the set up?

Ans. There being 7 pulleys in all, the load is supported by 7 strands of string. Hence it gets lifted only through $1/7^{th}$ of the distance through which the effort is applied. Hence velocity ratio of the set up has a value of 7.

Under ideal conditions,
$$\text{M.A.} = \text{V.R.} = 7$$

Hence, the maximum load that can be lifted under ideal conditions is:
$$\text{Effort} \times 7 = 1000 \text{ N} \times 7 = 7000 \text{ N.}$$

Under actual conditions,
$$\text{Load} = 6300 \text{ N}$$
and $$\text{Effort} = 1000 \text{ N}$$

Hence,
$$\text{Actual M.A.} = \frac{6300}{1000} = 6.3$$

Also $$\text{efficiency} = \frac{\text{M.A.}}{\text{V.R.}} \times 100\% = \frac{6.3}{7} \times 100\%$$
$$= 90\%.$$ **Ans.**

11. A boy lifts a load of 40 kgf through a vertical height of 2 m in 5 s by using a single fixed pulley when he applies an effort of 48 kgf. Calculate: (i) the mechanical advantage, and (ii) the efficiency of the pulley. Why is the efficiency of the pulley not 100%? (iii) the energy gained by the load in 5 s, and (iv) the power developed by the boy in raising the load.

Ans. Given: L = 40 kgf, d_L = 2 m, t = 5 s, E = 48 kgf.

(i) Mechanical advantage (M.A.)
$$= \frac{L}{E} = \frac{40}{48} = \frac{5}{6} = 0{\cdot}833$$ **Ans.**

(ii) If the effort moves a distance d downwards, the load also moves a distance d upwards.

So, velocity ratio (V.R.)
$$= d/d = 1$$
$$\text{Efficiency} = \frac{\text{M.A.}}{\text{V.R.}} \times 100\% = \frac{0{\cdot}833}{1} \times 100\%$$
$$= 83{\cdot}3\%.$$ **Ans.**

The efficiency of the pulley is not 100% because some energy is wasted in overcoming the friction in the pulley bearings.

(iii) The energy gained by the load in 5 s
$$= \text{Load} \times \text{displacement of load in 5 s}$$
$$= 40 \text{ kgf} \times 2 \text{ m} = 80 \text{ kgf} \times \text{m.}$$ **Ans.**

(iv) Power developed by the boy
$$= \frac{\text{Effort} \times \text{Displacement of effort}}{\text{Time}}$$
$$= \frac{48 \text{ kgf} \times 2 \text{ m}}{5 \text{ s}}$$
$$= 19{\cdot}2 \text{ kgf} \times \text{ms}^{-1}.$$ **Ans.**

Chapter 4. Refraction through lenses and Optical Instruments

1. A lens of focal length 20 cm forms an inverted image at a distance 60 cm from the lens.

(i) Identify the lens.

(ii) How far is the lens present in front of the object?

(iii) Calculate the magnification of the image.*

Ans. (i) Convex lens

(ii) f = + 20 cm, v = + 60 cm
$$\because \quad \frac{1}{f} = \frac{1}{v} - \frac{1}{u}$$
$$\Rightarrow \quad \frac{1}{20} = \frac{1}{60} - \frac{1}{u}$$
$$\Rightarrow \quad \frac{1}{u} = \frac{1}{60} - \frac{1}{20}$$

* Frequently asked previous years Board Exam Questions.

Long Numericals

$$= \frac{1-3}{60} = -\frac{2}{60} = -\frac{1}{30}$$

$\therefore \quad u = -30$ cm

The lens is at a distance 30 cm in front of the object.

(iii) $$m = \frac{v}{u} = \frac{+60}{-30} = -2$$

[– ve sign because image is real]

2. An object is placed at a distance 24 cm in front of a convex lens of focal length 8 cm.*

(i) What is the nature of the image so formed ?

(ii) Calculate the distance of the image from the lens.

(iii) Calculate the magnification of the image.

Ans. (i) A real, inverted and diminished image is formed.

(ii) Given : $u = -24$ cm, $f = +8$ cm

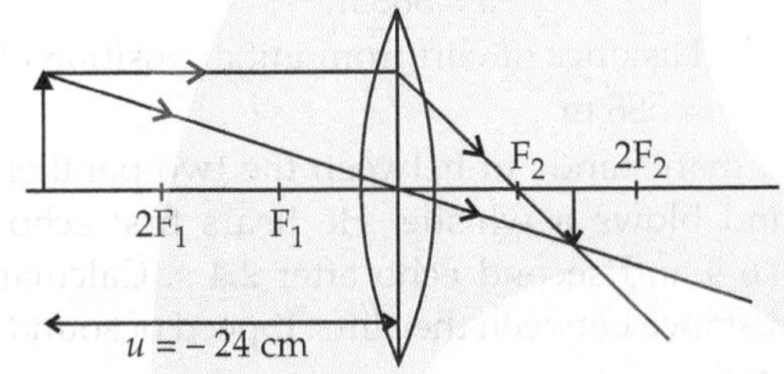

From the relation,

$$\frac{1}{v} - \frac{1}{u} = \frac{1}{f}$$

$$\Rightarrow \quad \frac{1}{v} = \frac{1}{u} + \frac{1}{f}$$

$$= \frac{1}{-24} + \frac{1}{8}$$

$$= \frac{-1+3}{24} = \frac{1}{12}$$

or $v = 12$ cm

Thus, the image is at a distance 12 cm behind the lens.

(iii) Magnification $(m) = \frac{v}{u} = \frac{12}{-24} = -\frac{1}{2}$

Negative sign signifies inverted image.

3. If there is a convex lens of focal length 75 cm and concave lens of focal length 40 cm, then calculate their combined power and combined focal length.

Ans. Power of convex lens $= \frac{100}{f(\text{cm})} = \frac{100}{75}$

$= +1.33$ D.

Power of concave lens $= \frac{100}{f(\text{cm})}$

$$= \frac{100}{-40} = -2.5 \text{ D.}$$

Their combined power

(P) = + 1.33 D + (–2.5 D)

P = – 1.17 D. **Ans.**

Their combined focal length

$$(f) = \frac{100}{P} = \frac{100}{-1.17}$$

$f = -85.47$ cm. **Ans.**

Negative sign indicates that the nature of their combination is concave.

4. The focal length of a camera lens is 40 cm. Find how far away from the film must the lens be set in order to photograph an object located at a distance of 80 cm from the lens.

Ans. Given: $f = +40$ cm, $u = -80$ cm

We know that,

$$\frac{1}{v} - \frac{1}{u} = \frac{1}{f}$$

$$\frac{1}{v} = \frac{1}{f} + \frac{1}{u}$$

$$\frac{1}{v} = \frac{1}{40} - \frac{1}{80}$$

$$\frac{1}{v} = \frac{2-1}{80} = \frac{1}{80}$$

or $v = 80$ cm

Thus, the lens must be set at a distance of 80 cm from the film towards the object.

5. If an object is placed at a distance of 12 cm from a convex lens of focal length 8 cm. Find the position of the image?

Ans. Given: $u = -12$ cm, $f = +8$ cm.

We know that,

$$\frac{1}{v} - \frac{1}{u} = \frac{1}{f}$$

$$\frac{1}{v} = \frac{1}{f} + \frac{1}{u}$$

$$\frac{1}{v} = \frac{1}{8} - \frac{1}{12}$$

$$\frac{1}{v} = \frac{3-2}{24} = \frac{1}{24}$$

$\Rightarrow \quad v = +24$ cm

Thus, the image is formed at a distance of 24 cm from the convex lens.

* **Frequently asked previous years Board Exam Questions.**

6. A convex lens of focal length 10 cm is placed at a distance of 12 cm from a wall. How far from the lens should an object be placed so as to form its real image on the wall?

Ans. Here, the real image is formed on the wall which is at a distance of 12 cm from the convex lens. This means that the distance of image from the convex lens or image distance will be 12 cm.

Now, $v = +12$ cm, $f = +10$ cm, $u = ?$

We know that,

$$\frac{1}{v} - \frac{1}{u} = \frac{1}{f}$$

$$\frac{1}{12} - \frac{1}{u} = \frac{1}{10}$$

$$\frac{1}{12} - \frac{1}{10} = \frac{1}{u}$$

$$\frac{5-6}{60} = \frac{1}{u}$$

$$\frac{-1}{60} = \frac{1}{u}$$

$$\Rightarrow \quad u = -60 \text{ cm}$$

Thus, the object should be placed at a distance of 60 cm in front of the convex lens.

Chapter 5. Echoes and Vibrations of Sound

1. (i) When a tuning fork [vibrating] is held close to ear, one hears a faint hum. The same [vibrating tuning fork] is held such that its stem is in contact with the table surface, then one hears a loud sound. Explain.*

(ii) A man standing in front of a vertical cliff fires a gun. He hears the echo after 3.5 seconds. On moving closer to the cliff by 84 m, he hears the echo after 3 seconds. Calculate the distance of the cliff from the initial position of the man.

Ans. (i) A tuning fork held close to ear, disturbs a small volume of air and hence, sound heard is faint, when the handle of the vibrating tuning fork is held against table, it sets up forced vibrators in the tabletop. As tabletop has a large surface area, large volume of air is set into vibration transmitting more energy thereby producing loud sound.

(ii) Let the initial position of man from the cliff be d metre

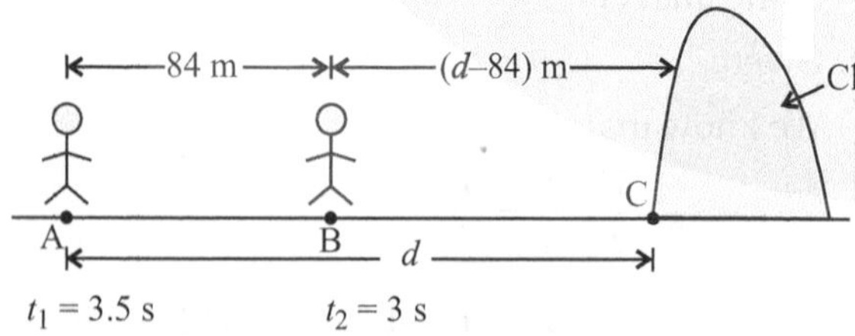

At Position	A At Position B
$d_1 = d$ metre	$d_2 = (d - 84)$ m
$t_1 = 3.5$ s	$t_2 = 3$ s

$$\text{Speed of sound} = \frac{2d}{t}$$

$$\Rightarrow \quad \frac{2d_1}{t_1} = \frac{2d_2}{t_2}$$

$$\Rightarrow \quad \frac{2d}{3.5} = \frac{2(d-84)}{3}$$

$$\Rightarrow \quad 6d = 7d - 588$$

$$\Rightarrow \quad d = 588 \text{ m}$$

$\therefore$ Distance of cliff from initial position of man = 588 m.

2. A man stands in between the two parallel cliffs and blows a whistle. He hears first echo after 0.6 s and second echo after 2.4 s. Calculate the distance between the cliffs. [Speed of sound = 336 m/s]

Ans. Given : t_1 = Time after which first echo is heard = 0.6 s

t_2 = Time after which second echo is heard = 2.4 s

v = Velocity of sound = 336 m/s.

For an echo the distance travelled by the sound is $2d$.

$$d_1 = \frac{v \times t_1}{2}$$

or, $$d_1 = \frac{336 \times 0.6}{2} = 100.8 \text{ m}$$

and $$d_2 = \frac{v \times t_2}{2}$$

or, $$d_2 = \frac{336 \times 2.4}{2} = 403.2 \text{ m}$$

$\therefore$ Total distance between two cliffs

$= d_1 + d_2$

$= (100.8 + 403.2)$ m $= 504$ m. **Ans.**

3. A boy stands in front of a cliff on the other side of a river. He fires a gun and hears an echo after 6 seconds. The boy then moves backward by 170 m and again fires the gun. He hears an echo after 7 seconds. Calculate (i) width of river, (ii) speed of sound.

*** Frequently asked previous years Board Exam Questions.**

Ans. (i) Let the width of river be x m.

When the boy stands on the bank of the river.

Now, $v = \frac{2d}{t}$

$v = \frac{2x}{6}$ ...(i)

(ii) When the boy moves 170 m away from the bank.

Distance of boy from cliff

$= x + 170$ m

Now, $v = \frac{2d}{t}$

or $v = \frac{2(x+170)}{7}$ ...(ii)

Comparing (i) and (ii), we get

$\frac{2x}{6} = \frac{2x+340}{7}$

$14x = 12x + 2040$

$x = 1020$ m

∴ Width of the river

$= 1020$ m. **Ans.**

Substituting value of x in (i), we get

$v = \frac{2 \times 1020}{6}$

$v = 340$ ms^{-1}. **Ans.**

4. A person standing between two vertical cliffs and is 640 m away from the nearest cliff. He shouted and heard the first echo after 4 s and the second echo after further 3 s. Calculate the speed of sound in air and the distance between the cliffs.

Ans. Given: $d_1 = 640$ m, $t_1 = 4$ s

$t_2 = 4 + 3 = 7$ s

First echo is heard from the nearest cliff.

Total distance travelled

$= 2d_1 = 2 \times 640 = 1280$ m

$v = \frac{2d_1}{t_1} = \frac{1280}{4} = 320$ ms^{-1}

Second echo is heard from the second cliff.

$v = \frac{2d_2}{t_2}$

$\Rightarrow \quad 320 = \frac{2d_2}{7}$

$d_2 = \frac{320 \times 7}{2} = 1120$ m

Hence, the distance between two cliffs

$= d_1 + d_2$

$= 640 + 1120 = 1760$ m. **Ans.**

5. A man standing in front of a vertical cliff fires a gun. He hears an echo after 3 s. On moving close to the cliff by 82.5 m, he fires again. This time he hears the echo after 2.5 s. Calculate the speed of sound, distance and initial position of man from the cliff.

Ans. Given: $t = 3$ s, $d = x$ m, $t_1 = 2.5$ s, $d_1 = (x - 82.5)$ m

Now, $v = \frac{2d}{t} = \frac{2x}{3}$ ...(i)

and $v = \frac{2d_1}{t_1} = \frac{2x - 165}{2.5}$ ...(ii)

Comparing (i) and (ii), we get

$\frac{2x}{3} = \frac{2x - 165}{2.5}$

or $5x = 6x - 165 \times 3$

$x = 495$ m

Substituting this value in (i), we get

$v = \frac{2 \times 495}{3} = 330$ ms^{-1}. **Ans.**

6. Waves produced on the surface of water are formed to move with a velocity of 24 ms^{-1}. If the wavelength of these waves equals 20 cm. Find the

(i) frequency (number of waves produced per second)

(ii) the time period (*i.e.*, the time required to produce one wave) for these waves.

Ans. (i) Let the frequency of the waves be n.

We have,

Wave velocity = Frequency × Wavelength

$24 \text{ ms}^{-1} = n \times (20 \times 10^{-2} \text{ m})$

or $n = \frac{24}{20 \times 10^{-2}}$

= 120 waves per second. **Ans.**

(ii) Time period $= \frac{1}{\text{Frequency}} = \left(\frac{1}{120}\right)$ s

$= 0{\cdot}00833$ s. **Ans.**

7. A man standing 25 m away from a wall produces a sound and receives the reflected sound.

(i) Calculate the time after which he receives the reflected sound if the speed of sound in air is 350 ms^{-1}.

(ii) Will the man be able to hear a distinct echo? Give a reason for your answer.

Ans. (i) We know that,

$v = \frac{2d}{t}$

Long Numericals

$\Rightarrow \quad 350 = \dfrac{2 \times 25}{t}$

$\therefore \quad t = \dfrac{50}{350} = \dfrac{1}{7}$ s = 0·14 s. **Ans.**

(ii) Echo will be heard because the conditions required for the formation of echo are fulfilled.

(a) 17 m minimum distance should be there.

(b) Persistence of hearing is 0·1 s.

As both the conditions are satisfied. So echo will be heard.

8. (i) A man stands at a distance of 68 m from a cliff and fires a gun. After what time interval will he hear the echo, if the speed of sound in air is 340 ms^{-1}?

(ii) If the man had been standing at a distance of 12 m from the cliff would he have heard a clear echo?

Ans. (i) Given: d = 68 m, v = 340 ms^{-1}.

$\therefore$ Time taken $= \dfrac{\text{Total distance}}{\text{Speed}} = \dfrac{2 \times 68}{340}$

$= \dfrac{2}{5} = 0{\cdot}4$ s

(ii) If the man had been standing at a distance of 12 m then

$t = \dfrac{2d}{v} = \dfrac{2 \times 12}{340} = \dfrac{24}{340} = 0{\cdot}07$ s

which is less than 0·1 sec.

Hence, the man cannot hear a clear echo.

9. A sound wavelength of 0.332 m has a time period of 10^{-3} s. If the time period is decreased to 10^{-4} s, calculate the wavelength and frequency of the new wave.

Ans. Wavelength, λ = 0·332 m

Time period, T = 10^{-3} s

Frequency $= \dfrac{1}{T} = \dfrac{1}{10^{-3}} = 10^3$ Hz

Now the time period is decreased and is now 10^{-4} s.

$\therefore$ New frequency $= \dfrac{1}{10^{-4}} = 10^4$ Hz **Ans.**

Speed of sound, $v = f \times \lambda = 10^3 \times 0{\cdot}332$

$= 332\ ms^{-1}$

Since, speed remains the same,

$\therefore$ New wavelength, $\lambda = \dfrac{v}{f} = \dfrac{332}{10^4} = 0{\cdot}0332$ m.

Ans.

10. The wavelength of waves produced on the surface of water is 10 cm. If the wave velocity is 25 ms^{-1}, calculate: (i) the number of waves produced in one second, and (ii) the time in which one wave is produced.

Ans. Given: $\lambda = 10$ cm $= \dfrac{10}{100}$ m = 0·1 m, v = 25 ms^{-1}

(i) We have, $v = f\lambda$

$f = \dfrac{v}{\lambda} = \dfrac{25}{0{\cdot}1} = 250$ Hz

$\therefore$ Number of waves produced in one second

= 250

(ii) $T = \dfrac{1}{f} = \dfrac{1}{250} = 0{\cdot}004\ s = 4 \times 10^{-3}$ s **Ans.**

Chapter 6. Electricity

1. A 10 m long wire of a particular material is of resistance 5 Ω. What will be the resistance: (i) of 5 m long wire of same material and same thickness, (ii) of 10 m long wire of same material but the double radius, (iii) if the wire is doubled on itself, (iv) if the wire is stretched to double its length?

Ans. Given: The resistance of 10 m long wire = 5 Ω

(i) R ∝ l for same thickness material so if the length of wire becomes half of previous value, the resistance will also become half of its previous value.

or $\quad R = \dfrac{1}{2} \times 5\ \Omega = 2.5\ \Omega$ **Ans.**

(ii) For the wire of double the radius, the area of cross-section ($a = \pi r^2$) of wire will be four times that of previous value. Since resistance, $R \propto \dfrac{1}{a}$. So, the resistance of new wire will be one-fourth of the previous value, *i.e.*, it will be $\dfrac{1}{4} \times 5\ \Omega = 1.25\ \Omega$ **Ans.**

(iii) On doubling the wire on itself the length is reduced to half and area of cross-section is doubled so the resistance is reduced to one-fourth *i.e.*, the resistance of the wire will become :

$\dfrac{1}{4} \times 5\ \Omega = 1.25\ \Omega$ **Ans.**

(iv) On stretching the wire to double its length, the length is doubled and its area of cross-section is reduced to half. So the resistance

Long Numericals

becomes four times *i.e.*, the resistance of wire will become:

$4 \times 5\ \Omega = 20\ \Omega$. **Ans.**

2.

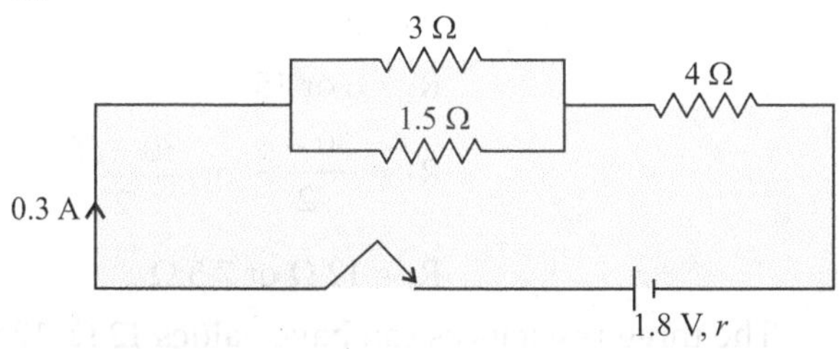

The diagram above shows three resistors connected across a cell of e.m.f. 1.8 V and internal resistance r. Calculate :

(i) Current through 3 Ω resistor.

(ii) The internal resistance *r*.

Ans. (i) Equivalent resistance of 3Ω and 1.5Ω in parallel,

$$R_1 = \frac{3 \times 1.5}{3 + 1.5}\Omega = 1\Omega$$

Potential difference, across R_1

$\Rightarrow \quad V_1 = IR_1$

$= 0.3 \times 1 = 0.3V$

∴ Current through 3Ω resistor,

$$I_1 = \frac{V_1}{3} = \frac{0.3}{3} = 0.1A$$

(ii) Total external resistance R = 1 + 4 = 5Ω

$\because \quad E = I(R + r)$

$\Rightarrow \quad 1.8 = 0.3(5 + r)$

$\Rightarrow \quad \frac{1.8}{0.3} = 5 + r \Rightarrow 6 = 5 + r$

$\therefore \quad r = 6 - 5 = 1\Omega$

3. A cell of e.m.f. 1·5 V and internal resistance 1·0 Ω is connected to two resistors of 4·0 Ω and 20·0 Ω in series.

Calculate the:

(i) Current in the circuit.

(ii) Potential difference across the 4·0 W resistor.

(iii) Voltage drop when the current is flowing.

(iv) Potential difference across the cell.

Ans. (i) Given: E.m.f. of cell = 1·5 V, Internal resistance = 1·0 Ω, $R_1 = 4{\cdot}0\ \Omega$, $R_2 = 20{\cdot}0\ \Omega$

External resistance (R)

$= R_1 + R_2 = 4 + 20 = 24\ \Omega$

$$\text{Current (I)} = \frac{E}{R + r}$$

or $$I = \frac{1{\cdot}5}{24 + 1} = \frac{1{\cdot}5}{25}$$

$= 0{\cdot}06$ A. **Ans.**

(ii) Potential difference across 4 Ω resistor

$= IR_1 = 0{\cdot}06 \times 4 = 0{\cdot}24$ V **Ans.**

(iii) Voltage drop $= Ir = 0{\cdot}06 \times 1 = 0{\cdot}06$ V **Ans.**

(iv) Potential difference across the cell

$= IR = 0{\cdot}06 \times 24$

$= 1{\cdot}44$ V. **Ans.**

4. When a resistance of 3 Ω is connected across a cell, the current flowing is 0.5 A. On changing the resistance to 7 Ω, the current becomes 0.25 A. Calculate the e.m.f. and the internal resistance of the cell.

Ans. We know that,

$$I = \frac{E}{(R + r)}$$

$$\therefore \quad 0.5 = \frac{E}{(3 + r)}$$

($\because$ I = 0.5 A when R = 3 Ω)

$\therefore \quad E = (3 + r) \times 0.5 \quad$ ...(i)

Also $$0{\cdot}25 = \frac{E}{7 + r}$$

($\because$ I = 0.25 A when R = 7 Ω)

$\therefore \quad E = (7 + r) \times 0.25 \quad$ ...(ii)

Comparing equations (i) and (ii), we get

$(3 + r) \times 0.5 = (7 + r) \times 0.25$

$\therefore \quad 1.5 + 0.5r = 1.75 + 0.25r$

$\therefore \quad 0.25r = 0.25$

Internal resistance,

$r = 1\ \Omega$

Now, $E = I(R + r)$

$\therefore \quad E = 0.5(3 + 1)$

($\because$ I = 0.5 A when R = 3 Ω, *r* = 1 Ω)

$\therefore \quad E = 1.5 + 0.5 = 2$ V

E.m.f. of the cell, E = 2 V **Ans.**

5. Three resistances of which two are equal when connected in series have an effective resistance of 30 ohms. When the three resistances are connected in parallel, the effective resistances is 3 ohms. Find the values of the individua resistances.

Ans. When connected in series,

$R = R_1 + R_1 + R_2$ (Two are equal)

$R = 2R_1 + R_2$

$\therefore \quad 2R_1 + R_2 = 30 \quad (\because R = 30\ \Omega)$

$\therefore \quad 2R_1 = 30 - R_2$

$$\therefore \quad R_1 = \frac{30 - R_2}{2} \quad \text{...(i)}$$

When connected in parallel,

$$\frac{1}{R} = \frac{1}{R_1} + \frac{1}{R_1} + \frac{1}{R_2}$$

Long Numericals

or $$\frac{1}{R} = \frac{1}{\frac{30-R_2}{2}} + \frac{1}{\frac{30-R_2}{2}} + \frac{1}{R_2}$$ [from equation (i)]

$$\therefore \quad \frac{1}{R} = \frac{2}{30-R_2} + \frac{2}{30-R_2} + \frac{1}{R_2}$$

$$\frac{1}{3} = \frac{4}{(30-R_2)} + \frac{1}{R_2} \quad (\because R = 3\ \Omega)$$

$$\therefore \quad \frac{1}{3} = \frac{4R_2 + 30 - R_2}{R_2(30-R_2)} = \frac{3R_2 + 30}{R_2(30-R_2)}$$

$$\therefore \quad R_2(30 - R_2) = 9R_2 + 90$$

$$30R_2 - R_2^2 = 9R_2 + 90$$

$$R_2^2 - 21R_2 + 90 = 0$$

$$R_2^2 - 6R_2 - 15R_2 + 90 = 0$$

$$\therefore \quad R_2 = 6 \text{ or } 15$$

$$\therefore \quad R_1 = \frac{30-6}{2} \text{ or } \frac{30-15}{2}$$

$$R_1 = 12\ \Omega \text{ or } 7.5\ \Omega.$$

The three resistances can have values 12 Ω, 12 Ω and 6 Ω or 15 Ω, 7.5 Ω and 7.5 Ω. **Ans.**

Chapter 7. Electric Power and Household Circuits

1. A bulb is connected to a battery of e.m.f. 6 V and internal resistance 2 Ω. A steady current of 0.5 A flows through the bulb. Calculate: (i) the total energy supplied by the battery in 10 minutes, (ii) the resistance of the bulb, (iii) the heat energy dissipated in the bulb in 10 minutes.

Ans. Given: E = 6 V, $r = 2\ \Omega$, $i = 0.5$ A, $t = 10$ minute = 10 × 60 s = 600 s.

(i) The total energy supplied by the battery

$$= Eit$$

$$= 6 \times 0.5 \times 600 = 1800 \text{ J.}$$

(ii) Let the resistance of bulb is R.

Then $$i = \frac{E}{R + r}$$

or $$0.5 = \frac{6}{R + 2}$$

or 0.5 (R + 2) = 6

or 0.5 R + 1 = 6

or 0.5R = 6 – 1 = 5

or $$R = \frac{5}{0.5} = 10\ \Omega$$

(iii) The heat energy dissipated in the bulb

$$= i^2Rt$$

$$= (0.5)^2 \times 10 \times 600 = 1500 \text{ J.}$$ **Ans.**

2. An immersion rod having resistance of 50 Ω is connected to 220 V main supply. Assuming that all the energy generated spend in heating the water, calculate the time taken to heat 5 kg water from 30°C to 100°C. [Take: $c = 4200$ J kg^{-1} K^{-1}]

Ans. We know that V = IR

$$\therefore \quad I = \frac{V}{R} = \frac{220}{50} = 4.4 \text{ A}$$

Let time taken be 't' second.

Heat produced in an immersion rod,

$$H = I^2Rt$$

$$H = (4.4)^2 \times 50 \times t$$

$$H = 968t \text{ joules.}$$

Heat needed to raise temperature of water

$$= mc\theta$$

$$= 5 \times 4200 \times (100 - 30) \text{ J}$$

$$= 1470000 \text{ J}$$

But Heat lost = Heat gained

$$968t = 1470000$$

$$t = \frac{1470000}{968}$$

$$t = 1518.59 \text{ s}$$

$$t = 25.31 \text{ minutes}$$ **Ans.**

3. An electric bulb is rated as 100 W, 250 V. What information does it convey? Calculate:

(i) the resistance of its filament while glowing, and

(ii) the current through the filament of the bulb.

Ans. The rating 100 W, 250 V on the bulb means that if the bulb is lighted on a 250 V supply, it consumes 100 W electrical power *i.e.,* 100 J electrical energy in 1 second.

Given: P = 100 W, V = 250 V.

(i) The resistance of filament,

$$R = \frac{V^2}{P} = \frac{(250)^2}{100} = 625\ \Omega.$$ **Ans.**

(ii) The current through the filament,

$$I = \frac{P}{V} = \frac{100}{250} = 0.4 \text{ A.}$$ **Ans.**

4. A battery of e.m.f. 12 V and internal resistance 1.6 Ω is connected to two resistors of 4 Ω and 6 Ω connected in parallel. Calculate: (i) the current drawn from the battery, (ii) the power dissipated in each resistor, (iii) the total power supplied by the battery.

Ans. Given : E = 12 V, $r = 1.6\ \Omega$, $R_1 = 4\ \Omega$, $R_2 = 6\ \Omega$

Equivalent resistance in parallel,

$$R = \frac{R_1R_2}{R_1 + R_2} = \frac{4 \times 6}{4 + 6} = 2.4\ \Omega$$

Total resistance of the circuit

$$= R + r = 2.4 + 1.6 = 4.0\ \Omega$$

(i) The current drawn from the battery,

$$I = \frac{E}{R + r} = \frac{12}{4.0} = 3.0 \text{ A.}$$ **Ans.**

(ii) The voltage drop in internal resistance,

$$= Ir = 3.0 \times 1.6 = 4.8 \text{ V}$$

Voltage across each resistor (connected in parallel),

$$V = E - Ir$$
$$= 12 - 4.8 = 7.2 \text{ V}$$

Power dissipated in resistance R_1,

$$P_1 = \frac{V^2}{R_1} = \frac{(7.2)^2}{4} = 12.96 \text{ W}$$

Power dissipated in resistance R_2,

$$P_2 = \frac{V^2}{R_2} = \frac{(7.2)^2}{6} = 8.64 \text{ W}$$

(iii) Total power supplied by the battery,

$$P = EI = 12 \times 3 = 36 \text{ W.}$$ **Ans.**

5. Two resistances $R_1 = 4\Omega$ and $R_2 = 6\Omega$ are connected in series. The combination is connected with a battery of e.m.f. 6 V and negligible internal resistance. Calculate: (i) the heat produced per minute in each resistor, (ii) the power supplied by the battery.

Ans. Given: $R_1 = 4\ \Omega$, $R_2 = 6\ \Omega$, E = 6 V, t = 1 minute = 60 s.

Total resistance of the circuit,

$$R = R_1 + R_2 = 4 + 6 = 10\ \Omega$$

Current through the battery,

$$I = \frac{E}{R} = \frac{6}{10} = 0.6 \text{ A}$$

The same current flows through each resistor.

(i) The heat produced in resistor R_1 is H_1

$$= I^2 R_1 t$$
$$= (0.6)^2 \times 4 \times 60 = 86.4 \text{ J}$$

The heat produced in resistor R_2 is H_2

$$= I^2 R_2 t$$
$$= (0.6)^2 \times 6 \times 60 = 129.6 \text{ J.}$$ **Ans.**

(ii) The power supplied by the battery

$$= E \times I$$
$$= 6 \times 0.6 = 3.6 \text{ W.}$$ **Ans.**

6. A bulb is marked 100 W, 220 V and an electric heater is marked 2000 W, 220 V.

(i) What is the ratio between the resistances of these two devices?

(ii) How does the power-voltage rating of a device help us to decide about the type of leads (connecting wires) to be used for it?

(iii) In which of the above two devices, a thicker lead (connecting wire) is required?

Ans. (i) For bulb: Let R_1 be the resistance of its filament wire.

$$P = \frac{V^2}{R}$$

$$\therefore \quad R_1 = \frac{V^2}{P_1}$$

$$= \frac{220 \times 220}{100}\ \Omega = 484\ \Omega$$

For heater: Let its resistance be R_2.

$$\therefore \quad R_2 = \frac{V^2}{P_2} = \frac{220 \times 220}{2000}\ \Omega = 24{\cdot}2\ \Omega.$$

i.e., $R_1 : R_2 = 484 : 24{\cdot}2$

$= 20 : 1$ **Ans.**

(ii) In the case of bulb,

$$\text{Current, } I_1 = \frac{P_1}{V} = \frac{100}{220} = \frac{5}{11} \text{ A} = 0.45 \text{ A}$$

In the case of heater,

$$\text{Current, } I_2 = \frac{P_2}{V} = \frac{2000}{220} = 9.09 \text{ A.}$$ **Ans.**

Hence power-voltage rating help us in this case. The current through the bulb is only 0.45 A while through the heater, it is 9.09 A. Hence a heavy lead (to avoid power loss due to heating effect) is needed for the heater while for a bulb, an ordinary thin connecting wire is required.

(iii) Electric heater requires a thicker wire of lead.

7. A house is provided with 15 bulbs of 40 W, 5 bulbs of 100 W, 5 fans of 80 W and one heater of 1.0 kW. Each day bulbs are used for 4 h, fans for 10 h and heater for 2 h. The voltage of mains is 220 V. Calculate:

(i) maximum power of the circuit in the house, (ii) maximum current capacity of the main fuse in the house, (iii) the electrical energy consumed in a week, (iv) cost of electricity consumed at 1.25 per kWh.

Ans. (i) Maximum power of the circuit

$$= (40 \times 15) + (100 \times 5) + (80 \times 5) + (1000 \times 1)$$
$$= 600 + 500 + 400 + 1000$$
$$= 2500 \text{ W} = 2.5 \text{ kW.}$$ **Ans.**

(ii) Maximum current capacity

$$= \frac{\text{Total wattage}}{\text{Voltage of mains}} = \frac{2500}{220}$$

= 11.36 (or 11.4) A. **Ans.**

(iii) Electrical energy consumed in a week

$$\text{In 40 W bulbs} = \frac{(40 \times 15) \times (4 \times 7)}{1000}$$

$$= 16.8 \text{ kWh}$$

$$\text{In 100 W bulbs} = \frac{(100 \times 5) \times (4 \times 7)}{1000}$$

$$= 14 \text{ kWh}$$

$$\text{In 80 W fans} = \frac{(80 \times 5) \times (10 \times 7)}{1000}$$

$$= 28 \text{ kWh}$$

$$\text{In 1 kW heater} = \frac{(1000 \times 1) \times (2 \times 7)}{1000}$$

$$= 14 \text{ kWh}$$

∴ Total electrical energy consumed

$$= 16.8 + 14 + 28 + 14$$

$$= 72.8 \text{ kWh.}$$ Ans.

(iv) Cost of electricity = 72.8 × 1.25 = ₹ 91 Ans.

8. Two bulbs are rated : bulb A 100 W, 120 V, bulb B 10 W, 120 V. If both are connected across a 120 V supply, which bulb will consume more energy. (i) When in series? (ii) When in parallel? Also calculate the current through each bulb in above cases.

Ans. In case of bulb A,

$$\text{Power, } P_A = \frac{V^2}{R_A}$$

$$\Rightarrow \quad R_A = \frac{V^2}{P_A} = \frac{120 \times 120}{100} = 144\ \Omega$$

i.e., Resistance (R_A) of bulb A

$$= 144\ \Omega$$

Similarly resistance R_B of bulb B

$$= \frac{V^2}{P_B} = \frac{(120 \times 120)}{10} = 1440\ \Omega$$

(i) Case I: When the bulbs are connected in series:

$$R = R_A + R_B = 144 + 1440 = 1584\ \Omega$$

∴ Current in the circuit

$$= \frac{V}{R} = \frac{120}{1584} = 0.076 \text{ A.}$$

Being in series, same current passes through both the bulbs.

Power consumed in bulb A,

$$P_A = i^2 R_A = (0.076)^2 \times 144 = 0.8317 \text{ W}$$

Power consumed in bulb B,

$$P_B = i^2 R_B = (0.076)^2 \times 1440 = 8.317 \text{ W}$$

This clearly shows that $P_A < P_B$, *i.e.,* bulb B (10 W, 120 V) consumes more energy when these are connected in series.

(ii) Case II: When the bulbs are connected in parallel:

Voltage is same (120 V) across each bulb.

∴ The current in bulb A or I_A

$$= \frac{V}{R_A} = \frac{120}{144} = 0.83\text{A}$$

∴ The current in bulb B or I_B

$$= \frac{120}{1440} = 0.083 \text{ A}$$

Hence, power consumed in bulb A,

$$P_A = \frac{V^2}{R_A} = \frac{120 \times 120}{144}$$

$$= 100 \text{ W.}$$

and power consumed in bulb B,

$$P_B = \frac{V^2}{R_B}$$

$$= \frac{120 \times 120}{1440}$$

$$= 10 \text{ W.}$$

Hence, bulb A (100 W, 120 V) consumes more energy than bulb B, when both are connected in parallel.

9. The following table gives the electrical appliances used, their power and the average time for which they are used each day in a home. Estimate the monthly electricity bill if the rate is 60 paise per unit.

S. No.	Name	Nos.	Power rating	Time/ day
(i)	Bulb	4	100 W	7.5 hr
(ii)	Fans	2	50 W	10 hr
(iii)	T.V.	1	100 W	2 hr
(iv)	Iron	1	500 W	1 hr
(v)	Electric stove	1	750 W	2 hr

Ans. We first calculate the total kWh consumption of each appliance per day. This is done below:

(i) Bulbs: kWh consumption

$$= \frac{4 \times 100}{1000} \text{ kW} \times 7.5 \text{ h} = 3 \text{ kWh}$$

(ii) Fans: kWh consumption

$$= 2 \times \frac{50}{1000} \text{ kW} \times 10 \text{ h} = 1 \text{ kWh}$$

(iii) T. V: kWh consumption

$$= 1 \times \frac{100}{1000} \text{ kW} \times 2 \text{ h} = 0.2 \text{ kWh}$$

(iv) Iron: kWh consumption

$$= 1 \times \frac{500}{1000} \text{ kW} \times 1 \text{ h} = 0.5 \text{ kWh}$$

(v) Electric stove: kWh consumption

$$= 1 \times \frac{750}{1000} \text{ kW} \times 2 \text{ h} = 1.5 \text{ kWh}$$

∴ Total kWh consumption per day

= (3 + 1 + 0.2 + 0.5 + 1.5) kWh

= 6.2 kWh = 6.2 unit.

∴ No. of units consumed in one month

= 6.2 × 30 unit = 186 units

∴ Monthly electricity bill

= 186 × 60 paise = ₹ 111.60. **Ans.**

10. A consumer uses 4 lamps of 60 watt, 2 lamps of 40 watt, and 2 lamps of 100 watt. All these are used for 6 hours daily. Find the total bill for 30 days when the rate of energy is 75 paise per unit and the meter rent is ₹ 1.

Ans. Total power consumption of 4 lamps of 60 watt each

= 60 × 4 = 240 W

Total power consumption of 2 lamps of 40 watt each

= 40 × 2 = 80 W

Total power consumption of 2 lamps of 100 watt each

= 100 × 2 = 200 W.

∴ Total power consumption

= (240 + 80 + 200) = 520 W

∴ Total watt hour for 6 hours a day for 30 days

= 520 × 6 × 30 = 93600 watt hour

$$\text{Hence, kilowatt hour} = \frac{93600}{1000} = 93.6 \text{ kWh.}$$

Cost of electric energy at 75 paise per kWh

$$= \frac{93.6 \times 75}{100} = ₹\ 70.20$$

∴ Total cost = ₹ 70.20 + ₹ 1.0 (Meter rent)

= ₹ 71.20. **Ans.**

11. Find the cost of operating an electric toaster for two hours if it draws 8 A current on a 110 volt circuit. The cost of electrical energy is ₹ 2·50 per kWh.

Ans. Given : Time = 2 hrs, I = 8 A, V = 110 V

Power = IV

= 8 × 110 W

= 880 W

Electrical energy (in kWh)

$$= \frac{\text{Power (in Watt)} \times \text{Time (in hrs)}}{1000}$$

$$= \frac{880 \times 2}{1000}$$

= 1·76 kWh

Cost of 1 kWh energy = ₹ 2·50

∴ Cost of 1·76 kWh energy = 2·50 × 1·76

= ₹ 4·40 **Ans.**

12. A bulb is rated at 100 W, 250 V and another one at 60 W, 250 V. What is the current flowing in the circuit if the two bulbs are put in series across a 220 V mains supply?

Ans. Resistance of first bulb $= \frac{V^2}{P_1} = \frac{250 \times 250}{100}\ \Omega$

= 625 Ω

Resistance of second bulb $= \frac{V^2}{P_2} = \frac{250 \times 250}{60}\ \Omega$

$$= \frac{3125}{3}\ \Omega$$

∴ Resistance of the circuit containing the two bulbs in series

$$= \left(625 + \frac{3125}{3}\right)\Omega = \frac{5000}{3}\ \Omega$$

∴ Current flowing $= 220 \div \left(\frac{5000}{3}\right)$ A

$$= \frac{220 \times 3}{5000} \text{ A}$$

$$= \frac{220 \times 3}{5000} \times 1000 \text{ mA}$$

= 132 mA. **Ans.**

13. The resistance of filament of an electric heater is 500 Ω. It is operated at 200 V for 1 hour daily. Calculate the current drawn by the heater and the energy consumed: (i) in joule (ii) in kWh, by the heater in a month of 30 days.

Ans. Given: R = 500 Ω, V = 200 V.

Total time, t = 1 × 30 h = 30 h = 30 × 60 × 60 s = 1,08,000 s.

Current drawn by the heater,

$$I = \frac{V}{R} = \frac{200}{500} = 0.4 \text{ A.}$$

(i) The energy consumed (in joule)

$$= \frac{V^2}{R} t$$

$$= \frac{(200)^2}{500} \times 1{,}08{,}000$$

= 86,40,000 J. **Ans.**

(ii) The energy consumed (in kWh)

$$= \frac{\text{Voltage} \times \text{Current}}{1000} \times \text{Time}$$

$$= \frac{200 \times 0.4}{1000} \times 30 = 2.4 \text{ kWh.}$$ **Ans.**

(iii) Cost of electrical energy

= ₹ 2·50 × 90

= ₹ 225. **Ans.**

14. An electrical appliance is rated 1500 W, 250 V. This appliance is connected to 250 V mains. Calculate :

(i) the current drawn,

(ii) the electrical energy consumed in 60 hours,

(iii) the cost of electrical energy consumed at ₹ 2.50 per kWh.

Ans. (i) Given: P = 1500 W, V = 250 V

$$\text{Current (I)} = \frac{P}{V}$$

$$= \frac{1500}{250} \text{ A}$$

= 6 A **Ans.**

(ii) $$\text{Energy consumed} = \frac{\text{Power} \times \text{Time}}{1000} \text{ kWh}$$

$$= \frac{1500 \times 60}{1000} \text{ kWh}$$

= 90 kWh. **Ans.**

15. An electric appliance is rated at 1000 W, 250 V. Calculate:

(i) The electrical energy consumed by the appliance in 12 hour.

(ii) The cost of energy consumed at ₹ 1.20 per kWh.

How long would this appliance take to use 1 kWh energy when operated at 250 V?

Ans. Given: P = 1000 W, V = 250 V, t = 12 h.

(i) Electrical energy consumed

$$= \frac{\text{Power}}{1000} \times \text{Time (in hr.)}$$

$$= \frac{1000}{1000} \times 12 = 12 \text{ kWh.}$$ **Ans.**

(ii) Cost = 12 × 1.20 = ₹ 14.40. **Ans.**

If it takes t hour to consume 1 kWh electric energy, then:

$$1 = \frac{1000 \times t}{1000}$$

or t = 1 hr. **Ans.**

Chapter 8. Specific Heat Capacity and Latent Heat

1. A Bunsen burner raises the temperature of 500 g of water from 10°C to 100°C in 5 minutes. What heat is supplied per second?

Ans. Given: m = mass of water = 500 g, θ_1 = Initial temperature = 10°C

c = Specific heat capacity of water = 4.2 J/g°C, θ_2 = Final temperature = 100°C

Let Q be the heat supplied by Bunsen burner.

Heat lost by Bunsen burner

= Heat gained by water

(Law of calorimetry)

$$Q = mc\,(\theta_2 - \theta_1)$$

$$Q = 500 \times 4.2 \times (100° - 10°)$$

$$Q = 500 \times 4.2 \times 90°$$

$$Q = 189000 \text{ J}$$

Heat supplied by burner per second

$$= \frac{\text{Total Heat}}{\text{Time}} = \frac{189000}{300}$$

= 630 J/s

Heat supplied per second is 630 J/s. **Ans.**

2. A body 'A' of mass m_1 of a substance of a specific heat capacity c_1 at a temperature θ_1 is mixed with a body 'B' of mass m_2 of other substance of specific heat capacity c_2 at a lower temperature θ_2. Deduce the expression for the temperature of the mixture. State the assumption made, if any.

Ans. Let the temperature of mixture be θ.

Fall in temperature of body,

$$A = \theta_1 - \theta$$

∴ Heat lost by the body,

$$A = m_1 \times c_1 \times (\theta_1 - \theta)$$

Rise in temperature of body,

$$B = \theta - \theta_2$$

∴ Heat gained by the body,

$$B = m_2 \times c_2 \times (\theta - \theta_2)$$

Assuming that there is no heat loss,

Heat lost by hot body A

= Heat gained by cold body B

i.e., $m_1c_1(\theta_1 - \theta) = m_2c_2(\theta - \theta_2)$

or $m_1c_1\theta_1 + m_2c_2\theta_2 = \theta\,(m_2c_2 + m_1c_1)$

or $$\theta = \frac{m_1c_1\theta_1 + m_2c_2\theta_2}{m_1c_1 + m_2c_2}$$

Assumption: It is assumed that there is no heat loss in this process.

3. A mass of 40 g of brass of specific heat capacity 0.85 Jg^{-1} K^{-1} is heated in an oven and then quickly transferred into 240 g of water at 30°C in a calorimeter of mass 60 g and specific heat capacity 0.4 $Jg^{-1}K^{-1}$. If the final temperature is 50°C. What was the temperature of the oven?

Ans.

Hot Body (Brass)	Cold Body (Water)	Copper Calori meter	Note
$m_1 = 40$ g	$m_2 =$ 240 g	$m_3 = 60$ g	(i) Temperature of oven is same as hot body
$\theta_1 = ?$ $c_1 = 0.85$ J $g^{-1}K^{-1}$	$\theta_2 = 30°C$ $c_2 = 4.2$ J $g^{-1}K^{-1}$	$\theta_2 = 30°C$ $c_3 = 0.4$ J $g^{-1}K^{-1}$	(ii) Temperature of calorimeter is same as cold body.

Final temperature, $\theta = 50°C$

$\therefore$ Fall in temperature of hot body

$$= \theta_1 - \theta = (\theta_1 - 50°)$$

Rise in temperature of cold body

$$= \theta - \theta_2 = (50° - 30°) = 20°C$$

Heat lost by hot body

= Heat gained by cold body + Copper calorimeter

$$m_1c_1 (\theta_1 - \theta) = m_2c_2 (\theta - \theta_2) + m_3c_3 (\theta - \theta_2)$$

$$40 \times 0.85 \times (\theta - 50°)$$

$$= (240 \times 4.2 \times 20°) + (60 \times 0.4 \times 20°)$$

$$40 \times 0.85 (\theta - 50°) = (20160 + 480)$$

$$(\theta - 50°) = \frac{20640}{40 \times 0.85}$$

$$\therefore \quad \theta = 607.05 + 50 = 657.05°C.$$

Temperature of oven = 657.05°C **Ans.**

4. 40 g of ice at 0°C is used to bring down the temperature of a certain mass of water at 60°C to 10°C. Find the mass of water used.

[Specific heat capacity of water = 4200 J kg^{-1} $°C^{-1}$]

[Specific latent heat of fusion of ice = 336 × 10^3 J kg^{-1}]

Ans. Let mass of water used = m g

By principle of calorimetry,

Heat given = Heat taken

$$m_1c_1\Delta\theta_1 = mL + m_2c_2 \Delta\theta_2$$

$$m \times 4{\cdot}2 \times (60° - 10°) = 40 \times 336 + 40 \times 4{\cdot}2 \times (10° - 0°)$$

$$m \times 4{\cdot}2 \times 50° = 40 \times 336 + 1680$$

$$m = 72 \text{ g}$$ **Ans.**

5. 250 g of water at 30 °C is present in a copper vessel of mass 50 g. Calculate the mass of ice required to bring down the temperature of the vessel and its contents to 5 °C.

Specific latent heat of fusion of ice
= 336 × 10^3 J kg^{-1}

Specific heat capacity of copper vessel
= 400 J kg^{-1} $°C^{-1}$

Specific heat capacity of water = 4200 J kg^{-1} $°C^{-1}$.

Ans. Given, Mass of water = 250 g

Initial temperature = 30°C

Mass of vessel = 50 g.

Final temperature = 5°C

Let mass of ice be m gm.

By the principle of calorimetry,

Heat given = Heat taken

$$250 \times 4{\cdot}2 \times (30° - 5°) + 50 \times 0{\cdot}4 \times 25°$$

$$= m \times 336 + m \times 4{\cdot}2 \times 5°$$

$$26250 + 500 = 336m + 21m.$$

$$26750 = 357m.$$

$$m = 74{\cdot}9 \text{ g.}$$ **Ans.**

6. The temperature of 600 g of cold water rise by 15°C when 300 g of hot water at 50°C was added to it. What was the initial temperature?

Ans. Let the initial temperature be t°C.

Final temperature will be $(t + 15)$°C.

Hence, fall in the temperature of hot water

$$= [50° - (t + 15°)] = (35 - t)°C.$$

Heat given out by hot water

= Heat taken up by cold water

or $300 \times 4.2 \times (35° - t)$

$$= 600 \times 4.2 \times 15°$$

or $\quad 35° - t = 30°$ or $t = 5°C$

$\therefore$ Initial temperature = 5°C. **Ans.**

7. Find the final temperature when a mass of 80 g of water at 100°C is mixed with a mass of 40 g of water at 25°C.

Ans. Mass of hot body, $m_1 = 80$ g , Mass of cold body, $m_2 = 40$ g

Initial temperature of hot body, $\theta_1 = 100°C$, Initial temperature of cold body, $\theta_2 = 25°C$

Let final temperature of mixture = θ°C

Heat lost by hot body

= Heat gained by cold body (Law of calorimetry)

$$m_1 C (\theta_1 - \theta) = m_2 C (\theta - \theta_2)$$

$$80 \times 4.2 \times (100° - \theta) = 40 \times 4.2 \times (\theta - 25°)$$

$$100° - \theta = \frac{40}{80}(\theta - 25°) = \frac{\theta - 25°}{2}$$

$$200° - 2\theta = \theta - 25°$$

$$\therefore \quad 3\theta = 225°$$

$$\theta = \frac{225°}{3} = 75°C.$$

Final temperature is 75°C. **Ans.**

8. Derive an expression for finding out the specific heat capacity of a body (solid) from the readings of an experiment given below:

(i) Mass of empty calorimeter (with stirrer) $= m_1$ g.

(ii) Mass of the metal piece = M g.

(iii) Mass of calorimeter and water $= m_2$ g.

(iv) Initial temperature of water $= t_1$°C.

(v) Temperature of hot solid (metal piece) $= t_2$°C.

(vi) Final temperature of the mixture $= t$°C

(vii) Specific heat capacity of calorimeter $= 0.4 \text{ J g}^{-1} °\text{C}^{-1}$

Ans. Calculations: Let 's' $\text{Jg}^{-1}°\text{C}^{-1}$ be the specific heat of the solid.

(i) Fall in temperature of hot solid $= (t_2 - t)$°C.

(ii) Rise in the temperature of calorimeter as well as water $= (t - t_1)$°C

(iii) Heat taken up by calorimeter $= m_1 \times 0.4 \times (t - t_1)$°C.

(iv) Heat taken up by water $= (m_2 - m_1) \times 4.2 \times (t - t_1)$°C

(v) Heat given out by hot solid $= M \times s \times (t_2 - t)$°C.

Now, Heat given out by hot solid
= Heat taken up by calorimeter
+ Heat taken up by water.

or $\quad Ms(t_2 - t) = m_1 \times 0.4 (t - t_1) + (m_2 - m_1) \times 4.2 \times (t - t_1)$

From this heat equation, 's' can be calculated when all other values are known.

9. A copper calorimeter of mass 50 g contains 100 g of water at 20°C. A metallic piece of mass 250 g is heated to 100°C and is then dropped into the calorimeter. The contents of the calorimeter are well stirred and its final highest temperature is recorded to be 28°C. If the specific heat capacity of water is 4.2 J/g°C and of copper is 0.4 J/g°C, find: (i) the heat gained by water, (ii) the heat gained by calorimeter, (iii) total heat supplied by the metal piece, and (iv) the specific heat capacity of metal.

Ans. (i) Heat gained by water
= Mass of water
× Specific heat capacity of water
× Rise in temperature
$= 100 \times 4.2 \times (28° - 20°) = 3360$ J.

(ii) Heat gained by calorimeter
= Mass of calorimeter
× Specific heat capacity of copper
× Rise in temperature
$= 50 \times 0.4 \times (28° - 20°) = 160$ J

(iii) Total heat supplied by the metal piece
$= 3360 + 160 = 3520$ J. **Ans.**

(iv) If the specific heat capacity of metal is c, then heat supplied by the metal piece
= Mass × Specific heat capacity
× Fall in temperature

or $\quad 3520 = 250 \times c \times (100° - 28°)$

or $\quad c = \frac{3250}{250 \times 72} = 0.195$ J/g°C. **Ans.**

10. 20 g of ice at 0°C is added to 200 g of water at 20°C. Calculate the drop in temperature ignoring the heat capacity of the container. (Specific latent heat of ice = 80 cal g^{-1}, Specific heat capacity of water = 1 cal $g^{-1}°C^{-1}$).

Ans. Let x be the drop in temperature of water.

Heat lost by water $= 200 \times 1 \times x = 200 \times x$

Heat gained by ice in melting
$= 20 \times 80 = 1600$ calories.

Heat gained by the water from ice in raising the temperature from 0°C to $(20 - x)$°C
$= 20 \times 1 \times (20 - x)$.

Heat gained = Heat lost

$$\therefore \quad 200 \times x = 20 \times 80 + 20(20 - x)$$

$$\text{or} \quad 200x + 20x = 400 + 1600$$

$$x = \frac{2000}{220} = 9.1°C.$$ **Ans.**

11. A piece of ice of mass 60 g is dropped into 140 g of water at 50°C.

Calculate the final temperature of water when all the ice has melted.

(Assume no heat is lost to the surrounding)

Specific heat capacity of water = 4.2 $Jg^{-1}k^{-1}$

Specific latent heat of fusion of ice = 336 Jg^{-1}

* Frequently asked previous years Board Exam Questions.

Ans. Let final temperature of water = x°C

Ice	Water
$m_1 = 60$ g	$m_2 = 140$ g
$T_1 = 0$°C	$T_1 = 50$°C
$T_2 = x$°C	$T_2 = x$°C
Rise in temp.	Fall in temp.
$(\Delta T) = (x - 0)$°C $= x$°C	$(\Delta T) = (50 - x)$°C

Heat gained by ice $= M_1L + m_1c\Delta T$

$= (60 \times 336 + 60 \times 4.2 \times x)$ J

Heat lost by water $= m_2c\Delta T$

$= 140 \times 4.2 \times (50 - x)$ J

Applying, the principle of mixtures,

$$140 \times 4.2 \times (50 - x) = 60 \times 336 + 60 \times 4.2 \times x$$

$$\Rightarrow 4.2\,(7000 - 140x - 60x) = 60 \times 336$$

$$\Rightarrow \quad 7000 - 200x = \frac{60 \times 336}{4.2}$$

$$= \frac{60 \times 336}{42} \times 10$$

$$= 4800$$

$$\Rightarrow \quad 200x = 2200$$

$$\Rightarrow \quad x = 11°C$$

$\Rightarrow$ Final temperature of water = 11°C

12. 104 g of water at 30°C is taken in a calorimeter made of copper of mass 42 g. When a certain mass of ice at 0°C is added to it, the final steady temperature of the mixture after the ice has melted, was found to be 10°C. Find the mass of ice added. [Specific heat capacity of water = 4.2 J g^{-1} °C^{-1}; Specific latent heat of fusion of ice = 336 J g^{-1}; Specific heat capacity of copper = 0.4 Jg^{-1}° C^{-1}]*

Ans. Given: $m_w = 104$ g $= T_w = 30$°C

$m_c = 42$ g, T = 10°C

$m_i = ?$

By calorimetry,

Heat lost = Heat gained

$$m_w S_w (T_w - T) + m_c S_c (T_w - T)$$

$$= m_i \alpha + m_i S_w (T - T_i)$$

$$\Rightarrow (104)(4.2)(30 - 10) + (42)(0.4)(30 - 10)$$

$$= m_i(336) + m_i(4.2)(10 - 0)$$

$$\Rightarrow \quad m_i = \frac{(104)(4.2)(20) + (42)(0.4)(20)}{(336 + 42)}$$

$$m_i = 24 \text{ g}$$

13. Calculate the amount of ice, which is sufficient to cool 200 g of water, contained in a vessel of mass 50 g at 30°C, such that the final temperature of the mixture is 10°C. (Specific heat capacity of vessel = 0·4 J g^{-1}°C^{-1}, Specific heat capacity of water = 4·2 J g^{-1}°C^{-1}, Specific latent heat of ice = 336 J g^{-1}).

Ans. Let the mass be x gram.

Heat gained by ice to form water at 0°C

$= x \times 336$ J

Heat gained by water formed from ice

$= x \times 4{\cdot}2 \times 10$ J

$= 42\,x$ J

Total heat gained $= 336x + 42x = 378x$ J

Heat lost by water at 30°C

$= 200 \times 4{\cdot}2 \times (30° - 10°)$ J

$= 16800$ J

Heat lost by vessel at 30°C

$= 50 \times 0{\cdot}4 \times (30° - 10°)$ J

$= 400$ J

Total heat lost $= (16800 + 400)$ J

$= 17200$ J

From the principle of calorimetry,

Heat gained = Heat lost

$$378\,x = 17200$$

$$\therefore \quad x = \frac{17200}{378} \text{ g}$$

$= 45{\cdot}50$ g. **Ans.**

14. (i) A molten metal weighing 150 g is kept at its melting point 800°C. When it is allowed to solidify at the same temperature, it gives out 75000 J of heat. What is the specific latent heat of the metal?

(ii) If its specific heat capacity is 200 J kg^{-1} K^{-1}, how much additional heat will it give out in cooling to –50°C?

Ans. Let 'L' be the specific latent heat of metal.

(i) When the metal is allowed to solidify at its melting point, i.e., 800°C, it will give out its latent heat only.

Therefore, mL = 75000

or $\quad 0.150 \times L = 75000$

$$\therefore \quad L = \frac{75000}{0{\cdot}150} = 500000 \text{ Jkg}^{-1}$$ **Ans.**

(ii) Fall in temperature of the metal piece,

$\Delta t = [800° - (-50°)] = 850$°C

Heat given out by the metal piece in cooling down to –50°C can be calculated as

Heat given out $= m \times s \times \Delta t$

$= (0.150 \times 200 \times 850)$ J

$= 25500$ J. **Ans.**

15. Heat energy is supplied at a constant rate to 100 g of ice at 0°C. The ice is converted into water at 0°C in 2 minutes. How much time will be required to raise the temperature of water from 0°C to 20°C? [Given: specific heat capacity of water = 4.2 g^{-1}°C^{-1}, specific latent heat of ice = 336 Jg^{-1}]

Ans. $m_{ice} = 100\text{ g} = 0.1\text{ kg}.$

Heat energy required to raise the temperature of water from 0°C to 20°C, Q

$$Q = mc\theta = 0.1 \times 4200 \times 20$$
$$= 420 \times 20 = 8400\text{ J}$$

Heat energy required for conversion of ice into water at 0°C,

$$Q' = mL$$
$$= 0.1 \times 336000 = 33600\text{ J}$$

$$\text{Power, } P = \frac{Q'}{t}$$

$$P = \frac{33600}{120} = \frac{3360}{12}$$
$$= 280\text{ W}$$

$$\text{Time, } t = \frac{Q}{P} = \frac{8400}{280} = 30\text{ s.}$$ **Ans.**

16. 50 g of ice at 0°C is added to 300 g of a liquid at 30°C. What will be the final temperature of the mixture when all the ice has melted? The specific heat capacity of the liquid is 2.65 J g^{-1}°C^{-1} while that of water is 4.2 J g^{-1} °C^{-1}. Specific latent heat of fusion of ice = 336 J g^{-1}.

Ans. Given: $m_1 = 50$ g, $t_1 = 0$°C, $m_2 = 300$ g, $t_2 = 30$°C.

Heat energy taken by ice to melt

$$= 50 \times 336\text{ J} = 16800\text{ J}$$

By law of conservation of energy,

Heat energy given by liquid

= Heat taken by water

$$300 \times 2{\cdot}65 \times (30 - t) = 16800 + 50 \times t \times 4{\cdot}2$$
$$\Rightarrow \quad 23850 - 795\,t = 16800 + 210\,t$$
$$\Rightarrow \quad -795\,t - 210\,t = 16800 - 23850$$
$$\Rightarrow \quad -1005\,t = -7050$$
$$\Rightarrow \quad t = \frac{7050}{1005} = 7{\cdot}014\text{°C}$$

17. Calculate the total amount of heat required to convert 100 g of ice at 0°C to water at 100°C. (Specific latent heat of fusion of ice = 336 J/g, specific heat capacity of water = 4.2 J/g°C).

Ans. Heat required to convert 100 g of ice at 0°C to water at 0°C

= Mass × Specific latent heat of fusion of ice

= 100 × 336 = 33600 J

Heat required to raise the temperature of 100 g water from 0°C to 100°C

= Mass × Specific heat capacity of water × Rise in temperature

= 100 × 4.2 × (100 – 0) = 42000 J

Total heat required = (33600 + 42000) J

= 75600 J. **Ans.**

18. A metal ball of mass 500 g and at 400°C is placed on a block of ice, 238 g of ice melts. If the specific heat capacity of the metal ball is 0·4 J g^{-1}°C^{-1}, calculate specific latent heat of ice.

Ans. Let specific latent heat of ice = L Jg^{-1}

Heat gained by ice to melt = mL

= 238 × L joule

Heat lost by the metal ball

$$= mc\theta$$
$$= 500 \times 0{\cdot}4 \times (400° - 0°)$$
$$= 80000\text{ J}$$

From the principle of calorimetry,

Heat gained = Heat lost

$$238 \times L = 80000$$
$$L = \frac{80000}{238}\text{ Jg}^{-1}$$
$$= 336{\cdot}13\text{ Jg}^{-1}.$$ **Ans.**

19. What will be the result of mixing 500 g of copper chips at 500°C with 500 g of crushed ice at 0°C? (Specific heat capacity of copper = 0·4 J g^{-1} K^{-1} and specific latent heat of fusion of ice = 336 J g^{-1}).

Ans. Heat lost by copper chips in cooling from 500°C to 0°C,

$$Q_1 = 500 \times 0{\cdot}4 \times (500° - 0°)\text{ J}$$
$$= 100000\text{ J}$$

Heat required to melt 500 g ice at 0°C,

$$Q_2 = 500 \times 336 = 168000\text{ J}$$

Since, $Q_2 > Q_1$, therefore all ice will not melt.

Let x gram of ice melt, then

$$x \times 336 = 100000$$
$$\therefore \quad x = \frac{100000}{336}\text{ g} = 297{\cdot}62\text{ g}$$

Thus, only 297·62 g ice will melt and temperature will remain 0°C. **Ans.**

20. Heat energy is supplied at a constant rate to 200 g of ice at 0°C. The ice is converted into water at 0°C in 4 minutes. How much time will be required to raise the temperature of water from 0°C to 90°C? (Specific latent heat of ice = 336 Jg^{-1}, specific heat capacity of water = 4·2 J g^{-1} K^{-1}).

Ans. Heat energy required to melt 200 g of ice at 0°C,

$$Q_1 = mL$$
$$= 200 \times 336\text{ J}$$
$$= 67200\text{ J}$$

Time taken = 4 minutes

= 4 × 60 s = 240 s

∴ Heat energy supplied per second

$$= \frac{67200}{240}\text{ J s}^{-1} = 280\text{ Js}^{-1}$$

Heat energy required to raise the temperature of 200 g water from 0°C to 90°C,

$$Q_2 = 200 \times 4{\cdot}2 \times (90° - 0°)\text{ J}$$
$$= 75600\text{ J}$$

Let the time required for 200 g water to raise its temperature from 0°C to 90°C be t second.

$$\therefore \quad 280 \times t = 75600$$

$$\text{Time required } (t) = \frac{75600}{280}\text{ s}$$
$$= 270\text{ s}$$
$$= 4 \text{ minutes } 30 \text{ second.}$$ **Ans.**

Chapter 9. Radioactivity and Nuclear Energy

1. A certain nucleus X has a mass number 15 and atomic number 7. Find the number of neutrons. The nucleus X loses (i) one proton (ii) one β-particle, (iii) one α-particle. Express each change and write the symbol of the new nucleus in each case.

Ans. Mass number = Number of protons + Number of neutrons.

Atomic number = Number of protons

∴ Number of neutrons = Mass number − Atomic number

= 15 − 7 = 8

Element X can be written as ${}^{15}{}_{7}X$.

(i) After the loss of 1 proton, the mass number and atomic number will decrease by 1. The change can be written as :

$${}^{15}_{7}X \rightarrow {}^{14}_{6}Y + {}^{1}_{1}p$$

(ii) After the loss of one β-particle, the mass number will remain the same, the atomic number will increase by 1. The element X changes to Z.

$${}^{15}_{7}X \rightarrow {}^{15}_{8}Z + {}^{0}_{-1}e$$

(iii) After the loss of one α-particle, the mass number is decreased by 4 and atomic number is decreased by 2. The element X changes to S.

$${}^{15}_{7}X \rightarrow {}^{11}_{5}X + {}^{4}_{2}He$$

2. An element ${}_{92}U^{238}$ loses one alpha particle, and then one β-particle and then again one β-particle to get the final nucleus ${}_{b}Q^{a}$.

(i) Write down the values of a and b.

(ii) State the relationship between the nuclei P and Q.

Ans. (i) Calculation:

$${}_{92}P^{238} \xrightarrow{\alpha\text{-particle}} {}_{90}X^{234} \xrightarrow{\beta\text{-particle}}$$

$${}_{91}Y^{234} \xrightarrow{\beta\text{-particle}} {}_{92}Q^{234}$$

${}_{b}Q^{a}$ and ${}_{92}Q^{234}$ being same. Hence, a = 234 and b = 92

(ii) On examining P and Q, we find that their mass number are different, *i.e.*, 238 and 234, but both have the same atomic number, *i.e.*, 92, hence P and Q are isotopes.

3. An isotope of uranium is ${}_{92}Y^{235}$, *i.e.*, it has mass number 235 and atomic number 92.

(i) Find the number of electrons in the neutral atom of this isotope.

(ii) Find the number of protons and neutrons in its nucleus.

(iii) Do all isotopes have the same number of neutrons?

(iv) What is the number of protons in ${}_{92}U^{238}$?

(v) Mention the relationship between ${}_{92}U^{235}$ and ${}_{92}U^{238}$.

Ans. (i) Since it is a neutral atom, the number of protons will be equal to the number of electons.

Number of protons = Atomic number = 92

∴ Number of electrons = 92

(ii) Number of protons = 92

Atomic mass = Number of protons + Number of neutrons.

or 235 = 92 + Number of neutrons

∴ Number of neutrons = 235 − 92 = 143

(iii) No. Isotopes of an element have the same atomic number but they differ in their atomic masses, and hence in such isotopes, the number of neutrons will be different.

(iv) In ${}_{92}U^{238}$, the atomic number is 92 and so the number of protons will also be 92.

(v) In ${}_{92}U^{238}$ and ${}_{92}U^{235}$, the difference is in their atomic masses but their atomic number is the same *i.e.*, both have the same atomic number 92. The elements, that differ in their atomic masses but have same atomic number are called isotopes, hence ${}_{92}U^{238}$ and ${}_{92}U^{235}$ are isotopes of the element uranium.

4. If 200 MeV energy is released due to fission of each nucleus of U-235, what mass of U-235 undergoes fission per hour in a reactor of power 600 MW? (Take 1 a.m.u. = $1{\cdot}66 \times 10^{-27}$ kg)

Ans. Given :

Power of reactor = 600 MW = 600×10^6 W

Energy obtained in one hour (or 3600 s) from the reactor

$$= \text{Power} \times \text{Time}$$
$$= 600 \times 10^6 \times 3600$$
$$= 2{\cdot}16 \times 10^{12} \text{ J} \quad \ldots\text{(i)}$$

Energy released due to fission of one U-235 nucleus

$$= 200 \text{ MeV}$$
$$= 200 \times 1{\cdot}6 \times 10^{-13} \text{ J}$$
$$= 3{\cdot}2 \times 10^{-11} \text{ J} \quad \ldots\text{(ii)}$$

From equations (i) and (ii),

Number of U-235 nuclei fissioned per hour

$$= \frac{2{\cdot}16 \times 10^{12}}{3{\cdot}2 \times 10^{-11}}$$
$$= 6{\cdot}75 \times 10^{22}$$

Since mass of $6{\cdot}02 \times 10^{23}$ U-235 nuclei is 235 g.

∴ Mass of $6{\cdot}75 \times 10^{22}$ nuclei

$$= \frac{235}{6{\cdot}02 \times 10^{23}} \times 6{\cdot}75 \times 10^{22}$$
$$= 26{\cdot}34 \text{ g.}$$ **Ans.**

5. If in nuclear fission of a piece of uranium, 0·5 g mass is lost, how much energy in kWh is obtained?

Ans. Given : Loss of mass,

$$\Delta m = 0{\cdot}5 \text{ g} = 0{\cdot}5 \times 10^{-3} \text{ kg}$$

The energy released in fission

$$E = (\Delta m)\, c^2$$
$$= (0{\cdot}5 \times 10^{-3}) \times (3 \times 10^8)^2 \text{ J}$$
$$= 4{\cdot}5 \times 10^{13} \text{ J}$$

Since 1 kWh = $3{\cdot}6 \times 10^6$ joule

$$\therefore \quad E = \frac{4{\cdot}5 \times 10^{13}}{3{\cdot}6 \times 10^6} \text{ kWh}$$
$$= 1{\cdot}25 \times 10^7 \text{ kWh.}$$ **Ans.**

Long Answers

Set 7

Chapter 1. Force, Work, Power and Energy

1. Why is it easier to open a door by applying the force at the free end of it?

Ans. If we open or shut a door, we need to apply less force at the free end because according to the formula of moment of force,

τ = Force × Perpendicular distance of force from axis of rotation

Magnitude of force is inversely proportional to the perpendicular distance of force from axis of rotation. Thus, as the distance between the point of force application and hinge will increase, the amount of force required to open the door will decrease.

2. What is meant by equilibrium and state the conditions of equilibrium of a body.

Ans. Equilibrium: When a number of forces acting on a body produce no change in its state of rest or of uniform motion, the body is said to be in equilibrium.

The two conditions of equilibrium of a body:

(i) The resultant of all the forces acting on the body should be equal to zero.

(ii) The algebraic sum of moments of all the forces acting on the body about a point should be zero.

3. What do you mean by dynamic and static equilibrium? Give one example of each.

Ans. When the body remains in the state of rest under the influence of applied forces, it is in static equilibrium and if it remains in the state of uniform motion, it is in dynamic equilibrium.

For example: In a beam balance, when the beam is balanced in the horizontal position, the clockwise and anticlockwise moments balance each other and the beam has no rotational motion *i.e.*, it is in static equilibrium.

A body moving with uniform velocity in a straight line over a frictionless surface is in dynamic equilibrium.

4. Explain the term 'centre of gravity' of a body.

Ans. A body can be considered to be made up of a number of particles, each particle has weight 'W'. Weight of all these particles act vertically downward and can be replaced by a single resultant force. This force passes through a fixed point 'G' which is a point where algebraic sum of moments of weight of all the particles is zero, and is called centre of gravity of the body.

Thus we can define 'C.G.' as the point through which the weight of the body acts irrespective of the position of the body.

5. State two methods of increasing the stability of the body.

Ans. The stability of a body is increased if:

(i) The centre of gravity is as low as possible. It should be above the base and near the geometric centre of the body. This is the case when mass of the body is large at the bottom.

(ii) The area of base of the body is large so that the vertical line through the C.G. of the body passes through its base. If the body is slightly tilted, its C.G. rises higher and it has a tendency to regain its original position to lower the C.G.

6. State three differences between work and power.

Ans.

Work	Power
(i) Work done does not depend on time.	Power spent depends on the time in which work is done.
(ii) Work done by a force is equal to the product of force and displacement in the direction of force.	Power of a source is the rate of work done by it.
(iii) S.I. unit of work is joule (J).	S.I. unit of power is watt (W).

7. Name the three forms of kinetic energy and give one example of each.

Ans. The three forms of kinetic energy are:

(i) **Translational kinetic energy**, example, a boy running in a straight path has translational kinetic energy.

(ii) **Rotational kinetic energy**, example, a rotating wheel has rotational kinetic energy.

(iii) **Vibrational kinetic energy, example**, a vibrating tuning fork has vibrational kinetic energy.

8. What is meant by gravitational potential energy? Derive expression for it.

Ans. The potential energy possessed by a body due to force of attraction of the earth is called its gravitational potential energy.

Work done (W) on a body in lifting it to a height (h) is,

W = Force of gravity (mg)

× displacement in direction of force (h) = mgh

The work done in lifting the body up, is stored in the body in the form of its gravitational potential energy when it is at a height 'h'.

Thus, gravitational potential energy (U) = mgh.

9. State three differences between potential energy and kinetic energy.

Ans.

Potential energy	Kinetic energy
(i) It is the energy possessed by a body due to its changed position or its changed configuration.	It is the energy possessed by a body due to its state of motion.
(ii) It is equal to the work done in bringing the body to its changed position or changed configuration.	It is equal to the work that a moving body can do before coming to rest.
(iii) It can change only in the form of kinetic energy.	It can change into my other form.

Chapter 2. Simple Machines

1. How does a machine work?

Ans. A machine does some useful work when energy is supplied to it.

Energy supplied to machine (or input)

= Effort × Displacement of the point of application of effort

Work obtained from machine (or output)

= Load × Displacement of the point of application of load

For an ideal machine, Work output = Work input.

2. Give an example for each of the following use of a machine:

(i) To change the point of application of the force.

(ii) To obtain gain in force.

(iii) To change the direction of force.

Ans. (i) Rotating the wheel of a cycle by applying force on the paddle with the help of chain.

(ii) Lifting a car with a jack.

(iii) Lifting a bucket full of water from a well using a single fixed pulley.

3. State four ways in which machines are useful to us.

Ans. (i) A machine can be used to obtain a gain in speed.

(ii) A machine can be used to lift a load in one direction by applying an effort in a convenient direction.

(iii) A machine can be used to lift a heavy load by applying a less effort.

(iv) A machine can be used to overcome a load at some point by applying an effort at a convenient point.

4. What is efficiency of a machine? Establish a relation between efficiency, M.A. and V.R.

Ans. The ratio between work output and work input is called efficiency.

$$\text{Efficiency} = \frac{\text{Work output}}{\text{Work input}}$$

Relation between M.A., V.R. and Efficiency: Let a machine overcome a load L by applying an effort E. In time t, the displacement of effort is d_E and displacement of load is d_L. Then

Work input = Effort × Displacement of effort

$= E \times d_E$

Work output = Load × Displacement of load

$= L \times d_L$

By definition,

$$\text{Efficiency} = \frac{\text{Work output}}{\text{Work input}}$$

$$= \frac{L \times d_L}{E \times d_E} = \frac{L/E}{d_E / d_L}$$

But L/E = M.A. and d_E/d_L = V.R.

$$\therefore \quad \text{Efficiency} = \frac{\text{M.A.}}{\text{V.R.}}$$

or M.A. = V.R. × Efficiency.

5. What is the relation between the mechanical advantage and the number of strands of string used to support the load, in a 'block and tackle' setup?

Ans. The mechanical advantage is equal to the number of strands of string, used to support the load, only if the 'block and tackle set up' can be assumed to be ideal (absence of friction, etc.) and the lower blocks of pulleys have a negligible weight. Both these conditions cannot be satisfied in actual practice. Hence, in practical situations, the M.A. of a 'block and tackle set up' is always less than the number of strands of string used to support the load.

6. Shears, used for cutting metals and scissors used for cutting clothes are both examples of levers of first order. However, the shears always have short blades and long handles whereas the scissors often have blades much longer than the handles. Explain, why this is so?

Ans. The mechanical advantage of lever equals the ratio between the effort arm to the load arm.

For the shears (used for cutting metals), the 'load' is really a formidable one. Therefore, we need a large mechanical advantage to keep the applied effort within reasonable limits. To ensure this, the shears are made to have short blades (small load arm) and long handles (long effort arm).

For the scissors (used for cutting clothes), the 'load' is almost negligible one. Therefore, mechanical advantage, may be more or less than one. Therefore, the scissors often have blades much longer than the handles.

7. The following are example of levers. State the class of lever to which each one belongs giving the relative positions of Load (L), Effort (E), Fulcrum (F):
(i) Scissors (ii) Sugar tongs (iii) Nut cracker (iv) Pliers.

S. No.	Name of Lever	Lever Class	Fulcrum	Position of Effort	Load
(i)	Scissors	First	Between E and L	At one end	At other end
(ii)	Sugar tongs	Third	At one end	Between F and L	At other end
(iii)	Nut cracker	Second	At one end	At other end	Between F and E
(iv)	Pliers	First	Between E and L	At one end	At other end

8. What is a block and tackle system of pulleys? What precautions would you observe while setting the string so that the effort is applied in the downward direction?

Ans. A block and tackle system of pulleys consist of two blocks of pulleys, each block having one or more than one pulleys. The upper block of pulleys is fixed to a rigid support and the lower block of pulleys is movable. The number of pulleys in the movable block is either equal or one less than the number of pulleys in the fixed block.

While setting the string, its one end is attached to the hook of movable block if the number of pulleys in the fixed block is more than that in the movable block or to the hook of the fixed block if the number of pulleys is the same in both the blocks.

Chapter 3. Refraction of Light

1. The refractive index of air with respect to glass is defined as :

$${}_g\mu_a = \sin i / \sin r$$

(i) Write down a similar expression for ${}_a\mu_g$ in terms of angle i and r.

(ii) If $r = 90°$, what is the corresponding angle i called?

Ans. (i) Given: ${}_g\mu_a = \dfrac{\sin i}{\sin r}$

Applying principle of reversibility:

$${}_a\mu_g = \frac{1}{{}_g\mu_a} = \frac{1}{\frac{\sin i}{\sin r}} \quad \left[\because {}_g\mu_a = \frac{\sin i}{\sin r}\right]$$

$${}_a\mu_g = \frac{\sin r}{\sin i}$$

[for the same value of $\angle i$ and $\angle r$ for a given pair of media]

(ii) If $r = 90°$

then ${}_a\mu_g = \dfrac{\sin 90°}{\sin i} = \dfrac{1}{\sin i}$

The corresponding angle i is called the critical angle for the given pair of media.

Chapter 4. Refraction through lenses and Optical Instruments

1. Why do we say that the '2F' and 'F' points of a convex lens can be regarded as a sort of 'turning points' as far as the nature of the image formed by it is concerned?

Ans. We know that :

(i) When an object is beyond the 2F point of a convex lens, the image formed by it is real, inverted and diminished in size.

(ii) When the object is in between the 2F and F points of the convex lens, the image formed by it is real, inverted and enlarged.

(iii) When object is between the F point and optical centre of the convex lens, the image formed by it is virtual, erect and enlarged.

We thus notice that there is a sudden change in the nature of the image formed when an object approaching from infinity, crosses the 2F and F points of a convex lens. It is for this reason that we regard these two points as a sort of turning points as far as the nature of the image formed by a convex lens is concerned.

Chapter 5. Spectrum

1. A prism causes dispersion of white light while a rectangular glass block does not. Explain.

Ans. In a prism, the refraction of light takes place at the two slant surfaces. The dispersion of white light occurs at the first surface of prism where its constituent colours are deviated through different angles. At the second surface, these splitted colours suffer only refraction and they get further separated. But in a rectangular glass block, the refraction of light takes place at the two parallel surfaces. At the first surface, although the white light splits into its constituent colours on refraction, but these splitted colours on suffering from refraction at the second surface emerge out in the form of a parallel beam, which give an impression of white light.

Chapter 6. Echoes and Vibrations of Sound

1. When a sound is made from a distance of 18 m, in front of a tall building, it is found to get repeated again. Which phenomenon is associated with this observation? Give a brief explanation of this phenomenon?

Ans. The associated phenomenon is known as Echo. We get echoes because of the reflection of sound. The reflected sound reaches the observer after a time interval of $2d/v$, where d is the distance of the reflecting wall from the source. For the human ear, this time must be at least 0.1 second to ensure that the reflected sound does not get mixed up with the direct sound. Since the velocity of sound in air is nearly 330 ms^{-1}, we must have

$$\frac{2d}{330} > 0{\cdot}1$$

or $$d > 16{\cdot}5 \text{ m}$$

Since the building in this case is at a distance of 18 m, we can hear the reflected sound clearly. We thus get a clear echo in this case.

2. Why is an echo not heard when distance between the source of sound and reflection body is less than 10 m?

Ans. The impression of sound lasts on the eardrum for $\frac{1}{10}$ of a second. In $\frac{1}{10}$ of a second the sound travels a distance of $\left(332\frac{m}{s} \times \frac{1}{10}s\right) = 33{\cdot}2$ m.

Thus, the minimum distance between the source of sound and the reflecting body should be $\frac{33{\cdot}2}{2}$ = 16·6 m.

As the distance 10 m is far less than 16·6 m, therefore the ear cannot make out when the original sound had died and echo has been received. Thus, no echo is heard.

3. What do you understand by (i) free (or natural) and (ii) forced vibrations?

Ans. (i) Any object when slightly disturbed from its position of rest or equilibrium vibrates with its own period irrespective of its size or shape, is said to make free or natural vibrations. The time period of said vibrations is called natural time period and the frequency is called natural frequency. In the absence of any resistance such as air, the amplitude of free vibration remains constant.

(ii) Sometimes to keep a body vibrating, a periodic force is applied on it. In such a case, the body does not vibrate with its

Long Answers

own natural frequency but gradually starts vibrating with the frequency of the applied periodic force. Such vibrations produced in a body are called forced vibrations, *e.g.*, in a string of musical instrument, when the vibrations produce a loud sound.

4. The rear view mirror of a motor bike starts vibrating violently at some particular speed of the motor bike.

(i) Explain why does this happen.

(ii) Name the phenomenon taking place.

(iii) Suggest one way by which these vibrations can be stopped.

Ans. (i) The frame of a motor cycle along with the rear view mirror is found to vibrate violently when it is driven to a particular speed. This happens when the frame has the natural frequency of vibration equal to that of the piston when the engine is driven at that particular speed.

(ii) This phenomenon is called 'resonance'.

(iii) To stop these violent vibrations, the speed of the motor bike should be changed.

5. What is meant by the 'loudness' of a sound? How is it related to the intensity?

Ans. Loudness is the characteristic by virtue of which a loud sound can be distinguished from a faint one, both having the same pitch and quality.

The loudness of sound depends on its intensity, but loudness is not same as intensity. Intensity is a measurable quantity, while loudness is a sensation. Experimentally, Weber and Fechner established a relationship between the loudness L and intensity I which is given as:

$$\boxed{L = K \log_{10} I}$$

where K is a constant of proportionality. Obviously, loudness increases with the increase in intensity, but not in the same proportion.

6. State three harmful effects of noise pollution.

Ans. (i) A long exposure to noise pollution may result in the loss of hearing.

(ii) It causes anger, tension and interferes with the sleep pattern of individuals.

(iii) It reduces concentration and results in the loss of work efficiency.

7. What is meant by the 'pitch' of a note ? Does it depend on the loudness or 'quality' of the note? How is pitch related to frequency?

Ans. The pitch of a note is determined solely by its frequency. The pitch in fact, is a subjective sensation in the ear depending only upon the frequency of the musical sound and is quite independent of its loudness or quality. Pitch is thus a subjective phenomenon and cannot be physically measured. However, the relation between pitch and frequency is linear to a very close approximation. Generally notes of high frequency or pitch are shrill and sharp whereas those of low frequency or pitch are flat and dull.

Chapter 7. Electricity

1. What is a resistance? Define it with respect to Ohm's law.

Ans. In the flow of current in a conductor, electrons are drifted. During this drift or flow, electrons mutually collide against each other and the atoms of the conductor. These collision give rise to some obstruction to the flow of electrons. This opposition or obstruction offered by a conductor to the flow of electrons is called 'Resistance'. According to Ohm's law, the ratio between potential difference V and the current I, is a constant provided the temperature and physical conditions of the conductor remain unaltered.

The ratio $\frac{V}{I}$ is called the resistance, *i.e.*, 'Ohmic resistance' of the conductor.

2. State and explain the laws of resistance.

Ans. The following are the main laws of resistance:

(i) Resistance of a conductor is directly proportional to its length, provided temperature and other physical conditions remain unchanged.

It means that $R \propto l$, *i.e.*, if the length increases, the resistance also increases and if length decreases, its resistance also decreases.

(ii) Resistance of a conductor is inversely proportional to its area of cross-section provided other conditions remaining the same.

If A is the area of cross-section, then:

$$R \propto \frac{1}{A}$$

or $$R \propto \frac{1}{\pi r^2}$$

where r is the radius of the wire.

Keeping the length same, if the radius of the wire is doubled then :

$$R' \propto \frac{1}{\pi(2r)^2} \propto \frac{1}{4\pi r^2}$$

or $R' \propto \frac{1}{4} \cdot \frac{1}{\pi r^2}$

$$R' \propto \frac{R}{4}$$

Therefore, R becomes one fourth.

Similarly if r is made half, then:

$$R'' \propto \frac{1}{\pi(r/2)^2}$$

$$R'' \propto \frac{4}{\pi r^2} \propto 4\left(\frac{1}{\pi r^2}\right)$$

$$R'' \propto 4R$$

Therefore, R becomes 4 times.

This shows that R is inversely proportional to A.

(iii) R depends on the nature of the material of the conductor. It means, if we take equal lengths of wires of copper, aluminium and iron and all of the same cross-sectional area, their resistance are different from each other since they are of different materials.

Chapter 8. Electric Power and Household Circuits

1. An electric heater is rated 220 V, 550 W.

(i) What does this statement mean?

(ii) Calculate the electrical energy consumed in 3 hours.

(iii) Electrical energy is also measured in kWh. What do you understand by a kilowatt-hour?

Ans. (i) The statement means that if the heater is connected across 220 V supply, it would consume energy at the rate of 550 W.

(ii) Energy consumed in 3 hrs $= \frac{550 \times 3}{1000}$

$= 1.65$ kWh $= 1.65 \times 3600 = 5940$ kJ.

(iii) It is the energy consumed in a resistor when the potential difference is 1 V and 1 A current flows for 1000 hrs.

2. An electric filament lamp is connected to a supply of voltage higher than the recommended value. Give reasons, why compared with a normal performance, the lamp emits a brighter light and have a short life ?

Ans. The lamp will have the filament of a particular resistance. If it is connected to a supply of higher voltage, the heat energy produced in it will be more $\left[\because H = \frac{V^2}{R} \times t\right]$ *i.e.,* the filament will get more heated so it will glow more and the lamp will emit a brighter light. On the other hand due to excess heating, the filament may vaporise and so its life may shorten.

3. At what voltage is the electric power generated at the generating station? Explain the transmission of this power to your house.

Ans. At the generating station, the electric power is generated at 11,000 volt. This is an alternating voltage of frequency 50 Hz. To transmit this power to our house the alternating voltage generated at the generating station is first stepped up from 11,000 V to 1,32,000 V. It is then transmitted to the main substation through the high tension line wires, where it is stepped down to 33,000 V and then transmitted to the intermediate substation where it is stepped down to 11,000 V and from there it is transmitted to the city substation. At the city substation, the voltage is further stepped down to 220 V for supply to the houses through the overhead line wires.

4. How is the amount of heat produced calculated due to passage of current in a metallic conductor? Derive an expression for it.

Ans. It is calculated by finding the amount of work done in sending the current through the conductor.

Let a current of I ampere flows in conductor having a potential difference of V volt across its ends for t second.

If Q be the amount of charge that flows in the conductor in t second.

Then $Q = I \times t$ coulomb

Since the amount of work done in moving a charge Q across the ends of a conductor is W.

Then $W = Q \times V$ joule

i.e., $= I \times t \times V = VIt$ joule

or, the amount of heat produced,

$$H = V\,It \text{ joule.}$$

5. (i) Two sets A and B, of three bulbs each, are glowing in two separate rooms. When one of the bulbs in set A is fused, the other two bulbs also cease to glow. But in set B, when one bulb fuses, the other two bulbs continue to glow. Explain why this phenomenon occurs.

(ii) Why do we prefer arrangements of set B for house circuiting?

Ans. (i) **In set A:** Bulbs are in series so if one gets fused, others are also affected. In series, potential difference is variable, and current is constant.

In set B: Bulbs are in parallel so if one goes off others continue to glow. In parallel, potential difference is constant and current varies inversely with resistance.

(ii) We prefer set B for household circuits as they are in parallel. Potential difference is constant (220 V) supplied by electricity department and current varies according to the value of resistors or appliances used. If one circuit goes off, others continue to function.

6. You are required to connect a bulb, a fan and a socket outlet to the mains in one set A and an air conditioner and a refrigerator to the mains in other set B. Will you recommend the wire of same thickness and same insulation in both ? Explain your answer.

Ans. No. The wire used in set B must be thicker and of better insulation than that used in set A. The reason is that in set B, the appliances will draw heavy current (=15A) when they are in use whereas in set A, the appliances will draw ordinary current (= 5A) when they are in use. Therefore, the resistance of wire used in set B must be less than that of wire used in set A. A thick wire has less resistance than a thin wire.

7. Why is fuse wire fitted in a porcelain casing?

Ans. The fuse wire is stretched between two terminals in a porcelain holder. This porcelain holder is tightly fit into a porcelain socket which is connected to the live wire in the circuit. Porcelain is a very good insulator and the part of the holder we touch is far from the live wire and so there are no dangers of getting any electric shock.

8. What is an 'electric fuse'? State two characteristics of electric fuse.

Ans. The electric fuse is a safety device in electric circuits used to limit the current and thus it acts as a safeguard for the household circuit and electric appliances connected in the circuit from being damaged. Its characteristics are:

(i) It has a low melting point of about 200°C and is usually made from an alloy of 50% tin and 50% lead.

(ii) It has a high resistance so that its temperature rises rapidly and soon exceeds its m.p. and thus breaks it, when current passing through it exceeds the permissible limits. In such a case, it acts as a switch.

9. State relative advantages and disadvantages of ring system of wiring over the tree system.

Ans. Advantages:

(i) The tree system of wiring requires the plugs and sockets of different sizes for different current capacities while they are all of the same size in ring system of wiring.

(ii) The tree system of wiring has one fuse in one distribution line. If the fuse in one distribution line blows, it disconnects all the appliances in that distribution line. But in the ring system, a separate fuse is connected with each appliances. So if one fuse blows, it does not affect the other appliances.

(iii) The tree system of wiring is expensive while the ring system of wiring is cheaper.

(iv) It is expensive and inconvenient to install a new appliance in the tree system of wiring since it requires to put the new leads up to the distribution box, while it is easy to install a new appliance in the ring system since the appliance can be directly connected to ring mains in the room.

Disadvantages:

The only disadvantage of ring system of wiring over the tree system is that while installing a new appliance, care has to be taken that the total load on the ring circuit does not exceed the main fuse rating (*i.e.*, generally 30 amp).

10. What do you understand by 'earthing'? What are the advantages of earthing in a household electric circuit ? Explain, how it is done?

Ans. By earthing, we mean that the metallic body of an electric appliance is connected to thick copper wire, which is buried deep in the earth and at its end is a copper plate surrounded by a mixture of charcoal and common salt.

Advantages: It is a kind of safety device which saves us from an electric shock in case when the metal casing of the faulty appliance comes in direct contact of the live wire due to short circuiting or leakage of electric current.

Whenever an appliance which is earthed, get short circuited, the current from the metal casing of the appliance flows into the earth which acts as an 'electric sink', *i.e.*, its potential always remains zero. Due to the flow of heavy current the fuse in that circuit melts and disconnects the appliance from the circuit. So, the user who is about to touch the appliance is protected from receiving any electric shock.

Another advantage is that the house wiring system is saved from being damaged and in the same time from being burnt out.

Procedure: A three core cord having three wires coated with insulation of brown, light blue and green colour is used for connecting the appliance to the mains for drawing current from the mains. At one end of the cord, brown is connected to the pin marked L (live), the light blue to the pin marked N (neutral) and green to thick pin of the plug marked E (earth). The three wires at the other end of the cord are connected to the appliance in such a way that live and neutral wires are connected to the element and the earth wire is connected to the metal body of the appliance so as to earth it.

Once the plug is put in the socket, the current flows through the appliance as soon as the switch is pressed. The live wire gets connected to the live wire of the mains, neutral wire is connected to the neutral of the mains and the earth wire gets connected to the earth in the mains.

11. How does earthing prevent us from getting an electric shock?

Ans. Local earthing provides us safety from electric shock in the case of short circuit. when excessive current flows through the live wire, it will pass to the earth through the earth wire. In the absence of local earthing, it may cause fire due to over heating of the live wire.

In the case of an electric appliance, if its outer metallic body is not earthed and by chance, if it happens to touch live wire, then metallic casing acquires the potential of the live wire and so any person touching it, is likely to get a fatal shock. But if earthed, a heavy current flows and the fuse connected to the appliance blows off. Thus the person touching it does not get any shock.

12. Give the role of each of the three terminals of a three way pin plug.

Ans. The live terminal of the three pin plug helps us to connect our appliance to the live (or high voltage wire) of the mains. The neutral terminal connects the other end of the appliance to the ground potential wire of the mains. The earth terminal helps us to connect the body of the appliance to a wire whose other end is buried deep in moist earth. This earthing of the body of the appliance acts as a safety measure and saves the user from dangerous shocks.

13. State the reason why, in a three pin plug, the earth pin is longer and thicker than the other two.

Ans. The earth pin being thicker can never enter live or neutral sockets thus a given electric appliance gets connected only in one proper electrical position, where the live wire, neutral and earthings are connected correctly.

The earth pins being longer, get connected to earth terminal first, thus ensuring that user will not get electric shock, because the current will melt the fuse in case the electrical appliance is short circuited.

Chapter 9. Magnetic Effect of Current

1. State important properties (or characteristics) of magnetic lines of force around a straight conductor.

Ans. The main properties are:

(i) The magnetic lines of force are in the form of concentric circles around the conductor but as their distance from the conductor increases, the curvature of lines of force decreases.

(ii) The direction of the magnetic lines of force reverses, if the direction of flow of current in the conductor reverses.

(iii) With the increase of current in the conductor, the magnetic lines of force become denser showing that the magnetic field strength increases.

2. State the factors on which the magnitude of the force acting on current carrying conductor depends, when it is kept in a magnetic field perpendicular to it.

Ans. Experimentally, it is found that the magnitude of the force acting on a current carrying conductor kept in a magnetic field in a direction perpendicular to it depends on the following factors:

(i) The F is directly proportional to the length of the current carrying conductor in the magnetic field, *i.e.*, $F \propto l$.

(ii) F is directly proportional to the strength of the magnetic field, *i.e.*, $F \propto B$.

(iii) F is directly proportional to the strength of the current flowing through the conductor, *i.e.*, $F \propto I$.

Combining the above factors we find that :

$$F \propto I B l$$

$$F = KIBl$$

3. A current carrying solenoid is suspended so as to freely rotate in a horizontal plane. State in what direction will it rest. Give reason to your answer.

Or

When a current carrying solenoid is freely suspended, it comes to rest along a particular direction. Why does this happen?

Ans. The current carrying solenoid when suspended freely, will rest along the N-S direction. The reason is that it will behave like a bar magnet, with N polarity at one face of solenoid and S polarity at the other face, depending upon the direction of current in the solenoid.

Current carrying solenoid behaves like a bar magnet. At end in which direction of current is anticlockwise, behaves as north pole and where current is in clockwise direction behaves as south pole. Therefore on suspending the current carrying solenoid freely, it sets itself in N-S direction as bar magnet does.

4. State the factors on which the strength of the magnetic field of an electromagnet depends.

Ans. Factors on which the strength of the magnetic field of an electromagnet depends are:

(i) The number of turns in the solenoid. If the number of turns in the solenoid increases or decreases, the magnetic intensity also increases or decreases.

(ii) It is directly proportional to the strength of the current that flows in the turns of the solenoid.

(iii) It is found to increase further if a soft iron piece, called core, is placed along the axis of the solenoid inside it.

5. Point out three differences between an electromagnet and a bar (or permanent) magnet.

Ans.

S. No.	Electromagnet	Bar magnet
(i)	An electromagnet exhibits a much stronger magnetic field.	A bar magnet is not able to exhibit a strong magnetic field.
(ii)	Its polarity can be reversed by reversing the direction of current flowing in the turns of the coil (or solenoid)	Its polarity is fixed.
(iii)	It is a temporary magnet, it behaves as a magnet only when the current flows through the solenoid.	It is a permanent magnet and cannot be readily demagnetized.

6. Why soft iron is preferred to be used as the core of the electromagnet of an electric bell?

Ans. Soft iron is chosen to make the core of the electromagnet of an electric bell on account of the following two reasons:

(i) On account of high magnetic permeability of soft iron, it can be easily magnetized and demagnetized. So it makes a strong magnet when current flows through the solenoid but it is almost demagnetized when no current flows through the solenoid. This fact helps in the working of the electric bell.

(ii) The soft iron intensifies the magnetic field of the solenoid and so it helps in making a strong magnet.

7. State Fleming's left hand rule.

Or

Name and state the law used to find the direction of force on a current carrying conductor kept in a magnetic field.

Ans. **Fleming's Left Hand Rule:** According to it, if we stretch forefinger, central finger and the thumb of our left hand mutually perpendicular to each other such that the forefinger indicates the direction of the magnetic field and the central finger indicates the direction of current, then the thumb will indicate the direction of force on the conductor.

8. The diagram below shows a magnetic needle kept just below the conductor AB which is kept in North-South direction.

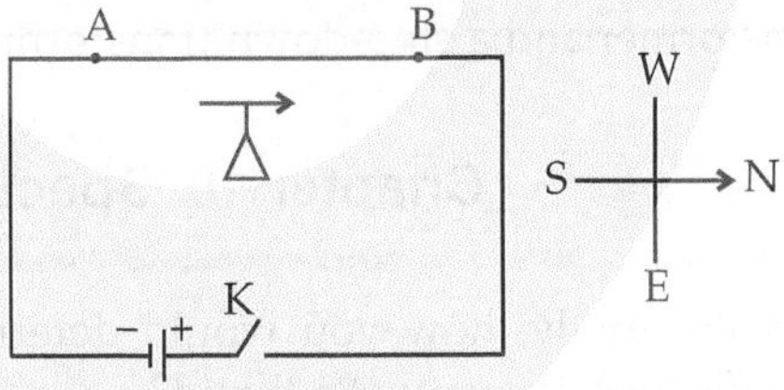

(i) In which direction will the needle deflect when the key is closed ?

(ii) Why is the deflection produced ?

(iii) What will be the change in the deflection if the magnetic needle is taken just above the conductor AB ?

(iv) Name one device which works on this principle.*

Ans. (i) Needle deflects towards the east.

(ii) On passing current in the wire AB, a magnetic field is produced around it and the magnetic needle experiences a torque in this magnetic field, so it deflects to align itself in the direction of magnetic field at that point.

(iii) Needle will deflect towards the west.

(iv) Electric motor.

* Frequently asked previous years Board Exam Questions.

9. State Faraday's laws of electromagnetic induction.

Ans. Faraday's laws of electromagnetic induction are:

(i) Whenever there is a change in magnetic flux linked with a coil, an e.m.f. is induced.

(ii) The magnitude of the induced e.m.f. is directly proportional to the rate of change of magnetic flux linked with the coil.

10. Name and state the law which determines the direction of induced current.

Or

State Fleming's right hand rule.

Ans. Fleming's Right Hand Rule: It states that if we stretch the thumb, the forefinger and the central finger of our right hand mutually perpendicular to each other such that the forefinger indicates the direction of magnetic field and the thumb indicates the direction of motion of the conductor, then the central finger will indicate the direction of induced current.

11. State three differences between a step-up transformer and a step-down transformer.

Ans.

S. No.	Step-up Transformater	Step-down Transformer
(i)	It has more number of turns in the secondary coil than in the primary coil.	It has less number of turns in the secondary coil than in the primary oil.
(ii)	It increases the a.c. voltage and decreases the current.	It decreases the a.c. voltage and increases the current.
(iii)	The wire of primary coil is thicker than that of the secondary coil.	The wire of the secondary coil is thicker than that of the primary coil.

12. Name three losses of energy in a transformer. How are they minimised?

Ans. (i) Heat losses in the windings of the transformer. They are minimised by taking a thicker wire in primary coil than in the secondary coil in a step up transformer, whereas a thicker wire in secondary coil than in primary coil in a step-down transformer.

(ii) Hysteresis losses in the core of the transformer. They are minimised by using soft iron core.

(iii) Eddy current losses in the core of the transformer. They are minimised by taking the core laminated instead of taking it as one solid piece.

Chapter 10. Specific Heat Capacity and Latent Heat

1. Explain briefly, how you would determine the specific heat capacity of a liquid?

Ans. In this case, take any solid of known specific heat and a liquid, whose specific heat capacity is to be determined. The solid chosen should be insoluble and non reactive with the liquid. The observations are to be noted as below:

(i) Mass of empty calorimeter with stirrer = m_1 g.

(ii) Mass of calorimeter + Liquid (of unknown specific heat capacity) = m_2 g.

(iii) Mass of solid = M g.

(iv) Specific heat of solid (J/g/°C) = s_1

(v) Initial temperature of liquid (as well as calorimeter) = t_1°C

(vi) Temperature of hot solid = t_2°C

(vii) Final temperature of the mixture = t°C.

Calculation: Let 's' be the specific heat capacity of the liquid.

(i) Fall in temperature of the hot solid
$= (t_2 - t)$°C

(ii) Rise in temperature of the liquid
$= (t - t_1)$°C

(iii) Rise in temperature of calorimeter
$= (t - t_1)$°C

(iv) Heat given out by hot solid
$= M \times s_1 \times (t_2 - t)$

(v) Heat taken up by liquid
$= (m_2 - m_1) \times s \times (t - t_1)$

(vi) Heat taken up by calorimeter
$= m_1 \times 0.4 \times (t - t_1)$

Heat taken up by liquid
+ Heat taken up by calorimeter
= Heat given out by solid

or $(m_2 - m_1) \times s \times (t - t_1) + m_1 \times 0.4 \times (t - t_1)$
$= M \times s_1 \times (t_2 - t)$

From this heat equation, we get

$$s = \frac{M \times s_1 \times (t_2 - t) - m_1 \times 0.4(t - t_1)}{(m_2 - m_1)(t - t_1)} \text{ J/g/°C.}$$

2. Describe a method to determine the specific heat capacity of a solid (say, a piece of copper).

Ans. First we weigh the given piece of solid and note its mass m_1. Then it is heated by suspending it inside a heater. Then a known mass (say m_2) of water is taken in a thin glass beaker and its temperature t_1 is recorded with a thermometer. When the given piece of solid becomes heated, its temperature t_2 is noted and it is quickly dropped into the water contained in the beaker, such that no water splashes out. The contents of beaker are well stirred and the final highest temperature t_3 is noted.

Assuming that the heat capacity of beaker is negligible and there is no heat loss to the surroundings,

Heat lost by the solid
= Heat gained by the water

∴ Mass of solid × Specific heat capacity of solid × Fall in temperature of solid
= Mass of water × Specific heat capacity of water × Rise in temperature of water

or $m_1 \times c \times (t_2 - t_3) = m_2 \times 4.2 \times (t_3 - t_1)$

or $$c = \frac{m_2 \times 4.2 \times (t_3 - t_1)}{m_1 \times (t_2 - t_3)} \text{ J/g°C.}$$

Here we have assumed that specific heat capacity of water is 4.2 J/g°C. **Ans.**

3. Discuss how high specific heat capacity of water helps in formation of land and sea breeze.

Ans. Specific heat capacity of land is five times less as compared to water. Thus, air above land becomes hot and light and rises up resulting in drop in pressure of land mass during day time. Thus, cool air from sea starts blowing towards the land forming sea breeze.

During night, land as well as sea radiates heat energy. However, temperature of land falls faster than sea water, because of high specific heat capacity of sea water. So, at night the temperature of sea water is more than land. Warm air above the sea rises and cold air from land starts blowing towards the sea resulting in land breeze.

4. State three differences between heat capacity and specific heat capacity.

Ans.

S. No.	Heat capacity	Specific heat capacity
(i)	It is the amount of heat energy required to raise the temperature of entire body by 1°C.	It is the amount of heat energy required to raise the temperature of unit mass of the body by 1°C.
(ii)	Its unit is JK^{-1}.	Its unit is $Jkg^{-1}K^{-1}$.
(iii)	It depends both on the material and mass of the body.	It does not depend on the mass of the body.

5. State and explain the principle of calorimetry.

Ans. The principle of calorimetry states that when a hot body is brought in contact with a cold body, heat passes from the hot body to the cold body till both attain the same temperature.

Heat lost by the hot body
= Heat gained by the cold body.

Let the mass of hot body be m_1, its specific heat capacity be s_1 and its temperature be t_1°C and let the mass of the cold body be m_2, its specific heat capacity be s_2 and its temperature be t_2 °C.

If t°C be the final temperature, then:

Fall in temperature of hot body
= $(t_1 - t)$°C and

Rise in temperature of cold body
= $(t - t_2)$°C

∴ Heat lost by hot body
= $m_1 s_1 (t_1 - t)$ and

Heat gained by cold body
= $m_2 s_2 (t - t_2)$

According to principle of calorimetry :

$$m_1 s_1 (t_1 - t) = m_2 s_2 (t - t_2)$$

6. State four precautions during the determination of specific heat capacity of a solid by the method of mixtures.

Ans. (i) The calorimeter should be polished, so as to minimise the heat loss due to radiation.

(ii) The calorimeter should be insulated by placing it in wooden jacket and surrounded by some poor conductor.

(iii) A very sensitive thermometer should be used.

(iv) Hot solid should be transferred to the calorimeter quickly and carefully without splashing any water out of the calorimeter.

7. Will the value of specific heat capacity and specific latent heat of a substance change if the scale is °F instead of °C?

Ans. The specific heat capacity will change if the scale is °F instead of °C. This is because it is the heat needed to raise the temperature of a unit mass of the substance by one degree. Since the size of the Fahrenheit degree is not the same as that of the Celsius degree, a change of scale will affect the specific heat capacity.

The specific latent heat will not change with a change of scale. This is because this quantity is simply the heat needed to change the state of 1 kg of a substance at its melting/boiling point without any change in temperature. It is thus heat per unit mass and does not depend upon the scale of temperature used.

8. Derive an expression for the amount of heat given out or taken up, when its temperature falls or rises by t°C.

Ans. Let the mass of the body be 'm' and its specific heat capacity be 's'. Let t°C be the rise in its temperature, then heat taken up by the body is calculated as below:

Since 's' is the specific heat capacity of the body which means that the unit mass of the body for a rise of 1°C takes up 's' amount of heat.

∴ Mass 'm' of the body for a rise of t°C will need heat

$= m \times s \times t$

= Mass of the body × Specific heat capacity × Rise in the temperature

= mst joule.

Chapter 11. Radioactivity and Nuclear Energy

1. Mention four properties of Becquerel rays.

Ans. The four properties of Becquerel rays are:

(i) They have penetrating power.

(ii) They affect a photographic plate.

(iii) They can ionize a gas or air through which they pass.

(iv) They are deflected by electric and magnetic fields.

2. State four properties of α-particles.

Ans. (i) An alpha particle consists of two protons and two neutrons.

(ii) The mass of an alpha particle is roughly four times the mass of a proton.

(iii) Alpha particles are positively charged, so they are deflected by the electric and magnetic fields.

(iv) Alpha particles strongly ionize the gas through which they pass.

3. An atomic nucleus A is composed of 84 protons and 128 neutrons. The nucleus A emits an alpha particle and is transformed into a nucleus B.

(i) What is the composition of B?

(ii) The nucleus B emits a beta particle and is transformed into a nucleus C. What is the composition of C?

(iii) What is mass number of the nucleus A?

(iv) Does the composition of C change if it emits gamma radiations?*

Ans. For nucleus A

Atomic no. (Z) = 84

Mass no. (A) = 84 + 128 = 212

(i) $_{84}A^{212} \xrightarrow{\alpha} {_{82}B^{208}}$ Nucleus B will have 82 protons and (208 – 82) = 126 neutrons.

(ii) $_{82}B^{208} \xrightarrow{\beta} {_{83}C^{208}}$ Nucleus C will have 83 protons and (208 – 83) = 125 neutrons

(iii) Mass number of nucleus, A = 212

(iv) No, the composition of nucleus C does not change due to emission of gamma radiations.

4. State four properties of β-particles.

Ans. (i) Beta particles are the fast moving electrons emitted from the nucleus of an atom.

(ii) Beta particles are negatively charged, so they are deflected by electric and magnetic fields.

(iii) Beta particles produce X-rays when they are stopped by the metals of high atomic number and high melting point (*eg.*, tungsten).

(iv) Beta particles cause fluorescence on striking a fluorescent material.

5. How many alpha and beta particles are emitted when Uranium nucleus $^{238}_{92}U$ decays to Lead $^{206}_{82}Pb$?

Ans. On emission of one alpha particle, the mass number decreases by 4 and the atomic number decreases by 2 but when a β-particle is emitted, the atomic number increases by 1.

$_{92}U^{238} \xrightarrow{-\alpha} {_{90}X^{234}}$, where X represents the new nucleus

$_{90}X^{234} \xrightarrow{-\beta} {_{91}X^{234}} \xrightarrow{-\alpha} {_{89}X^{230}} \xrightarrow{-\beta}$

$_{90}X^{230} \xrightarrow{-\alpha} {_{88}X^{226}} \xrightarrow{-\beta} {_{89}X^{226}} \xrightarrow{-\alpha}$

* **Frequently asked previous years Board Exam Questions.**

$$_{87}X^{222} \xrightarrow{-\beta} {}_{88}X^{222} \xrightarrow{-\alpha} {}_{86}X^{218} \xrightarrow{-\beta}$$

$$_{87}X^{218} \xrightarrow{-\alpha} {}_{85}X^{214} \xrightarrow{-\beta} {}_{86}X^{214} \xrightarrow{-\alpha}$$

$$_{84}X^{210} \xrightarrow{-\alpha} {}_{82}Pb^{206}$$

No. of α-particles emitted = 8

No. of β-particles emitted = 6

6. What do you mean by background radiations? Is it possible for us to keep ourselves away from it?

Ans. Background radiations are the radioactive radiations (such as α, β and γ) to which we all are exposed in the absence of an actual radioactive sources.

There are two sources of background radiations:

(i) Internal source such as carbon (C-14), potassium (K-40).

(ii) External source such as cosmic rays, naturally occurring radioactive elements etc.

It is not possible to keep ourselves away from the background radiations.

7. State three safety precautions that you would take while handling the radioactive substances.

Ans. (i) The people working with radioactive materials should use mechanical devices such as tongs, clamps, tweezers, etc., to help minimize exposure.

(ii) The people working with radioactive materials should wear special lead lined lab coats, gloves and safety glasses and remove them when leaving the work area or laboratory.

(iii) All radioactive materials should be stored in specially designed thick lead containers with a very narrow opening so as to stop radiations coming out from other directions.

8. What precautions must be taken for the safe disposal of nuclear waste?

Ans. The nuclear waste obtained from laboratories, hospitals, scientific establishments or power plants must be first kept in thick casks and then they should buried in the specially designed deep underground stores. The stores should be made quite far from the populated area and the casks should also be buried and sealed in useless mines after storing it.

9. What safety measures are needed in a nuclear power plant?*

Ans. (i) The nuclear power plant must be shielded with lead and steel walls so as to stop radiations from escaping out to the environment during its normal operations.

(ii) The nuclear power plant must be housed in an airtight building of strong concrete structure which can withstand earthquakes, fires and explosions.

(iii) There must be a back up of cooling system for the reactor core, so that in case of failure of one system, the other system could take its place.

10. (i) Differentiate between nuclear fusion and nuclear fission.

(ii) State one safety precaution in the disposal of nuclear waste.*

Ans. (i)

Nuclear Fission	Nuclear Fusion
(1) In nuclear fission when neutrons are bombarded on a heavy nucleus, it splits in two nearly equal light fragments.	In nuclear fusion, two light nuclei combine to form a heavy nucleus at a very high temperature and high pressure.
(2) It can be controlled.	It cannot be controlled.

(ii) For disposal of nuclear waste, they must be first kept in thick casks and then buried in specially constructed deep underground stores and also while handling nuclear waste, we should wear special lined lead aprons and lead gloves.

11. (i) Define nuclear fission.

(ii) Rewrite and complete the following nuclear reaction by filling in the atomic number of Ba and mass number of Kr:*

$$^{235}_{92}U + ^{1}_{0}n \rightarrow ^{144}_{\dots}Ba + ^{\dots}_{36}Kr + 3^{1}_{0}n + \text{Energy}$$

Ans. (i) Nuclear fission is the process in which a heavy nucleus is splitted into two light nuclei of nearly the same size, by bombarding it with slow neutrons.

(ii) $^{235}_{92}U + ^{1}_{0}n \rightarrow ^{144}_{56}Ba + ^{89}_{36}Kr + 3^{1}_{0}n + \text{Energy}$

* Frequently asked previous years Board Exam Questions.

ICSE Solved Paper 2020

(Two Hours)

Answer to this paper must be written on the paper provided separately.

*You will **not** be allowed to write during the first 15 minutes.*

This time is to be spent in reading the Question paper.

The time given at the head of this Paper is the time allowed for writing the answers.

***Section I** is compulsory. Attempt **any four** questions from Section II.*

The intended marks for questions or parts of questions are given in brackets [].

SECTION-I (40 Marks)

*Attempt **all** questions from this Section*

Question 1.

(a) (i) *Define moment of force.* [2]

(ii) *Write the relationship between the SI and CGS unit of moment of force.*

(b) *Define a kilowatt hour. How is it related to joule ?* [2]

(c) *A satellite revolves around a planet in a circular orbit. What is the work done by the satellite at any instant ? Give a reason.* [2]

(d) (i) *Identify the class of the lever shown in the diagram below :* [2]

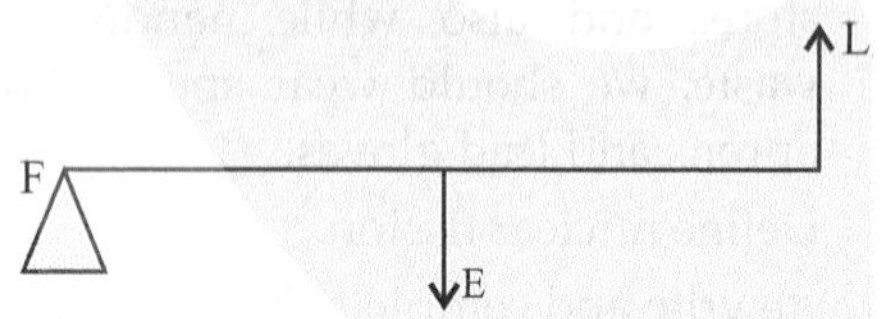

(ii) *How is it possible to increase the M. A. of the above lever without increasing its length ?*

(e) *Give one example of each when :* [2]

(i) *Chemical energy changes into electrical energy.*

(ii) *Electrical energy changes into sound energy.*

Answer 1.

(a) (i) The turning effect of force on the body about an axis is due to the moment of force applied on the body and is equal to the product of the magnitude of the force and the perpendicular distance of the line of action of the force from the axis of rotation.

(ii) 1 Nm = 10^7 dyn cm or 1 dyn cm = 10^{-7} Nm.

(b) One kilowatt hour (kWh) is the energy spent or work done by a source of power 1 kW in 1 hour.

$$1 \text{ kWh} = 3.6 \times 10^6 \text{ J or } 3{\cdot}6 \text{ MJ}$$

(c) The work done by the satellite at any instant is zero because the force required (centripetal force) to go around the planet is perpendicular to the displacement at any instant of its motion.

(d) (i) Class III lever.

(ii) Without increasing the length of the lever, its mechanical advantage can be increased by shifting the effort (E) towards the load (L), *i.e.*, by decreasing the load arm.

(e) (i) A dry cell in use.

(ii) Loud speaker.

Comments of Examiners

1. (a) (i) Majority of the candidates were able to answer this question correctly. Some errors which were observed are as follows:

- Instead of writing 'line' of action, they wrote 'point' of application. In many cases, candidates did not mention perpendicular distance. Some even mentioned displacement.
- Some merely wrote turning effect and did not mention the axis of rotation or fulcrum. Instead of turning effect of force, many wrote turning effect of a body.

(ii) Most of the candidates answered it correctly. But some made following errors:

A few candidates wrote 1 N *m* = 10^7 erg instead of dyn-cm.

Some wrote 1 J = 10^7 dyn.

Some related the SI and CGS unit of force, instead of relating the SI and CGS unit of moment of force.

(b) Most candidates could establish the relation between kW h and J. Some errors were as follows:

- Several candidates did not mention that it is the energy consumed / required; in some cases, they mentioned power.

- Some comprehended 1kW h as 1watt energy consumed in 1 hour.
- Some candidates used the concept of kilo and multiplied by 103 for writing the relation between kilowatt hour and joule.
- A few candidates committed errors in converting 1kW h to J, or while converting into MJ.

(c) Most candidates answered correctly. However, some candidates could not give a satisfactory reason for their answer. The mistake which they made was common i.e., they wrote displacement as zero. Candidates overlooked the words work at any instant.

(d) (i) Some candidates did not identify the class of lever correctly.

(ii) This part of the question was not answered by most candidates. Many candidates expressed the answer learnt by rote, *i.e.*, "By increasing effort arm". They did not pay attention to the statement without increasing the length of the lever. The following errors were also observed.

- Some candidates wrote by increasing load or by decreasing effort without realising that as load changes, effort also changes proportionately.
- A few candidates wrote - change the position of load and effort without realising that the class of lever would change.

(e) (i) Most of the candidates answered it correctly. Many candidates were confused between 'cell in use' and 'charging of a cell'. A few candidates wrote vague answers.

(ii) A number of candidates answered this part correctly but following errors were also observed:

- Many of them wrote microphone instead of loudspeaker.
- Some gave technically incorrect answers such as, bell instead of an electric bell.

Answering Tips

- ➢ The difference between distance and perpendicular distance. Also, the difference between the point of application of force and line of application of force should be well known.
- ➢ Explain the Draw diagrams corresponding to the two different situations and probe till the concept is clear.
- ➢ Write definition in a statement form instead of a formula.
- ➢ Explain the derivation between the kW h and joule. Also administer a few sums on conversions to verify the understanding of the topic.
- ➢ Comprehend a physical quantity before defining its unit.
- ➢ To develop the idea of correct identification of different classes of levers through diagrams.
- ➢ Understand the difference between a microphone and a speaker and then relate them with respect to the energy conversions.
- ➢ Practice the maximum possible types of energy conversions.

Question 2.

(a) *A crane 'A' lifts a heavy load in 5 seconds, whereas another crane 'B' does the same work in 2 seconds. Compare the power of crane 'A' to that of crane 'B'.* **[2]**

(b) *A ray of light falls normally on a rectangular glass slab.* **[2]**

Draw a ray diagram showing the path of the ray till it emerges out of the slab.

(c) *Complete the path of the monochromatic light ray AB incident on the surface PQ of the equilateral glass prism PQR till it emerges out of the prism due to refraction.* **[2]**

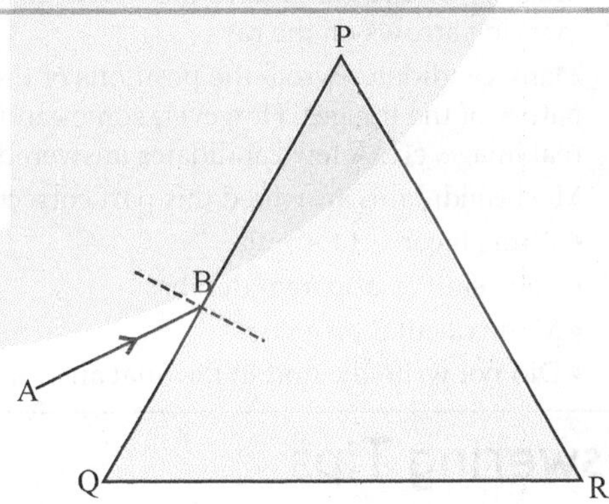

(d) *Where should an object be placed in front of a convex lens in order to get :* **[2]**

(i) *an enlarged real image*

(ii) *enlarged virtual image ?*

(e) *A pond appears to be 2.7 m deep. If the refractive index of water is $\frac{4}{3}$, find the actual depth of the pond.* **[2]**

Answer 2.

(a) Let the work done in both case be x joule.

Crane A	Crane B
$W_1 = x$ Joule	$W_2 = x$ Joule
$t_1 = 5\,s$	$t_2 = 2\,s$
$P_A = \frac{W_1}{t_1} = \frac{x}{5}$ W	$P_B = \frac{W_2}{t_2} = \frac{x}{2}$ W

$$\therefore \frac{P_A}{P_B} = \frac{x/5}{x/2} = \frac{x}{5} \times \frac{2}{x} = \frac{2}{5}$$

$\therefore$ Power of Crane A : Power of Crane B = 2 : 5

(b)

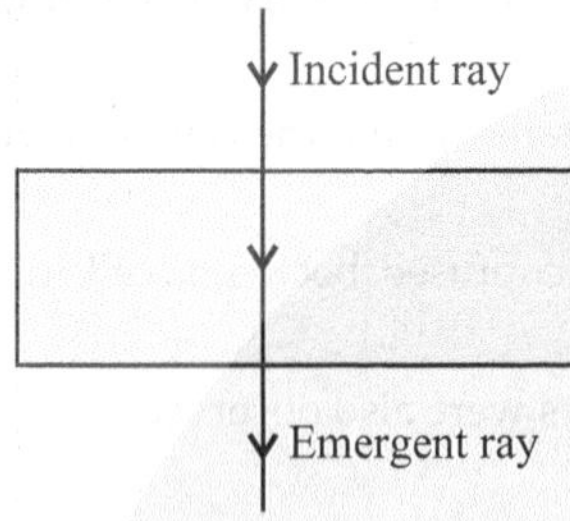

(c) AB → Incident Ray

BC → Refracted Ray

CD → Emergent Ray.

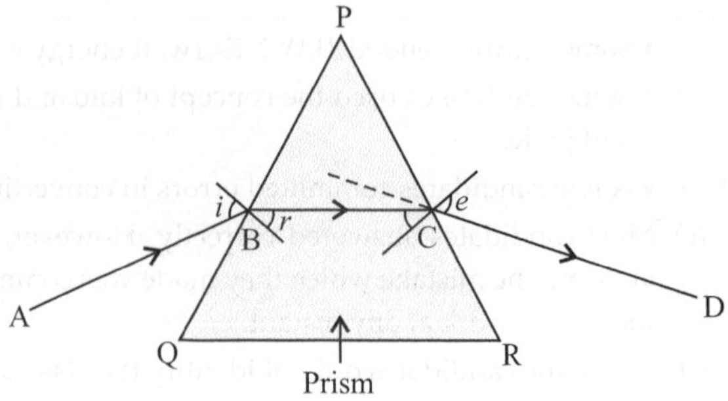

(d) (i) Object must be placed between first focal point (F_1) and the centre of curvature ($2F_1$) of the lens.

(ii) Object must be placed between the first focal point (F_1) and the lens.

(e) Given : Apparent depth = 2.7 m

$$\mu_w = \frac{4}{3}$$

$$\because \quad \mu_w = \frac{\text{Actual depth}}{\text{Apparent depth}}$$

$$\Rightarrow \quad \frac{4}{3} = \frac{\text{Actual depth}}{2.7}$$

$$\Rightarrow \quad \text{Actual depth} = \frac{4}{3} \times 2.7 \text{ m} = 3.6 \text{ m}$$

Comments of Examiners

2. (a) This part was well attempted by a number of candidates. Some did not express their answers in ratio form and left it in fractional form. A few did not get the inverse proportion between power and time for same work. Some candidates wrote their answer in the form of > and < and not in ratio form.

(b) Some candidates took the ray normal at one surface but after refraction, the ray shown was bent at the other surface. The angle with the second surface was marked as 900 by them. Several candidates did not mark arrows on the rays.

(c) Majority of the candidates answered correctly barring some who showed dispersion or some who missed marking arrows on the rays.

(d) Many candidates wrote the positions of the objects correctly in subparts (i) and (ii) corresponding to the asked nature of the images. However, some candidates were confused in F_1 and $2F_1$, F_2 and $2F_2$, between virtual and real image, etc. A few candidates answered by drawing ray diagram, which was not asked.

(e) Most candidates answered this part correctly. However, some made the following errors:

- Wrote incorrect formula.
- Took shift as apparent depth.
- Made calculation errors.
- Did not write the unit in the final answer.

Answering Tips

- Whenever comparison is asked, answer in ratio form if the data is available to calculate the ratio. If data is not available, it can be accepted in > or < form.
- Write calculated ratio, with the left-hand side of the expression to clarify which quantity is compared with which *i.e.*, whether the first quantity is compared with the second one or vice versa.
- Give adequate practice of drawing ray diagrams on a rectangular glass slab, taking the incident ray obliquely / normally.
- Understand the difference between monochromatic and poly chromatic light clearly.
- Understand how to draw normal to an inclined surface.
- Always mark arrows on the rays before and after refraction.

- ➤ If a descriptive question is asked, the answer cannot be in the form of a diagram.
- ➤ Point out the difference between real depth and apparent depth and shift clearly to students. More practice of these type of numerical problems to be done.
- ➤ Always write the data before substituting the values in the formula.
- ➤ Lay stress on writing the final answer with the unit.

Question 3.

(a) *The wave lengths for the light of red and blue colours are nearly 7.8×10^{-7} m and 4.8×10^{-7} m respectively.* **[2]**

(i) *Which colour has the greater speed in a vacuum ?*

(ii) *Which colour has a greater speed in glass ?*

(b) *Draw a graph between displacement from mean position and time for a body executing free vibration in a vacuum.* **[2]**

(c) *A sound wave travelling in water has wavelength 0.4 m.* **[2]**

Is this wave audible in air ? (The speed of sound in water = 1400 ms^{-1})

(d) *Why does stone lying in the sun get heated up much more than water lying for the same duration of time ?* **[2]**

(e) *Why is it not advisable to use a piece of copper wire as fuse wire in an electric circuit ?* **[2]**

Answer 3.

(a) (i) Both colours of light have same speed in vacuum.

(ii) In glass speed of red light is more than that of blue light.

(b)

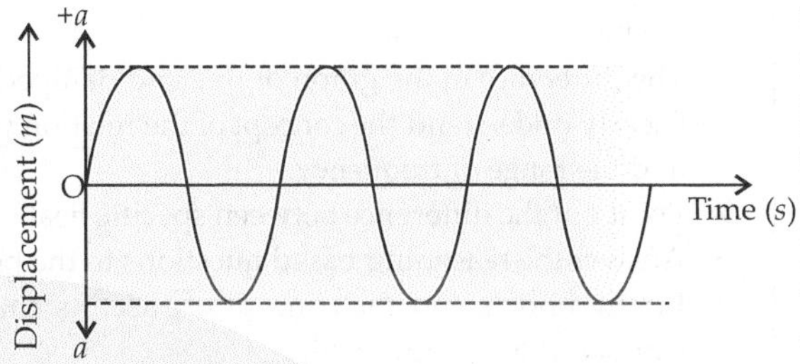

(c) Given : $\lambda = 0.4$ m

$V = 1400$ ms^{-1}

$$\because \quad V = f\lambda \Rightarrow f = \frac{V}{\lambda} = \frac{1400}{0.4} \text{ Hz}$$

$$= 3500 \text{ Hz}$$

$\because$ Frequency remains unchanged in air, the wave is audible in air because 3500 Hz falls in the audible range of frequency.

(d) The specific heat capacity of stone is much less than the specific heat capacity of water, hence for the same heat supplied the temperature of stone rises much more than water for the same duration and gets heated up much more.

(e) Copper wire should not be used as a fuse wire because it has high melting point and its specific resistance is much less than a normal fuse wire. (Normal fuse wire should have low melting point and high resistivity)

Comments of Examiners

3. (a) (i) Most candidates answered this part correctly but a few who did not know the concept wrote blue or red randomly. Some candidates treated wavelengths as speed.

(ii) Some candidates did not establish a relation between speed and wavelength in a medium and thus wrote incorrect answers. Several candidates just wrote blue randomly.

(b) Many candidates drew the free vibrations in vacuum graphically in a correct manner, but axes were not labelled. A few candidates did not take X-axis as the mean position. Some candidates drew a single curve while several drew a graph of minimum deviation and angle of incidence but labelled the axes as displacement and time. A few candidates drew the graph only in the positive quadrant of the graph. It did not have any negative amplitude. Many candidates drew the graph for damped vibrations.

(c) Many candidates committed the mistake of simply writing "audible in air" or "not audible" without calculating frequency. Some wrote an incorrect unit of frequency. A few used the velocity formula of echo. Several candidates were not aware of the audible range. Other candidates calculated 'time', treated it as time period to calculate frequency and finally wrote that the wave is not audible in air. A number of candidates used speed of sound in water as 336 m s^{-1} for calculation.

(d) Almost all the candidates gave partially correct answers. They related it to the difference in the absorption of heat or to heat capacity instead of specific heat capacity of the stone and water. Some candidates wrote that the stone takes less time since it is a good conductor instead of writing about the specific heat capacity.

(e) Majority of the candidates wrote about resistance instead of resistivity. Some candidates wrote the reason for not using a copper wire as a fuse wire was the low melting point of copper/copper melts quickly/low specific heat capacity of copper. Some wrote the characteristics of a fuse only without comparing it with copper.

Answering Tips

- Understand the concept that the velocity of light of all colours in vacuum or air remains the same and it changes only when the medium changes.
- Do the sufficient practice of graph-based questions. Also, understand the concept that in free vibrations, the amplitude remains the same, above and below the X-axis, that is, the wave is equally spaced about the mean position.
- The difference in the graph of free and damped vibrations should be clearly known.
- Clearly understand the concept of the relation velocity = frequency × wavelength, the unit of frequency and the audible range of frequency.
- Point out the difference between specific heat capacity and heat capacity.
- Answer the reasoning-based questions to the point.
- Briefly understand the concept of fuse, its characteristics and uses in detail. Also, how a fuse wire can be compared with copper wire.

Question 4.

(a) *Calculate the total resistance across AB :* **[2]**

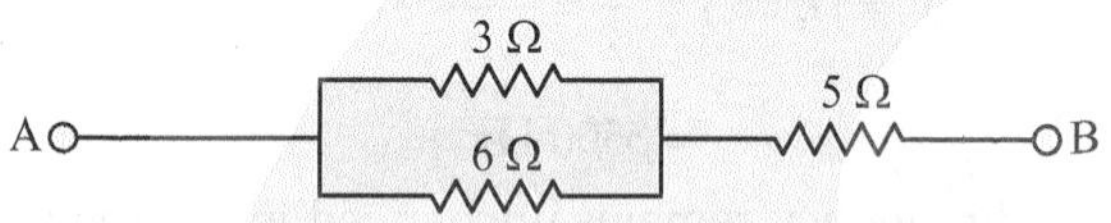

(b) *Two metallic blocks P and Q having masses in ratio 2 : 1 are supplied with the same amount of heat. If their temperatures rise by same degree, compare their specific heat capacities.* **[2]**

(c) *When a current carrying conductor is placed in a magnetic field, it experiences a mechanical force. What should be the angle between the magnetic field and the length of the conductor so that the force experienced is :* **[2]**

(i) *Zero*

(ii) *Maximum ?*

(d) *A nucleus $_{84}X^{202}$ of an element emits an alpha particle followed by a beta particle. The final nucleus is $_aY^b$. Find a and b.* **[2]**

(e) *The diagram below shows a loop of wire carrying current I :* **[2]**

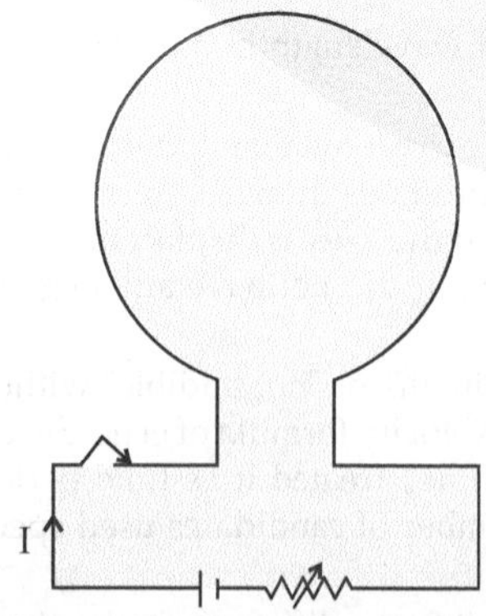

(i) What is the magnetic polarity of the loop that faces us ?

(ii) With respect to the diagram how can we increase the strength of the magnetic field produced by this loop ?

Answer 4.

(a) Equivalent resistance of 3 Ω and 6 Ω in parallel is,

$$R_1 = \frac{3 \times 6}{6+3} = \frac{18}{9} = 2\ \Omega$$

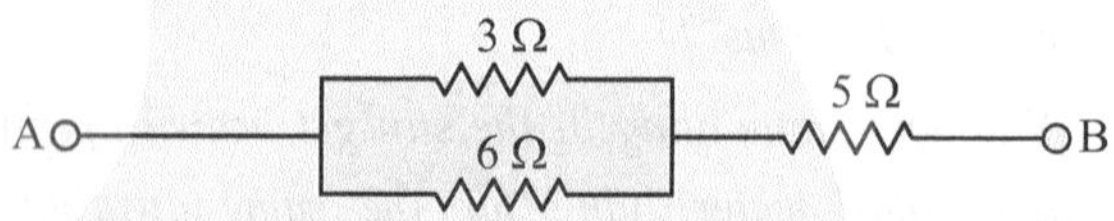

Total resistance across AB is $R_1 + R_2 = (5 + 2) = 7\ \Omega$.

(b) Let specific heat capacity of block P and block Q be C_P and C_Q respectively.

Metallic block P	**Metallic block B**
$m_1 = 2$ m	$m_2 =$ m
$c_1 = C_P$	$c_2 = C_Q$
$\Delta T = \Delta T$	$\Delta T = \Delta T$
$H_1 = m_1 c_1 \Delta T$	$H_2 = m_2 c_2 \Delta T$
$= 2m\, C_P\, \Delta T$	$= m\, C_Q\, \Delta T$

$\because \quad H_1 = H_2$

$\Rightarrow 2m\, C_P\, \Delta T = mC_Q\, \Delta T$

$$\therefore \quad \frac{C_P}{C_Q} = \frac{m \times \Delta T}{2m \times \Delta T} = \frac{1}{2}$$

∴ Ratio of specific heat capacities of block P and block Q is 1 : 2.

(c) **(i)** 0°

(ii) 90°

(d) $_{84}X^{202} \longrightarrow {}_{82}Z^{198} + {}_2He^4$

$_{82}Z^{198} \longrightarrow {}_{83}Y^{198} + {}_{-1}\beta^0$

$\therefore a = 83$, and $b = 198$

(e) **(i)** South

(ii) By increasing the strength of the current.

Comments of Examiners

4. (a) Most of the candidates answered this part correctly. However, the following errors were also observed:
- In the parallel combination formula, substitution was incorrect.
- Calculation errors were common.
- Units were not written in the final answer.

(b) Following errors were observed:
- Substitution for masses was incorrect.
- Some candidates compared heat capacities.
- Several candidates substituted for masses correctly but could not express the specific heat capacities in the ratio form.

(c) This was answered correctly by most candidates. However, a few candidates answered this part of the question by keeping a coil in mind instead of a straight conductor. Some candidates were not able to write the angles between the magnetic field and the length of the conductor for the asked forces.

(d) Most of the candidates answered this part correctly. However, a few candidates did not answer what is 'a' and 'b' in the final nucleus and left the answer as 83Y198, leaving it for the examiner to identify which is 'a' and which is 'b'. Some interchanged atomic and mass numbers.

(e) (i) Most of the candidates answered this part correctly, but some candidates wrote north instead of south. A few candidates wrote that the magnetic polarity of the loop that faces us as south to north or north to south, which was incorrect.

(ii) Some candidates did not pay attention to the loop shown in the diagram (the number of turns remains the same) and wrote that the strength of the magnetic field produced by this loop could be increased by increasing the number of turns of the coil. A few candidates wrote that the strength of the magnetic field can be increased by increasing the magnetic strength.

Answering Tips

- Give adequate practice of numerical problems or conceptual problems based on combination of resistors and specific heat capacity.
- Lay emphasis on correct mathematical operations.
- Revise conceptual questions frequently.
- Draw conclusion about the proportion of remaining quantities if some quantities are kept constant. Also, the comparison has to be expressed in ratio form only.
- For better understanding learn the concept that the force experienced by a current carrying straight conductor and by a current carrying coil.
- Find the atomic number and mass number of the products giving the reactants on emission of an α / β particle and vice-versa.
- Lay Stress on practicing balanced nuclear equations. Also understand the correct position of mass number and atomic number of an element.

SECTION-II (40 Marks)

*Attempt any **four** questions from this Section*

Question 5.

(a) *The figure below shows a simple pendulum of mass 200 g. It is displaced from the mean position A to the extreme position B. The potential energy at the position A is zero. At the position B the pendulum bob is raised by 5 m.* [3]

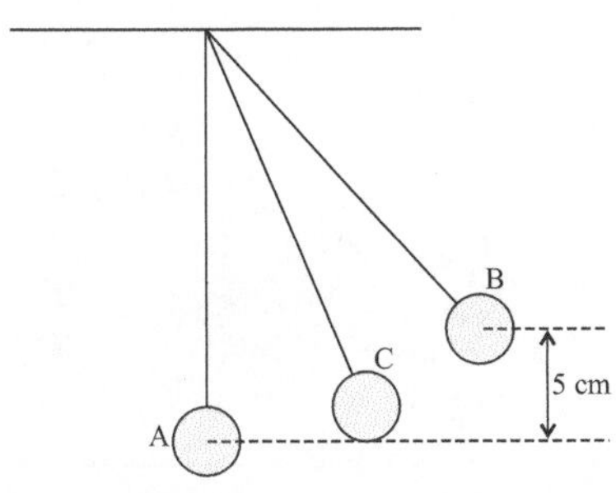

(i) *What is the potential energy of the pendulum at the position B ?*

(ii) *What is the total mechanical energy at point C ?*

(iii) *What is the speed of the bob at the position A when released from B ?*

(Take g = 10 ms^{-2} and there is no loss of energy.)

(b) **(i)** *With reference to the direction of action, how does a centripetal force differ from a centrifugal force during uniform circular motion ?* [3]

(ii) *Is centrifugal force the force of reaction of centripetal force ?*

(iii) *Compare the magnitudes of centripetal and centrifugal force.*

(c) *A block and tackle system of pulleys has velocity ratio 4.* [4]

(i) *Draw a neat, labelled diagram of the system indicating clearly the points of application and direction of load and effort.*

(ii) *What will be its V. R. if the weight of the movable block is doubled ?*

Answer 5.

(a) (i) Given :

$m = 200 \text{ g} = 0.2 \text{ kg}, g = 10 \text{ ms}^{-2}, h = 5 \text{ m}$

$\therefore$ Potential energy (U) = mgh

at position B = $0.2 \times 10 \times 5 = 10$ J

(ii) 10 J $\quad (\because K + U = \text{constant})$

(iii) At position A, kinetic energy (K) = 10 J

$$\frac{1}{2}mv^2 = 10$$

$$\Rightarrow \frac{1}{2} \times 0.2 \times v^2 = 10$$

$$\Rightarrow v^2 = \frac{10 \times 2}{0.2} = 100$$

$$\therefore v = \sqrt{100} = 10 \text{ ms}^{-1}$$

(b) (i) Centripetal force acts in a direction towards the centre of circular path whereas centrifugal force acts in a direction away from the centre of circular path.

(ii) No, centrifugal force is not the force of reaction of centripetal force because action and reaction do not act on the same body.

(iii) Magnitudes of centripetal and centrifugal forces are in the ratio 1 : 1.

(c) (i) Diagram of a block and tackle system having velocity ratio = 4

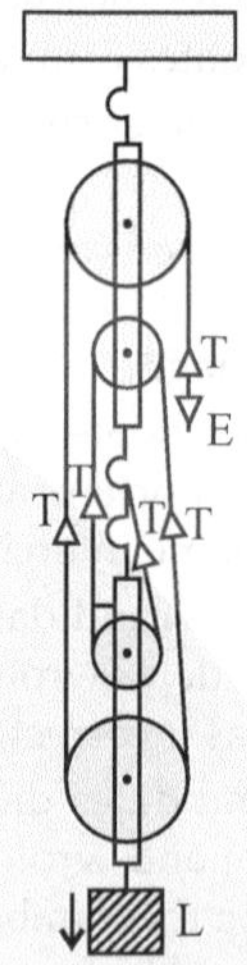

(ii) If weight of movable pulley is doubled there will be no change in the velocity ratio.

Comments of Examiners

5. (a) (i) Several candidates did not convert mass from g to kg, hence the answer was incorrect.

(ii) Many candidates did not realize the principle of conservation of energy involved in the process. Some got this incorrect because of errors made in the first part. Many candidates simply wrote total mechanical energy = P.E. + K.E. but did not write the exact value.

(iii) Many candidates wrote the unit of acceleration instead of writing the unit of velocity. A large number of candidates could not apply the principle of conservation of mechanical energy and thus, could not arrive at the correct answer.

Some wrote v^2 = 10 m/s. In this subpart, some candidates did not convert mass into the SI unit 'kg', however in the comparison of kinetic energy and potential energy, that is ½ m v^2 = m g h, as m was cancelled, they got the correct velocity, as the error got nullified.

(b) (i) Almost all candidates answered correctly. However, some only wrote inside or outside instead of writing radially inside or radially outside the circular path.

(ii) Barring a few, most candidates answered this part correctly.

(iii) Most candidates answered this part correctly. However, a few candidates wrote vague answers. They explained centrifugal and centripetal forces but did not compare their magnitudes. Some wrote that centripetal force is greater /more than centrifugal force.

(c) (i) Following errors were observed:

- No fixed support was shown.
- Strands drawn were loose.
- Connections shown were incorrect.
- Directions of either/both load and effort were not marked.
- Rough sketches of pulleys were drawn.

(ii) Most candidates wrote that V.R. becomes half/double.

Answering Tips

- Lay stress on conversion of units.
- Mention the principle involved while solving a numerical to arrive at the answer.
- Give adequate practice to solve diagram-based questions / conceptual questions in order to learn how to apply the concept of conservation of energy.
- Understand the concept of centripetal and centrifugal forces. The centripetal and centrifugal forces are equal in magnitude but opposite in direction.
- Draw pulley related diagrams and mark the tension force in the strings. Practice should be given with odd and even velocity ratios.
- Understand concept of velocity ratio for a given design of machine and effect of friction or weight of the parts of the machine on the velocity ratio.

Question 6.

(a) *A diver in water looks obliquely at an object AB in air.* **[3]**

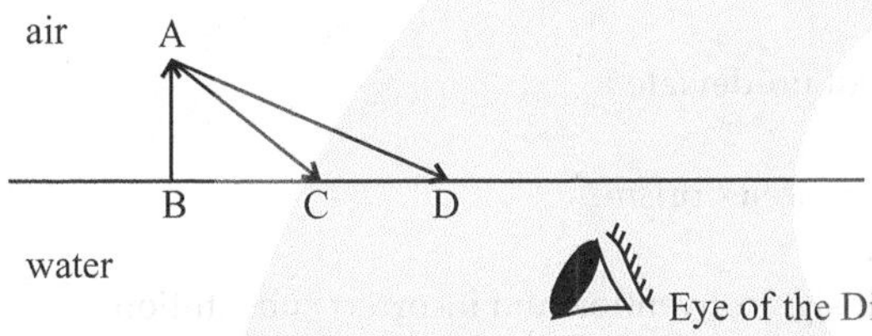

(i) *Does the object appear taller, shorter or of the same size to the diver ?*

(ii) *Show the path of two rays AC and AD starting from the tip of the object as it travels towards the diver in water and hence obtain the image of the object.*

(b) *Complete the path of the ray AB through the glass prism in PQR till it emerges out of the prism. Given the critical angle of the glass as 42°.* **[3]**

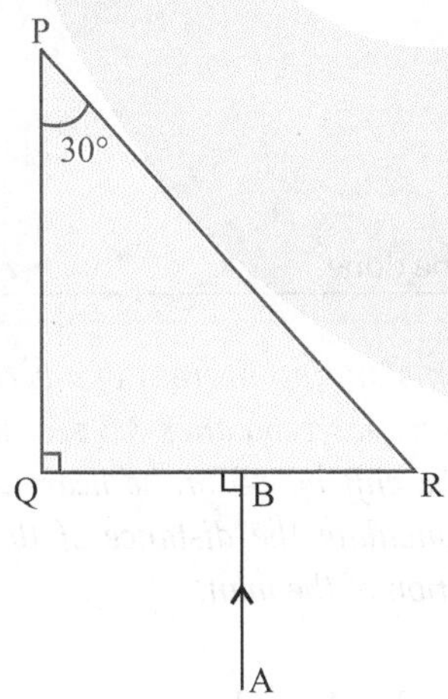

(c) *A lens of focal length 20 cm forms an inverted image at a distance 60 cm from the lens.* **[4]**

(i) *Identify the lens.*

(ii) *How far is the lens present in front of the object ?*

(iii) *Calculate the magnification of the image.*

Answer 6.

(a) (i) The object will appear taller.

(ii) A'B is the image formed of the object AB.

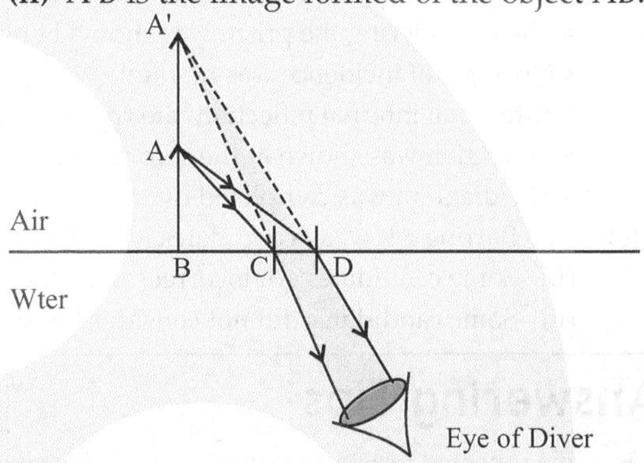

(b) The path of the ray till it emerges out of the prism is ABCDE.

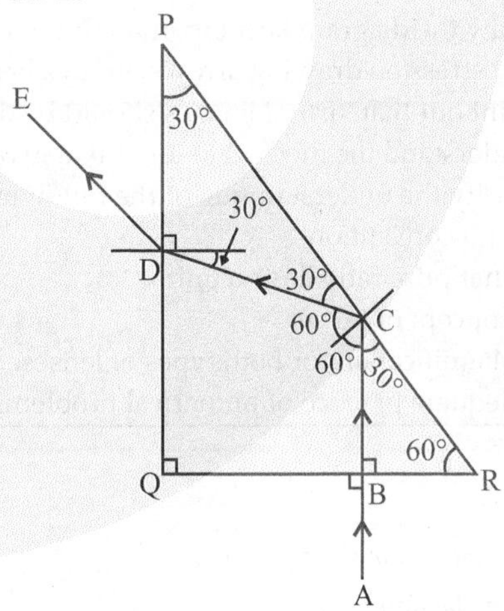

(c) (i) Convex lens

(ii) Given : $f = +20$ cm, $v = +60$ cm

$$\because \quad \frac{1}{f} = \frac{1}{v} - \frac{1}{u}$$

$$\Rightarrow \quad \frac{1}{20} = \frac{1}{60} - \frac{1}{u}$$

$$\Rightarrow \quad \frac{1}{u} = \frac{1}{60} - \frac{1}{20}$$

$$= \frac{1-3}{60} = -\frac{2}{60} = -\frac{1}{30}$$

$\therefore \quad u = -30$ cm

The lens is at a distance 30 cm in front of the object.

(iii) $m = \frac{v}{u} = \frac{+60}{-30} = -2$ [– ve sign because image is real]

Comments of Examiners

6. (a) (i) Many candidates answered the part correctly. A few candidates made mistakes as they could not establish the relation of between the size image and change in medium. Some candidates considered the light ray passing from denser medium to rarer medium. Some wrote that the object appears shorter while others wrote, of the same size.

(ii) Candidates made several mistakes like:

1. Not marking arrows on rays.
2. Showing partial reflection with refraction.
3. Not showing extension of rays and dotted image.
4. Showing incorrect image position and incorrect refraction.

(b) Following errors were observed in this part:

- The ray entering the prism was shown bending.
- No angle of incidence was marked.
- After total internal reflection, the ray was shown coming out un-deviated.
- Refraction was shown at the second surface.
- The diagram was completed by treating it as right-angled isosceles prism.

(c) (i) Barring a few, all candidates identified the lens correctly.

(ii) Some candidates got incorrect answers due to improper sign convention and incorrect substitution.

(iii) Some candidates did not consider the sign convention. A few candidates wrote unit for magnification.

Answering Tips

- Give repeated practice of drawing ray diagrams.
- Draw diagrams applying laws of refraction at every surface when the ray changes medium, rather than remembering the diagram by rote.
- Draw the diagram keeping in mind the refractive index of the medium, e.g. for water as 4/3.
- Lay stress on drawing arrows on rays before and after the refraction.
- Point out that virtual images should be dotted. Also, apparent intersection should be shown by dotted lines.
- Understand the method to identify a given lens from the description given in the question.
- Clarify the understanding of the following terms:
  - Sign convention
  - That pure ratio has no unit.
  - Concept of linear
  - Magnification for both types of lenses.
- Adequate practice of numerical problems on the lens formula should be done.

Question 7.

(a) *Give reasons for the following :* **[3]**

During the day :

(i) *Clouds appear white.*

(ii) *Sky appears blue.*

(b) (i) *Name the system which enables us to locate underwater objects by transmitting ultrasonic waves and detecting the reflecting impulse.* [3]

(ii) *What are acoustically measurable quantities related to pitch and loudness ?*

(c) (i) *When a tuning fork [vibrating] is held close to ear, one hears a faint hum. The same [vibrating tuning fork] is held such that its stem is in contact with the table surface, then one hears a loud sound. Explain.* **[4]**

(ii) *A man standing in front of a vertical cliff fires a gun. He hears the echo after 3.5 seconds. On moving closer to the cliff by 84 m, he hears the echo after 3 seconds. Calculate the distance of the cliff from the initial position of the man.*

Answer 7.

(a) (i) During the day, clouds appear white because they contain dust particles and aggregates of water molecules of size bigger than the wavelength of visible light, hence they scatter all colours of incident white light from the sun to the same extent.

(ii) During the day, sky appears blue because blue (or violet) light due to its short wavelength is scattered more as compared to the red light of long wavelength. Hence, the sky other than the direction of sun appears blue.

(b) (i) SONAR or Sound Navigation and Ranging.

(ii) The acoustically measurable quantities related to pitch is frequency or wavelength and for loudness, it is intensity of sound.

(c) (i) A tuning fork held close to ear, disturbs a small volume of air and hence, sound heard is faint, when the handle of the vibrating tuning fork is held against table, it sets up forced vibrators in the tabletop. As tabletop has a large surface area, large volume of air is set into vibration transmitting more energy thereby producing loud sound.

(ii) Let the initial position of man from the cliff be d metre

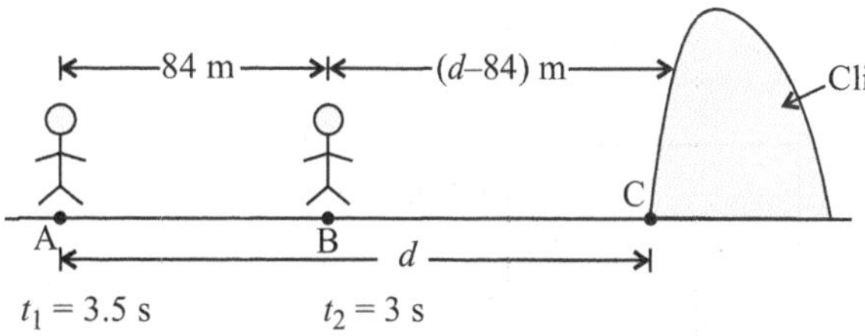

At Position A	**At Position B**
$d_1 = d$ metre	$d_2 = (d - 84)$ m
$t_1 = 3.5$ s	$t_2 = 3$ s

$$\text{Speed of sound} = \frac{2d}{t}$$

$$\Rightarrow \frac{2d_1}{t_1} = \frac{2d_2}{t_2}$$

$$\Rightarrow \frac{2d}{3.5} = \frac{2(d-84)}{3}$$

$$\Rightarrow 6d = 7d - 588$$

$$\Rightarrow d = 588 \text{ m}$$

$\therefore$ Distance of cliff from initial position of man = 588 m.

Comments of Examiners

7. (a) (i) Many candidates used the term reflection or dispersion instead of scattering. The size of the particles and uniform scattering was not mentioned by some candidates. Some candidates were confused between the white colour of the sky at noon and white colour of clouds. They mixed both concepts and wrote the answer.

(ii) Most of the candidates explained this part correctly. However, some candidates did not compare the wavelength. A few candidates gave incorrect reasons.

(b) (i) Most of the candidates gave the correct answer 'SONAR'. However, some candidates wrote the process as 'echo depth sounding'.

(ii) Some candidates wrote the unit for pitch while some gave the answer as quality. For loudness many candidates wrote amplitude which was incorrect.

(c) (i) Some candidates related it to resonance rather than surface area.

(ii) Most candidates were able to form correct equations but could not solve the same correctly.

Answering Tips

- Lay stress on key words like dispersion and scattering.
- Understand the dependence of extent of scattering with respect to wavelength of light.
- Give sufficient practice on conceptual or reasoning-based questions.
- Understand the difference between a process and a system.
- Understand the subjective nature of characteristics of sound and the objective nature of the physical quantities on which they depend, with relevant examples.
- Give adequate practice on echo based conceptual or numerical problems.
- Draw a rough diagram for better comprehension and for framing correct equations.

Question 8.

(a) *The diagram below shows the core of a transformer and its input and output connections.* **[3]**

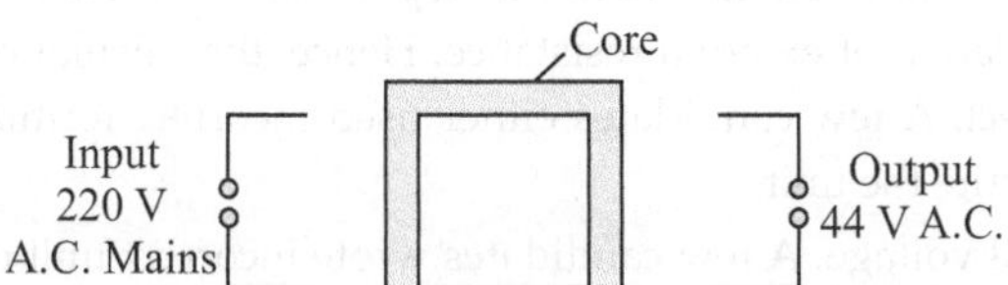

(i) *State the material used for the core.*

(ii) *Copy and complete the diagram of the transformer by drawing input and output coils.*

(b) (i) *What are superconductors ?* **[3]**

(ii) *Calculate the current drawn by an appliance rated 110 W, 220 V when connected across 220 V supply.*

(iii) *Name a substance whose resistance decreases with the increase in temperature.*

(c)

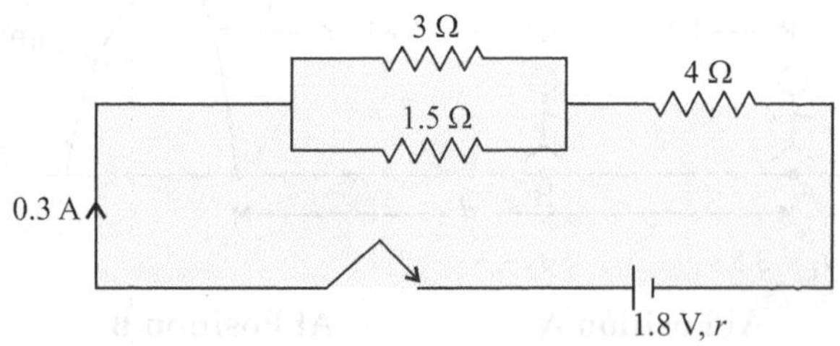

The diagram above shows three resistors connected across a cell of e.m.f. 1.8 V and internal resistance r. Calculate : [4]

(i) *Current through 3 Ω resistor.*

(ii) *The internal resistance r.*

Answer 8.

(a) (i) The material used for core is laminated soft iron.

(ii)

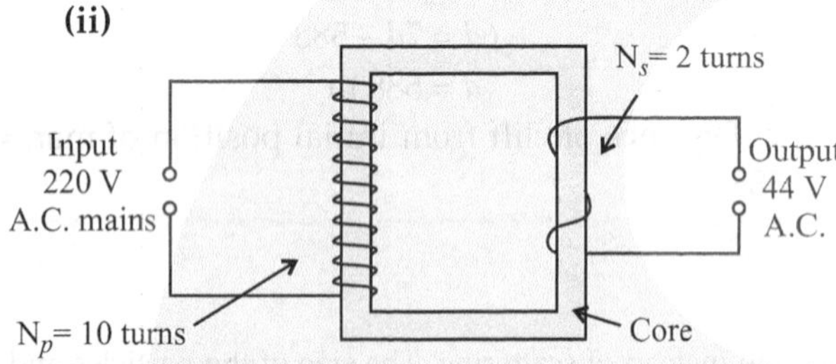

$$\text{Turns ratio} = \frac{N_s}{N_p} = \frac{E_s}{E_p} = \frac{44}{220} = \frac{1}{5}$$

Hence, number of turns in secondary coil : primary coil is equal to 1 : 5

(b) (i) Superconductors are substances of zero resistance at temperature closer to absolute zero.

(ii) P = 110 W V = 220 volt

$$\therefore \quad \text{Current I} = \frac{P}{V} = \frac{110}{220} \text{ A}$$

$$= 0.5 \text{ A}$$

(iii) For semiconductors such as silicon, germanium, or carbon resistance decreases with the increase in temperature.

(c) (i) Equivalent resistance of 3Ω and 1.5Ω in parallel,

$$R_1 = \frac{3 \times 1.5}{3 + 1.5} \Omega = 1\Omega$$

Potential difference, across R_1

$$\Rightarrow \quad V_1 = IR_1$$

$$= 0.3 \times 1 = 0.3\text{V}$$

$\therefore$ Current through 3Ω resistor,

$$I_1 = \frac{V_1}{3} = \frac{0.3}{3} = 0.1\text{A}$$

(ii) Total external resistance R = 1 + 4 = 5Ω

$$\because \quad E = I(R + r)$$

$$\Rightarrow \quad 1.8 = 0.3(5 + r)$$

$$\Rightarrow \quad \frac{1.8}{0.3} = 5 + r \Rightarrow 6 = 5 + r$$

$$\therefore \quad r = 6 - 5 = 1\Omega$$

Comments of Examiners

8. (a) (i) Most of the candidates answered correctly but a few candidates stated the wrong material to be used for the core.

(ii) Many candidates did not connect the coil to the terminals shown in the diagram. Some showed the coils in the incorrect arms or showed incorrect number of turns of the coil i.e. more turns for less voltage and vice-versa. A few candidates labelled it correctly.

(b) (i) Most of the candidates answered this part correctly but some candidates were confused between superconductors and semiconductors.

(ii) A large number of the candidates solved this problem correctly. However, following errors were noticed in many scripts:

- errors in copying the given data
- error in calculation
- unit not written
- incorrect unit given

(iii) Almost all candidates answered correctly, but some wrote random names such as, galvanized rubber, metals, copper, aluminium, tungsten, constantan, mangnanin, nichrome, which were incorrect.

(c) (i) Some candidates solved this question correctly. However, a few candidates took potential difference across 3 Ω as 1.8 V. Several candidates did incorrect calculation of external resistance. Hence, the calculation for current through the given resistor was also incorrect. A few candidates either used incorrect formula or made calculation errors. Some candidates did not write the unit.

(ii) Some candidates considered the emf as the terminal voltage. A few candidates wrote incorrect unit of the asked physical quantity.

Answering Tips

- Learn the process of extraction thoroughly.
- Construction and working of a transformer should be clearly known.
- Sufficient practice of drawing labelled diagrams should be done with clear explanation of the working.
- In the definition of superconductors, instead of 'very low resistance' it is advisable to write 'at temperatures closer to the absolute zero', as very low is vague. The concept of superconductivity with infinite conductance must be clearly understood.
- Give regular practice of solving numerical.
- Understand the variation of resistance with temperature in detail.
- The combination of resistors in series and parallel should be essentially known.
- Give adequate practice of conceptual or numerical problems based on combination of resistors.
- Comparison of the EMF, terminal voltage with the voltage drop should be known in detail.

Question 9.

(a) **(i)** *Define heat capacity of a substance.* [3]

(ii) *Write the SI unit of heat capacity.*

(iii) *What is the relationship between heat capacity and specific heat capacity of a substance ?*

(b) *The diagram below shows the change of phases of a substance on a temperature vs time graph on heating the substance at a constant rate.* [3]

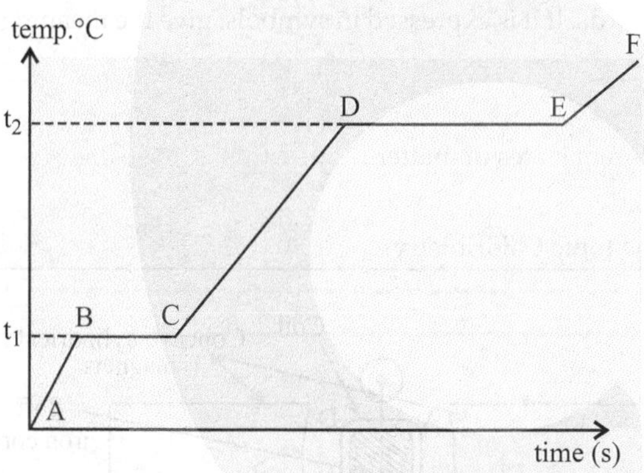

(i) *Why is the slope of CD less than slope of AB ?*

(ii) *What is the boiling and melting point of the substance ?*

(c) *A piece of ice of mass 60 g is dropped into 140 g of water at 50°C.* [4]

Calculate the final temperature of water when all the ice has melted.

(Assume no heat is lost to the surrounding)

Specific heat capacity of water = 4.2 $Jg^{-1}k^{-1}$

Specific latent heat of fusion of ice = 336 Jg^{-1}

Answer 9.

(a) **(i)** Heat capacity of a body is the amount of heat energy required to raise its temperature by 1 Kelvin.

(ii) SI Unit of heat capacity is joule per Kelvin (JK^{-1}).

(iii) Heat capacity = Mass × Specific heat capacity

(b) **(i)** The slope of CD is less than slope of AB because specific heat capacity of liquid phase of same material can be different from that of solid phase of same material.

(ii) Boiling point is t_2°C and melting point is t_1°C.

(c) Let final temperature of water = x°C

Ice	Water
$m_1 = 60$ g	$m_2 = 140$ g
$T_1 = 0°C$	$T_1 = 50°C$
$T_2 = x°C$	$T_2 = x°C$
Rise in temp.	fall in temp.
$(\Delta T) = (x - 0)°C = x°C$	$(\Delta T) = (50 - x)°C$

Heat gained by ice $= M_1L + m_1c\Delta T$

$= (60 \times 336 + 60 \times 4.2 \times x)$ J

Heat lost by water $= m_2c\Delta T$

$= 140 \times 4.2 \times (50 - x)$ J

Applying, the principle of mixtures,

$$140 \times 4.2 \times (50 - x) = 60 \times 336 + 60 \times 4.2 \times x$$

$$\Rightarrow \quad 4.2\,(7000 - 140x - 60x) = 60 \times 336$$

$$\Rightarrow \quad 7000 - 200x = \frac{60 \times 336}{4.2}$$

$$= \frac{60 \times 336}{42} \times 10$$

$$= 4800$$

$$\Rightarrow \quad 200x = 2200$$

$$\Rightarrow \quad x = 11°C$$

∴ Final temperature of water = 11°C

Comments of Examiners

9. (a) (i) Almost all candidates answered correctly barring a few, who wrote the definition of specific heat capacity. Instead some candidates wrote the definition assuming the substance to be water.
(ii) Most candidates answered correctly but a few wrote the unit of specific heat capacity.
(iii) Some candidates wrote the relationship between heat capacity and specific heat capacity of a substance in the form of symbols without giving the meaning of the symbols.
(b) (i) Many candidates could not answer this question correctly. They related it with the time.
(ii) Most of the candidates answered correctly but a few wrote vague values such as, 100°C, 90 °C, etc. Some also wrote BC, DE as answers. A few candidates interchanged the values of melting point and boiling point of the substance. Some answered assuming the substance to be water.
(c) Majority of the candidates attempted this numerical correctly. However, the following errors were observed in other scripts:
- Not taking into consideration the heat absorbed by ice after melting to reach equilibrium temperature.
- Not substituting the change in temperature correctly.
- Not writing the unit for the final answer.

Answering Tips

- Understand the difference between heat capacity and specific heat capacity.
- Insist on writing the SI unit of temperature kelvin, (K) for example, unit of heat capacity as J/K instead of J/C.
- Learn the correct SI unit of a given physical quantity.
- Write the relation between different physical quantities in words. If it is expressed in symbols, give the meaning of the symbols.
- Give more practice on graph-based questions.
- Point out the how specific heat capacity is different for different states of matter.
- Read the information from the graph and interpret it correctly.
- Give sufficient practice on numerical problems based on the topic Calorimetry.

Question 10.

(a) **(i)** *Draw a neat labeled diagram of a d.c. motor.* [3]

(ii) *Write any one use of a d.c motor.*

(b) **(i)** *Differentiate between nuclear fusion and nuclear fission.* [3]

(ii) *State one safety precaution in the disposal of nuclear waste.*

(c) *An atomic nucleus A is composed of 84 protons and 128 neutrons. The nucleus A emits an alpha particle and is transformed into a nucleus B.* [4]

(i) *What is the composition of B ?*

(ii) *The nucleus B emits a beta particle and is transformed into a nucleus C. What is the composition of C ?*

(iii) *What is mass number of the nucleus A ?*

(iv) *Does the composition of C change if it emits gamma radiations ?*

Answer 10.

(a) **(i)** Diagram of a d.c. motor.

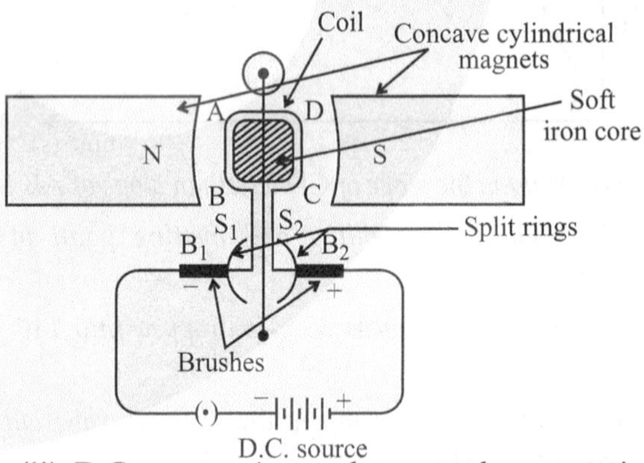

(ii) D.C. motor is used to produce rotational motion by using electricity in devices such as spinning and weaving machines, Fan etc.

(b) **(i)**

Nuclear Fission	Nuclear Fusion
1. In nuclear fission when neutrons are bombarded on a heavy nucleus, it splits in two nearly equal light fragments.	In nuclear fusion, two light nuclei combine to form a heavy nucleus at a very high temperature and high pressure.
2. It can be controlled.	It cannot be controlled.

(ii) For disposal of nuclear waste, they must be first kept in thick casks and then buried in specially constructed deep underground stores and also while handling nuclear waste, we should wear special lined lead aprons and lead gloves.

(c) For nucleus A

Atomic no. (Z) = 84

Mass no. (A) = 84 + 128 = 212

(i) ${}_{84}A^{212} \xrightarrow{\alpha} {}_{82}B^{208}$ Nucleus B will have 82 protons and (208 – 82) = 126 neutrons.

(ii) ${}_{82}B^{208} \xrightarrow{\beta} {}_{83}C^{208}$ Nucleus C will have 83 protons and (208 – 83) = 125 neutrons

(iii) Mass number of nucleus, A = 212

(iv) No, the composition of nucleus C does not change due to emission of gamma radiations.

Comments of Examiners

3. (a) (i) Many candidates drew the diagram of a d.c. motor correctly. However, in some scripts the following errors were observed:
   - drew diagram of AC generator.
   - did not label the diagram or labeled incorrectly.
   - drew slip rings instead of split rings.
   - drew the diagram with the load.
   - did not draw the field magnet.

   (ii) Most of the candidates answered it correctly, except for a few who wrote energy conversion as the answer.

(b) (i) A number of candidates answered correctly. Some candidates mixed up the definitions of fission and fusion. The words bigger and smaller were used in place of heavier and lighter, respectively. Vague statement was written by a few candidates such as, large energy is given out during nuclear fusion without comparing the quantity of nuclear material.

(ii) Some candidates gave precautions to be taken while handling the radioactive substances.

(c) This part of the question was well attempted by most candidates. Some of the common mistakes made in the subparts are as follows:

(i) A few candidates did not write number of neutrons. Some forgot to add the number of protons and neutrons to get the mass number. They subtracted 4 from the neutron number and 2 from the proton number instead of subtracting 4 from the mass number.

(ii) Some candidates used neutrons as atomic mass for beta emission. For candidates who made mistakes in the first part, the second part also went incorrect, since the second part was related to the first part.

(iii) Most of the candidates answered correctly but some forgot to add the number of protons and neutrons and some wrote only the number of protons as mass number.

(iv) A few candidates answered this subpart as Yes, which was incorrect.

Answering Tips

- Comparison of DC motor with AC generator should be done while understanding the diagrams.
- Differentiate between the function of a device and its usage clearly. For example, conversion of electrical to mechanical energy is the function of a DC motor but it is used in devices such as washing machine, mixer, toy-fans, etc.
- Stress on correct use of key words in definitions.
- Difference between safety precaution taken in handling of radioactive materials and safety precaution in disposal of nuclear waste should be known properly.
- Understand the effect on the nucleus during alpha, beta and gamma emission. Also understand it as the effect on number of protons and neutrons as well as atomic number and mass number.
- The composition of an element explicitly with atomic number and mass number should be clearly known.
- The change in atomic number and mass number during an alpha decay or beta decay from a nucleus, with reactions should be known.
- Understand the concept of when a nucleus emits a gamma radiation, there is no change in the mass number and atomic number.

